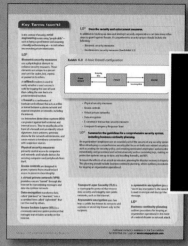 

COURSE TECHNOLOGY
CENGAGE Learning™

MIS2
Hossein Bidgoli

Executive Vice President and Publisher: Jonathan Hulbert

Executive Vice President of Editorial, Business: Jack W. Calhoun

Vice President of Marketing: Bill Hendee

Publisher: Joe Sabatino

Publisher/Director 4LTR Press: Neil Marquardt

Sr. Acquisitions Editor: Charles McCormick, Jr.

Sr. Product Manager: Kate Mason

Product Development Manager, 4LTR Press: Steven Joos

Sr. Marketing Communications Manager: Libby Shipp

Marketing Manager: Adam Marsh

Marketing Coordinator: Suellen Ruttkay

Editorial Assistant: Courtney Bavaro

Content Project Manager: Jennifer Feltri

Media Editor: Chris Valentine

Sr. Art Director: Stacy Jenkins Shirley

Print Buyer: Julio Esperas

Cover Designer: Joe Devine, Red Hangar

Cover Photos: ©Getty Images/Thinkstock

Compositor: Bill Smith Group

To so many fine memories of my mother, Ashraf, my father, Mohammad, and my brother, Mohsen, for their uncompromising belief in the power of education.
–Hossein Bidgoli

• This publication incorporates screen capture of the web page http://www.cert.org/certcc.html (c) 2010 Carnegie Mellon University (c) 2010 Carnegie Mellon University, with special permission from its Software Engineering Institute.
• ANY MATERIAL OF CARNEGIE MELLON UNIVERSITY AND/OR ITS SOFTWARE ENGINEERING INSTITUTE CONTAINED HEREIN IS FURNISHED ON AN "AS-IS" BASIS. CARNEGIE MELLON UNIVERSITY MAKES NO WARRANTIES OF ANY KIND, EITHER EXPRESSED OR IMPLIED, AS TO ANY MATTER INCLUDING, BUT NOT LIMITED TO, WARRANTY OF FITNESS FOR PURPOSE OR MERCHANTABILITY, EXCLUSIVITY, OR RESULTS OBTAINED FROM USE OF THE MATERIAL. CARNEGIE MELLON UNIVERSITY DOES NOT MAKE ANY WARRANTY OF ANY KIND WITH RESPECT TO FREEDOM FROM PATENT, TRADEMARK, OR COPYRIGHT INFRINGEMENT.
• This publication has not been reviewed nor is it endorsed by Carnegie Mellon University or its Software Engineering Institute.
• CERT® is a registered trademark of Carnegie Mellon University.

Library of Congress Control Number: 2010941999

Student Edition:
ISBN-13: 978-1-111-53396-0
ISBN-10: 1-111-53396-2

Instructor's Edition:
ISBN-13: 978-1-111-53399-1
ISBN-10: 1-111-53399-7

Course Technology
20 Channel Center Street
Boston, MA 02210
USA

Some of the product names and company names used in this book have been used for identification purposes only and may be trademarks or registered trademarks of their respective manufacturers and sellers.

Any fictional data related to persons or companies or URLs used throughout this book is intended for instructional purposes only. At the time this book was printed, any such data was fictional and not belonging to any real persons or companies.

Course Technology, a part of Cengage Learning, reserves the right to revise this publication and make changes from time to time in its content without notice.

Cengage Learning is a leading provider of customized learning solutions with office locations around the globe, including Singapore, the United Kingdom, Australia, Mexico, Brazil and Japan. Locate your local office at **www.cengage.com/global**

Cengage Learning products are represented in Canada by Nelson Education, Ltd.

To learn more about Course Technology, visit **www.cengage.com/ coursetechnology**

Purchase any of our products at your local college store or at our preferred online store **www.CengageBrain.com**

Printed in the United States of America
2 3 4 5 6 7 17 16 15 14 13 12 11

Brief Contents

Contents

7 The Internet, Intranets, and Extranets 118

8 E-Commerce 140

SPEAK UP!

MIS2 was built on a simple principle: to create a new teaching and learning solution that reflects the way today's faculty and students teach and learn. Through conversations, focus groups, surveys, and interviews, we collected data that drove the creation of the version of MIS2 that you are using today.

But it doesn't stop there – in order to make MIS2 an even better learning experience, we'd like you to SPEAK UP and tell us how MIS2 works for you.

What do you like about it? What would you change? Do you have additional ideas that would help us build a better product for next year's Management Information Systems students?

Speak Up! Go to CourseMate for MIS2. Access at login.cengagebrain.com.

INFORMATION SYSTEMS: AN OVERVIEW

t his chapter starts with an overview of common uses for computers and information systems, explains the difference between computer literacy and information literacy, and then reviews transaction processing systems as one of the earliest applications of information systems. Next, we discuss the components of a management information system (MIS), including data, databases, processes, and information, and see how information systems relate to information technologies. This chapter also covers the roles and applications of information systems and explains the Five Forces Model, used to develop strategies for gaining a competitive advantage. Finally, we review the IT job market and touch on the future of information systems.

Organizations use computers and information systems to reduce costs and gain a competitive advantage in the marketplace.

learning outcomes

After studying this chapter, you should be able to:

LO1 Discuss common applications of computers and information systems.

LO2 Explain the differences between computer literacy and information literacy.

LO3 Define transaction processing systems.

LO4 Define management information systems.

LO5 Describe the four major components of an information system.

LO6 Discuss the differences between data and information.

LO7 Explain the importance and applications of information systems in functional areas of a business.

LO8 Discuss how information technologies are used to gain a competitive advantage.

LO9 Explain the Five Forces Model and strategies for gaining a competitive advantage.

LO10 Review the IT job market.

LO11 Summarize the future outlook of information systems.

1 Computers and Information Systems in Daily Life

Organizations use computers and information systems to reduce costs and gain a competitive advantage in the marketplace. Throughout this book, you will study many information system applications. For now, let's look at some common applications used in your daily life.

Computers and information systems are all around you. As a student, you use computers and office suite software and might take online classes. Computers are often used to grade your exam answers and generate detailed reports comparing the performance of each student in your class. Computers and information systems also calculate grades and GPAs and can deliver this information to you.

Computers and information systems are commonly used in grocery and retail stores as well. For example, a point-of-sale (POS) system speeds up service by reading the universal product codes (UPCs) on items in your shopping cart (see Exhibit 1.1). This same system also manages store inventory, and some information systems can even reorder stock automatically. Banks, too, use computers and information systems for generating your monthly statement and running ATM machines for many banking activities.

Exhibit 1.1 *A point-of-sale system*

Many workers are now telecommuters who perform their jobs at home, and others often use their PDAs (personal digital assistants such as a Palm) to conduct business while on the go. The most common PDA is a smartphone (such as an iPhone, Droid, or a Blackberry). A typical PDA includes a calendar, address book, and task-listing programs; more advanced PDAs often allow for wireless connection to the Internet and have built-in MP3 players. Smartphones are mobile phones with advanced capabilities, much like a mini PC. They include e-mail and Web browsing features, and most have a built-in keyboard or an external USB keyboard (see Exhibit 1.2).

The Internet is used for all kinds of activities, from shopping to learning to working. Search engines and broadband communication bring information to your desktop in seconds. The Internet is also used for social purposes. With social networking sites, such as Facebook, Twitter, and Foursquare, you can connect with friends, family, and colleagues online and meet people with similar interests and hobbies. Twitter (*www.twitter.com*), for example, is a social networking and short-message service. Users can send and receive brief text updates, called "tweets." These posts are displayed on your profile page, and other users can sign up to have them delivered to their in-boxes.

Organizations also use social networking sites to give customers up-to-date information and even how-to support with videos. These sites can reduce organizations' costs by providing an inexpensive medium for targeting a large customer base.

In addition, people use video-sharing sites to watch news, sporting events, and entertainment videos. One of the most popular sites is YouTube (*www.youtube.com*). You can upload and share video clips via Web sites, mobile devices, blogs, and e-mail. Users upload most of the content on YouTube, although media corporations such as CBS, BBC, Sony Music Group, the Sundance Channel, and others also provide content. Anyone can watch videos on YouTube, but you must register to upload videos. Businesses are increasingly using YouTube to promote their products and services. The information box below highlights a few such companies.

Exhibit 1.2 *Examples of smartphones*

A New Era of Marketing: YouTube

Companies use newspapers, magazines, TV shows, and search engines to promote their products, services, and their brands. YouTube is a popular video-sharing service that can be used as a marketing tool. The videos on YouTube are very well indexed and organized. They are categorized and sorted by "channels." The channels range from film & animation to sports, short movies, and video blogging. Individual YouTube users have used this tool to share videos and stories. Corporations can also take advantage of this popular platform. YouTube represents a great opportunity for marketers to reach consumers who are searching for information about a brand or related products and services. It can also be used as a direct marketing tool. The following are examples of corporations that are using YouTube to promote their products and services:

Quiksilver—This manufacturer of apparel and accessories, including the Roxy brand, frequently posts new videos of its products, continually renewing its Web presence.

Ford Models—Since 2006, it has uploaded over 554 videos promoting its brand.

University of Phoenix Online—This site has hundreds of video testimonials, reviews, and documentaries that promote the university's degree programs.

The Home Depot—Here you'll find free content, including practical knowledge and money-saving tips for home improvements.

Nikefootball—Nike maintains several distinct YouTube channels that cater to specific audiences. Consumers can find content that is relevant to their needs without having to sift through everything.[1,2]

So what do all these examples mean to you? As a knowledge worker of the future, computers and information technology will help you perform more effectively and productively, no matter what profession you choose. In addition, you can connect to the rest of the world to share information, knowledge, videos, ideas, and almost anything else that can be digitized. Throughout this book, we will explore these opportunities as well as the power of computers and information systems.

As you read, keep in mind that the terms *information systems* and *information technologies* are used interchangeably.

Information systems are broader in scope than information technologies, but the two overlap in many areas.

In the future, computers and information technology will help you be a more effective and productive knowledge worker, no matter what profession you choose.

Both are used to help organizations be more competitive and to improve their overall efficiency and effectiveness. Information technologies offer many advantages for improving decision making but involve some challenges, too, such as security and privacy issues. The information box on this page describes one of the potential challenges.

2 Computer Literacy and Information Literacy

In the 21st century, knowledge workers need two types of knowledge to be competitive in the workplace: computer literacy and information literacy. **Computer literacy** is skill in using productivity software, such as word processors, spreadsheets, database management systems, and presentation software, as well as having a basic knowledge of hardware and software, the Internet, and collaboration tools and technologies. **Information literacy**, on the other hand, is understanding the role of information in generating and using business intelligence. **Business intelligence (BI)** is more than just information. It provides historical, current, and predictive views of business operations and environments and gives organizations a competitive advantage in the marketplace. (BI is discussed in more detail in Chapter 3.) To summarize, knowledge workers should know the following:

Computer literacy is skill in using productivity software, such as word processors, spreadsheets, database management systems, and presentation software, as well as having a basic knowledge of hardware and software, the Internet, and collaboration tools and technologies.

Information literacy is understanding the role of information in generating and using business intelligence.

Business intelligence (BI) provides historical, current, and predictive views of business operations and environments and gives organizations a competitive advantage in the marketplace.

Transaction processing systems (TPS) focus on data collection and processing; the major reason for using them is cost reduction.

Social Networking and the Vulnerability of Personal Information

The popularity of social networking sites such as Facebook, Twitter, and Foursquare is on the rise. As of July 2010, Facebook had more than 500,000 registered users, and the number is increasing on a daily basis.[3] But so is the potential risk. According to a Consumer Reports study published on May 4, 2010, over half of all users of social networks in this country are putting themselves at risk by posting information that could be misused by cybercriminals. Many social networkers post their full birth dates, their home addresses, photos of themselves and their families, and the times when they will be away from home. This information could be used by cybercriminals for malicious purposes. According to the report, 9% of the 2000 people who participated in the study had experienced some kind of computer-related trouble, such as malware infections, scams, identity theft, or harassment. To reduce risk and improve the privacy of your personal information, the study offers several tips[4]:

- *Always use the privacy controls offered by the social networking sites.*
- *Use passwords that mix upper- and lowercase letters with numbers and symbols.*
- *Do not post a phone number or a full address.*
- *Do not post children's names, even in photo tags or captions.*
- *Do not be specific when posting information about vacations or business trips.*

- Internal and external sources of data
- How data is collected
- Why data is collected
- What type of data should be collected
- How data is converted to information and eventually to business intelligence
- How data should be indexed and updated
- How data and information should be used to gain a competitive advantage

3 The Beginning: Transaction Processing Systems

For past 60 years, **transaction processing systems (TPSs)** have been applied to structured tasks such as record keeping, simple clerical operations, and inventory control. Payroll, for example, was one of the first applications to be automated. TPSs focus on data collection and processing, and have provided enormous reductions in costs.

Computers are most beneficial in transaction processing operations. These operations are repetitive, such as printing numerous checks, or involve enormous volumes of data, such as inventory control in a multinational textile company. When these systems are automated, human involvement is minimal. For example, in an automated payroll system, there's little need for managerial judgment in the task of printing and sending checks, which reduces personnel costs.

4 Management Information Systems

 management information system (MIS) is an organized integration of hardware and software technologies, data, processes, and human elements designed to produce timely, integrated, relevant, accurate, and useful information for decision-making purposes.

The hardware components, discussed in more detail in Chapter 2, include input, output, and memory devices and vary depending on the application and the organization. MIS software, also covered in Chapter 2, can include commercial programs, software developed in-house, or both. The application or organization determines the type of software used. Processes are usually methods for performing a task in an MIS application. The human element includes users, programmers, systems analysts, and other technical personnel. This book emphasizes users of MISs.

In designing an MIS, the first task is to define the system's objectives clearly. Second, data must be collected and analyzed. Finally, information must be provided in a useful format for decision-making purposes.

Many MIS applications are used in both the private and public sectors. For example, an MIS for inventory

A **management information system (MIS)** is an organized integration of hardware and software technologies, data, processes, and human elements designed to produce timely, integrated, relevant, accurate, and useful information for decision-making purposes.

control provides data, such as how much of each product is on hand, what items have been ordered, and what items are back-ordered. Another MIS might forecast sales volume for the next fiscal period. This type of system uses recent historical data and mathematical or statistical models to generate the most accurate forecast, and sales managers can use this information for planning purposes. In the public sector, an MIS for a police department, for example, could provide information such as crime statistics, crime forecasts, and allocation of police units. Management can examine these statistics to spot increases and decreases in crime rates or types of crimes and analyze this data to determine future deployment of law enforcement personnel.

As you'll see in this book, many organizations use information systems to gain a competitive advantage. The information box on Hertz is one example of this use. (*Note:* MISs are often referred to as just *information systems,* and these terms are used interchangeably in this book.)

5 Major Components of an Information System

n addition to hardware, software, and human elements, an information system includes four major components, discussed in the following sections: data, a database, process, and information, as shown in Exhibit 1.3.[6]

Information Technology at The Hertz Corporation

Executives in the car rental business must be able to electronically sift through important information on a wide array of topics, such as cities, climates, holidays, business cycles, tourist activity, past promotions, and market forecasts. Examining this information helps executives make effective marketing decisions so they can compete in the car rental business.

To gain a competitive edge, Hertz uses a mainframe-based decision support system (DSS) and an executive information system (EIS) that includes tools for analyzing the massive amount of demographic data to make real-time marketing decisions. (DSSs and EISs are discussed in Chapter 12.)

With the EIS, Hertz executives can now analyze essential information from both external and internal sources. Internal sources include rental agreements, fleet purchases, computer reservation system reports, and airport reports comparing revenues for Hertz and other car-rental companies. In addition, Hertz managers can manipulate and refine data to make it more meaningful and use data for a variety of what-if analyses. According to Hertz executive Scott H. Meadow, using an EIS doesn't ensure prosperity, but "how you use it" does have an impact.[5]

If an organization has defined its strategic goals, objectives, and critical success factors, then structuring the data component to define what type of data is collected and in what form is usually easy.

5.1 Data

The **data** component of an information system is considered the input to the system. The information that users need affects the type of data that is collected and used. Generally, there are two sources of data: external and internal. An information system should collect data from both sources, although organizational objectives and the type of application also determine what sources to use. Internal data includes sales records, personnel records, and so forth. The following list shows some examples of external data sources:

> **Data** consists of raw facts and is a component of an information system.
> A **database** is a collection of all relevant data organized in a series of integrated files.

- Customers, competitors, and suppliers
- Government agencies and financial institutions
- Labor and population statistics
- Economic conditions

Typically, data has a time orientation, too. For example, past data is collected for performance reports, and current data is collected for operational reports. In addition, future data is predicted for budgets or cash flow reports. Data can also be collected in different forms, such as aggregated (reporting subtotals for categories of information, for example) or disaggregated (itemized lists, for instance). An organization might want disaggregated data to analyze sales by product, territory, or salesperson. Aggregated data can be useful for reporting overall performance during a particular sales quarter, for instance, but it limits the ability of decision makers to focus on specific factors.

If an organization has defined its strategic goals, objectives, and critical success factors, then structuring the data component to define what type of data is collected and in what form is usually easy. On the other hand, if there are conflicting goals and objectives or the company isn't aware of critical success factors, many problems in data collection can occur, which affects an information system's reliability and effectiveness.

5.2 Database

A **database**, the heart of an information system, is a collection of all relevant data organized in a series of integrated files. (You learn more about databases in Chapter 3.) A comprehensive database is essential for the success of any information system. To create,

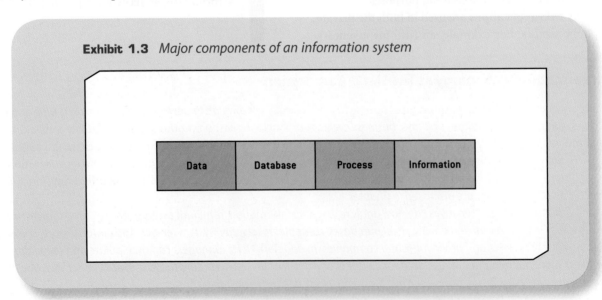

Exhibit 1.3 *Major components of an information system*

| Data | Database | Process | Information |

organize, and manage databases, a database management system (DBMS) is used, such as Microsoft Access or FileMaker Pro for home or small-office use. In a large organization, a DBMS such as Oracle or IBM DB2 might be used.

Databases are also important for reducing personnel time needed to gather, process, and interpret data manually. With a computerized database and a DBMS, data can be treated as a common resource that's easy to access and use.

5.3 Process

The purpose of an information system's **process** component is generating the most useful type of information for making decisions. This component generally includes transaction processing reports and models for decision analysis that can be built into the system or accessed from external sources.

An information system can include a wide range of models to support all levels of decision making. Users should be able to query an information system and generate a variety of reports. In addition, an information system should be able to grow with the organization so that users can redefine and restructure models and incorporate new information into their analyses.

5.4 Information

Although they might seem the same, data and information are different. Data consists of raw facts and by itself is difficult to use for making decisions. **Information**—the output of an information system—consists of facts that have been analyzed by the process component and, therefore, are more useful to the MIS user. For example, XYZ Company's total sales last month was $5,000,000. This number is data, because it doesn't tell you how the company performed. Did it meet the sales goal? Did sales increase or decrease from the previous month? How did the company perform against its top competitors? These questions and more can be answered by the information that an information system provides.

The quality of information is determined by its usefulness to users, and its usefulness determines the success of an information system. Information is useful if it enables decision makers to make the right decision in a timely manner. To be useful, information must have the following qualities:

- Timeliness
- Integration with other data and information
- Consistency and accuracy
- Relevance

If information lacks any of these qualities, the results are incorrect decisions, misallocation of resources, and overlooked windows of opportunity. If the system can't give users a minimum level of confidence in its reliability, it won't be used or users might dismiss the reports it generates. Information must provide either a base for users to explore different options or insight into tasks.

Another factor affecting the usefulness of information is the information system's user interface. Because this interface must be flexible and easy to use, most information systems make use of graphical user interfaces (GUIs), with features such as menus and buttons. To be useful, information systems should also produce information in different formats, including graphics (pie charts and bar graphs, for example), tables, and exception reports, which highlight data that is outside a specified range. Supplying information in a variety of formats increases the likelihood of users understanding and being able to use the information. Note that, in addition to the formal information that an information system generates, users need to be able to make use of informal information, such as rumors, unconfirmed reports, and stories, when solving problems.

The ultimate goal of an information system is to generate business intelligence (BI), described earlier in this chapter. As you'll learn throughout this book, many different tools, techniques, and types of information system technologies are used to generate BI.

5.5 Examples of Information Systems

To better understand the four main components of an information system, take a look at the following examples.

Example 1 A state university stores all student data in a database. The collected data includes each student's first name, last name, age, gender, major, nationality, and so forth. The process component of the information system performs all sorts of analysis on this data. For example, the university's DBMS has a built-in query capability that can generate the following information:

The **process** component of an information system generates the most useful type of information for decision making, including transaction processing reports and models for decision analysis.

Information consists of facts that have been analyzed by the process component and is an output of an information system.

© Blend Images/Jupiter Images

component of the information system conducts analysis on the data to provide the following information about the preceding month:

- Which salesperson generated the highest sales?
- Which product generated the highest sales? The lowest sales?
- Which region generated the highest sales?

Again, forecasting models can be used to generate predictions for the next sales period, and these predictions can be broken down by product, region, and salesperson. Based on this information, many decisions could be made, such as allocating the advertising budget to different products and regions.

- How many students are in each major?
- Which major is the fastest growing?
- What is the average student age?
- Among the international students, which country represents the highest number of students?
- What is the ratio of male to female students in each major?

Many other types of analysis can be done. A forecasting model (part of the process component) could be used to generate the estimated number of students for 2015, for instance. In addition, predictions could be made or improved, based on information this system provides. For example, knowing which major is the fastest growing can help with decisions on hiring faculty, and knowing the estimated number of students for 2015 can help with planning facilities.

Example 2 Teletech, an international textile company, uses a database to store data on products, suppliers, sales personnel, costs, and so forth. The process

Information technologies support information systems and use the Internet, computer networks, database systems, POS systems, and radio-frequency-identification (RFID) tags.

6 Using Information Systems and Information Technologies

nformation systems are designed to collect data, process the collected data, and deliver timely, relevant, and useful information that can be used for making decisions. To achieve this goal, an information system might use many different **information technologies**. For example, organizations often use the Internet as a worldwide

Information Technologies at The Home Depot

The Home Depot revolutionized the do-it-yourself home-improvement industry in the United States. Its stores use a POS system[7] for fast customer service and improved inventory management and a wireless network for efficient in-store communication. The Home Depot also has a Web site to communicate with customers and increase sales with online orders. It also uses RFID tags to better manage inventory and improve the efficiency of its supply chain network.

The Home Depot maintains a high-speed network connecting its stores throughout the United States and Canada and uses a data warehousing application to analyze variables affecting its success—customers, competitors, products, and so forth.[8] The information system gives The Home Depot a competitive advantage by gathering, analyzing, and using information to better serve customers and plan for customers' needs.

> Information systems are designed to collect data, process the collected data, and deliver timely, relevant, and useful information that can be used for making decisions.

network to communicate with one another. Computer networks (wired and wireless), database systems, POS systems, and radio-frequency-identification (RFID) tags are just a few examples of information technologies used to support information systems. The information box on The Home Depot gives you an idea of how companies use information technologies to stay competitive.

© Miodrag Gajic/Shutterstock.com

6.1 The Importance of Information Systems

Information is the second most important resource (after the human element) in any organization. Timely, relevant, and accurate information is a critical tool for enhancing a company's competitive position in the marketplace and managing the four *Ms* of resources: manpower, machinery, materials, and money.

To manage these resources, different types of information systems have been developed. Although all have the major components shown in Exhibit 1.3, they vary in the kind of data they collect and the analyses they perform. This section discusses some major types of information systems, focusing on the types of data and analysis used in each.

A personnel information system (PIS) or human resource information system (HRIS) is designed to provide information that helps decision makers in personnel carry out their tasks more effectively. Web technologies have played a major role in improving the efficiency and effectiveness of HR departments. For example, intranets are often used to provide

basic HR functions, such as employees checking how much vacation time they have left or looking up how much they have in their 401(k) plans. Intranets reduce personnel costs and speed up responses to common employee requests. As we will discuss in Chapter 7, an intranet is a network within an organization that

Information Technologies at UPS

Established in 1907, United Parcel Service (UPS; www.ups.com/content/us/en/about/index.html) is now a $49.7 billion global company and uses a sophisticated information system to manage the delivery of more than 14 million packages every day.[9] UPS uses several types of networks in its operations, particularly GPS and wireless networks.

To better serve customers, UPS has developed UPS Delivery Intercept, a Web-based service that allows customers to intercept and reroute packages before they are delivered, thus avoiding potentially costly mistakes and wasted time and costs. UPS calls the technology behind this service Package Flow Technology, which is also used to map efficient routes for drivers and mark packages for special handling. Kurt Kuehn, senior vice president of worldwide sales and marketing, says, "Innovations like Package Flow Technology and services like UPS Delivery Intercept are key components of UPS's drive to treat each of our millions of customers as if they're our only customer. We're constantly working on new and innovative ways to harness technology to help our customers meet their unique needs."[10]

uses Internet protocols and technologies for collecting, storing, and disseminating useful information that supports business activities such as sales, customer service, human resources, and marketing. The main difference between an intranet and the Internet is that intranets are private and the Internet is public. A PIS/HRIS supports the following actions, among others:

- Choosing the best job candidate
- Scheduling and assigning employees
- Predicting the organization's future personnel needs
- Providing reports and statistics on employee demographics
- Allocating human and financial resources

A logistics information system (LIS) is designed to reduce the cost of transporting materials while maintaining safe and reliable delivery. The following are a few examples of decisions supported by an LIS:

- Improving routing and delivery schedules
- Selecting the best modes of transportation
- Improving transportation budgeting
- Improving shipment planning

The information box on UPS shows uses of information systems and information technologies, particularly logistics information systems.

A manufacturing information system (MFIS) is used to manage manufacturing resources so that companies can reduce manufacturing costs, increase product quality, and make better inventory decisions. MFISs can perform many types of analysis with a high degree of timeliness and accuracy. For example, managers could use an MFIS to assess the effect on final product costs of a 7% increase in raw materials or determine how many assembly-line workers are needed to produce 200 automobiles in the next 3 weeks. Here are some decisions that an MFIS supports:

- Ordering decisions
- Product cost calculations
- Space utilization
- The bid evaluation process used with vendors and suppliers
- Analysis of price changes and discounts

The goal of a financial information system (FIS) is to provide information to financial executives in a timely manner. An FIS is used to support the following decisions, among others:

- Improving budget allocation
- Minimizing capital investment risks
- Monitoring cost trends
- Managing cash flows
- Determining portfolio structures

In addition, marketing information systems (MKISs) are used to improve marketing decisions. An effective MKIS should provide timely, accurate, and integrated information about the marketing mix (price, promotion, place, and product). Here are some decisions an MKIS supports:

- Analyzing market share, sales, and sales personnel
- Sales forecasting
- Price and cost analysis of items sold

Wal-Mart Stores, Inc.

Wal-Mart, the largest retailer in the world (http://walmartstores.com), built the Wal-Mart Satellite Network, which is the largest private satellite communication system in the United States. It links branch stores with the home office in Bentonville, Arkansas, by using two-way voice and data and one-way video communication. In addition to the POS systems used for many years, Wal-Mart uses the following information technologies to gain a competitive advantage:

- *Telecommunication is used to link stores with the central computer system and then to suppliers' computers. This system creates a seamless connection among all parties.*
- *Network technologies are used to manage inventory and implement a just-in-time inventory system. As a result, products and services can be offered at the lowest possible prices.*
- *Wal-Mart uses an extranet, called RetailLink, to communicate with suppliers. Suppliers can use this extranet to review product sales records in all stores and track current sales figures and inventory levels.[11] (Extranets are discussed in Chapter 7.)*
- *Electronic data interchange (EDI, discussed in Chapter 11) is used to streamline the order-invoice-payment cycle, reduce paperwork, and improve accuracy.*
- *Wal-Mart is a major user of RFID technologies, which have improved its supply chain and inventory management systems.*

6.2 Using Information Technologies for a Competitive Advantage

Michael Porter, a professor at Harvard Business School, identified three strategies for competing in the marketplace successfully:[12]

- Overall cost leadership
- Differentiation
- Focus

Information systems can help organizations reduce the cost of products and services and, if designed correctly, assist with differentiation and focus strategies, too. Throughout this book, you'll see many case examples of the cost savings that organizations have achieved with information systems and technologies. For example, Wal-Mart has been using overall cost leadership strategies successfully (see the information box).

Information technologies can help bottom-line and top-line strategies. The focus of a bottom-line strategy is improving efficiency by reducing overall costs. A top-line strategy focuses on generating new revenue by offering new products and services to customers or increasing revenue by selling existing products and services to new customers. For example, e-commerce businesses are adapting business models to reduce distribution costs dramatically. A good example is antivirus vendors using the Internet to distribute software. For a subscription fee of around $30, you can download the software and receive updates for a year. Without the Internet for easy, inexpensive distribution, vendors couldn't afford to offer software at such a low price.

As we will discuss in Chapter 11, many organizations use enterprise systems, such as supply chain management (SCM), customer relationship management (CRM), enterprise resource planning (ERP), and collaboration software, to reduce costs and improve customer service. The goal of these systems is to use information technologies to create the most efficient, effective link between suppliers and consumers. A successful CRM program, for example, helps improve customer service and create a long-term relationship between an organization and its customers.

For differentiation strategies, organizations try to make their products and services different from their competitors. Apple has been successful with this strategy by designing its computers to look much different from PCs and focusing on its computers' ease of use. As another example, Amazon.com has differentiated its Web site by using certain information technologies, such as personalization technologies (covered in more detail in Chapter 11) to recommend products to customers based on their previous purchases. Amazon.com also uses the one-click system for fast checkout. With this system, customers can enter credit card numbers and addresses once, and in subsequent visits simply click once to make a purchase, without having to enter information again.

With focus strategies, organizations concentrate on a specific market segment to achieve a cost or differentiation advantage. Apple has also used this strategy to target iPhones to consumer users rather than business users. Similarly, Macintosh computers are heavily marketed to creative professionals such as designers, photographers, and writers. As another example, Abercrombie & Fitch targets high-end clothing to low-income customers, such as teenagers and young adults, while Nordstrom targets their high-end clothing to high-income customers. Information technologies could assist these companies in reaching their target market segments more cost effectively.

Remember that focus and differentiation strategies work only up to a certain point. Customers are often willing to pay more for a unique product or service or one with a specific focus. However, cost still plays a major role. If a product or service becomes too expensive, customers might not be willing to purchase it.

Certain information technology tools, such as the Internet, have evened the playing field by giving customers more access to all sorts of data, such as being able to compare prices.

Exhibit 1.4 *The Five Forces Model*

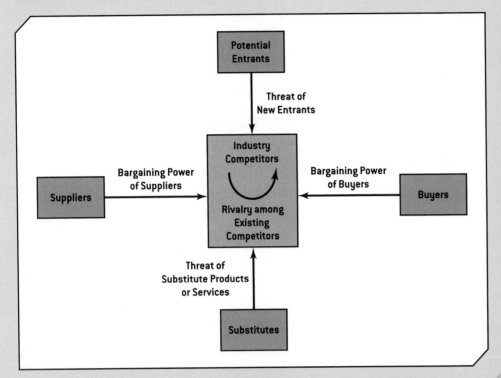

more competitive.[12] The five forces are as follows (see Exhibit 1.4):

- Buyer power
- Supplier power
- Threat of substitute products or services
- Threat of new entrants
- Rivalry among existing competitors

Buyer power is high when customers have many choices, low when they have few choices. Typically, organizations try to limit buyers' choices by offering services that make it difficult for customers to switch, which is essentially using a differentiation strategy. For example, Dell Computer was among the first to offer computer customization options to customers, and other computer manufacturers followed suit. Grocery stores, such as Sams Club, offer club cards that encourage customers

6.3 Porter's Five Forces Model: Understanding the Business Environment

Harvard Business School's Michael Porter created a comprehensive framework called the **Five Forces Model** for analyzing an organization, its position in the marketplace, and how information systems could be used to make it

> Michael Porter's **Five Forces Model** analyzes an organization, its position in the marketplace, and how information systems could be used to make it more competitive. The five forces include buyer power, supplier power, threat of substitute products or services, threat of new entrants, and rivalry among existing competitors.

Information Technology at Boeing

Boeing is the world's leading aerospace company and the largest manufacturer of commercial jetliners and military aircraft (www.boeing.com/companyoffices/aboutus/). Additionally, Boeing designs and manufactures electronic and defense systems, missiles, satellites, launch vehicles, and advanced information and communication systems.

Boeing uses intelligent information systems (discussed in Chapter 13) to stay ahead of the competition. These systems include artificial intelligence technologies such as neural networks, natural language processing, and expert systems that perform tasks usually carried out by human experts. Boeing uses these systems to develop and provide technology solutions and applications in information management, collaborative technologies, knowledge management, data and text mining, and natural language processing. For example, one solution is a tool for communication between vehicle diagnostic systems and technical support databases that is used to collect information sent to airplane mechanics. Boeing also uses these systems internally to support its 787 program, commercial aviation services, the Multi-Mission Maritime Aircraft program, and many military aircraft and space programs.[13]

to shop by giving them big discounts, an example of overall cost leadership strategies. Similarly, airlines and hotels offer free mileage and points when customers use their services. Information systems can make managing these strategies easier and more cost effective.

By using these strategies, organizations try to combat the threat of new entrants or substitute products by increasing customer loyalty. However, certain information technology tools, such as the Internet, have evened the playing field by giving customers more access to all sorts of data, such as being able to compare prices. This increased access to data increases buyers' bargaining power and decreases supplier power, discussed next.

Supplier power—the opposite of buyer power—is high when customers have fewer options and low when customers have more options. Organizations might use information systems to make their products and services cheaper or offer more services to distinguish themselves from competitors (again, another use of a differentiation strategy). Boeing, for example, uses information technologies to offer unique products and services, which increases its power in the marketplace (for a use of these focus strategies, see the information box on Boeing). In addition to information systems and technologies, organizations have other tools for increasing their power. For example, drug companies get patents for their products to reduce competition.

Threat of substitute products or services is high when many alternatives to an organization's products or services are available. Some organizations add services to make them more distinct in the marketplace, such as Amazon.com's personalized recommendations. Other organizations add fees to discourage customers from switching to a competitor, such as cell phone companies adding charges for switching to another provider before the customer contract is up.

Threat of new entrants is low when duplicating a company's product or service is difficult. Organizations often use focus strategies to ensure that this threat remains low. For example, developing a search engine that could compete successfully with Google would be difficult. In addition, organizations use information technologies to increase customer loyalty, as mentioned previously, which reduces the threat of new entrants. For instance, banks offer free bill paying to attract customers and keep them from switching to another bank; setting up a bill-paying service at another bank takes time that most customers don't want to spend. Similarly, after customizing their home pages with options offered by sites such as Yahoo! and Google, many users don't want to repeat this process at a new site.

Rivalry among existing competitors is high when many competitors occupy the same marketplace position; it is low when there are few competitors. For example, online brokerage firms operate in a highly competitive environment, so they use information technologies to make their services more unique.

7 The IT Job Market

uring the past decade, the IT job market has been one of the fastest growing segments in the economy, and it continues to be so. Even during the economic downturn, certain segments of the IT job market, such as Web design, infrastructure, and computer and network security, have shown growth compared to the rest of the job market. Currently, cloud computing-related jobs are in high demand (discussed in Chapter 14).[14] Broadly speaking, IT jobs fall into the following categories:

- Operations and help desk
- Programming
- Systems design
- Web design and Web hosting
- Network design and maintenance
- Database design and maintenance
- Robotics and artificial intelligence

Popular jobs in the information systems field are described in the following sections.[15]

7.1 CTO/CIO

The top information systems job belongs to either the chief technology officer (CTO) or the chief information officer (CIO). This person oversees long-range planning and keeps an eye on new developments in the field that can affect a company's success. Some organizations also have a chief privacy officer (CPO). This executive position includes responsibility for managing the risks and business impacts of privacy laws and policies.

7.2 Manager of Information Systems Services

This person is responsible for managing all the hardware, software, and personnel within the information systems department.

7.3 Systems Analyst

This person is responsible for the design and implementation of information systems. In addition to computer knowledge and an information systems background, this position also requires a thorough understanding of business systems and functional areas within a business organization.

7.4 Network Administrator

This person oversees a company's internal and external network systems, designing and implementing network systems that deliver correct information to the right decision maker in a timely manner. Providing network and cybersecurity is part of this position's responsibility.

7.5 Database Administrator

A database administrator (DBA) is responsible for database design and implementation. Additionally, a database administrator should have knowledge and understanding of data warehouses and data mining tools.

7.6 Computer Programmer

A computer programmer writes computer programs or software segments that allow the information system to perform a specific task. There are many computer languages available, and each one requires a specific knowledge suitable for a specific application.

7.7 Webmaster

A webmaster designs and maintains the organization's Web site. Because of the popularity of e-commerce applications, webmasters have been in high demand.

The educational backgrounds for an IT position can include an AA, BA, BS, MS, or MBA in information systems and related fields. The salaries vary based on educational background, experience, and the job's location. Salaries range from $52,000 for a programmer to over $180,000 for a CIO.

8 Future Outlooks

by examining various factors related to designing, implementing, and using information systems, the following predictions can be made:

- Hardware and software costs will continue to decline, so processing information will be less expensive. These cost savings should make information systems affordable for any organization, regardless of its size and financial status

Industry Connection

Microsoft Corporation*

Microsoft, founded in 1975, is the world's largest software company and is involved in all aspects of desktop computing. It's best known for the Disk Operating System (DOS), Windows operating systems, and office software suites such as Office. Here are some of the products and services Microsoft offers:

- Windows—The most popular operating system for PCs and PC-compatible computers
- Windows XP, Windows Vista, and Windows 7—Three widely used OSs for PCs
- Windows Server 2003 and Server 2008—Widely used server operating systems used in network environments
- Office—The most widely used office suite; includes Word, Excel, Access, and PowerPoint
- Internet Explorer—A popular Web browser
- Expression Web (replacing FrontPage)—An HTML editor and Web design program for developing Web pages and other HTML applications
- MSN—An Internet portal combining Web services and free Web-based e-mail (Hotmail)

- SharePoint Server—Microsoft's groupware for facilitating information sharing and content management
- SQL Server 2008—A widely used database management system
- Xbox—A video game system
- Visual Studio—An integrated development environment (IDE) that can be used to program applications in a number of different languages (such as C++, Java, Visual Basic, and C#); used for console or GUI applications as well as Web applications
- Zune—A portable media player and software; includes Zune Marketplace, which provides online music, video, and podcast downloads
- Windows Live ID—A single sign-on service for multiple Web sites

*This information has been gathered from the Microsoft Web site and other promotional materials. For detailed information and updates, visit *www.microsoft.com.*

- Artificial intelligence and related technologies will continue to improve and expand, which will have an impact on information systems. For example, further development in natural language processing should make information systems easier to use.

- The computer literacy of typical information system users will improve, as computer basics are taught more in elementary schools.

- Networking technology will improve, so connecting computers will be easier, and sending information from one location to another will be faster. Compatibility issues between networks will become more manageable, and integrating voice, data, and images on the same transmission medium will improve communication quality and information delivery.

- Personal computers will continue to improve in power and quality, so that most information system software will be able to run on them without problems. This trend should make information systems more affordable, easier to maintain, and more appealing to organizations.

- Internet growth will continue, which will put small and large organizations on the same footing, regardless of their financial status. Internet growth will also make e-collaboration easier, despite geographical distances.

- Computer criminals will become more sophisticated, and protecting personal information will become more difficult.

The Industry Connection highlights Microsoft and its products and services.

9 Chapter Summary

In this chapter, you have seen examples of how computers and information systems are used today, you've learned the difference between computer literacy and information literacy, and you've learned that transaction processing systems are one of the earliest applications of information technology. You have learned what a management information system (MIS) is and its major components: data, a database, process, and information. You have also been given an overview of how information systems and information technologies are used in different areas of business and learned how companies can use Michael Porter's three competitive strategies and his Five Forces Model to gain a competitive advantage. Finally, you learned about IT jobs and the future of information systems.

Key Terms

business intelligence (BI) (6)

computer literacy (6)

data (8)

database (8)

Five Forces Model (14)

information (9)

information literacy (6)

information technologies (10)

management information system (MIS) (7)

process (9)

transaction processing system (TPS) (6)

Problems, Activities, and Discussions

1. What are some examples of information technologies used for telecommuting?

2. List two examples of decisions that could be improved by a personnel/human resources information system.

3. List some examples of external data used by an information system.

4. What are some examples of job titles in the information systems field?

5. What are some IT jobs that are in high demand even during economic downturns?

6. What should you do in order to reduce risk and improve the privacy of your personal data when using social networks?

7. After reading the information at the following links and from other sources, write a one-page paper that explains how YouTube can be used as a marketing tool for promoting a company's products and services.

   www.itbusinessedge.com/cm/blogs/all/companies-putting-youtube-to-work/?cs=11503

   http://videos.webpronews.com/2007/05/17/answers-matt-youtube/

   http://nybw.businessweek.com/smallbiz/tips/archives/2008/10/using_youtube_to_promote_your_business.html

8. Log on to the IBM Web site at *www-01.ibm.com/software/success/cssdb.nsf/CategoryL1ViewFM?ReadForm&Site=dmmain_industryL1VW&cty=en_us.* This Web site lists several information systems and information management case studies used in industries and government agencies. Review some of the industries in which you're interested in seeking employment. Do you see similarities among these information systems?

9. Buyer power is high when customers have many choices and low when they have few choices. True or False?

10. A typical information system includes which of the following components? (Choose all that apply.)

    a. Data entry system

    b. Database

    c. Process

    d. Information

casestudy

USING INFORMATION TECHNOLOGIES AT FEDERAL EXPRESS

Federal Express, founded in 1971, handles an average of 3 million package-tracking requests every day (*http://about.fedex.designcdt.com*). To stay ahead in a highly competitive industry, the company focuses on customer service by maintaining a comprehensive Web site, *www.FedEx.com*, to assist customers and reduce costs. For example, every request for information that is handled at the Web site instead of by the call center saves an estimated $1.87. Federal Express has reported that customer calls have decreased by 83,000 per day since 2000, which saves the company $57.56 million per year. And since each package-tracking request costs Federal Express 3 cents, costs have been reduced from more than $1.36 billion per year to $21.6 million per year by customers using the Web site instead of the call center.

Another technology that improves customer service is Ship Manager, an application installed on customers' sites so that users can weigh packages, determine shipping charges, and print shipping labels. Customers can also link their invoicing, billing, accounting, and inventory systems to Ship Manager.[16]

© Denise Kappa/Shutterstock.com

However, Federal Express still spends almost $326 million per year on its call center to reduce customers' frustration when the Web site is down or when customers have difficulty using it. Federal Express uses customer relationship management software called Clarify in its call centers to make customer service representatives' jobs easier and more efficient and to speed up response time.[17]

Answer the following questions:

1. Is technology by itself enough to ensure high-quality customer service?

2. What are Federal Express's estimated annual savings from using information technology?

3. What are a couple of examples of information technologies used by Federal Express?

REVIEW!

HE DID

MIS2 puts a multitude of study aids at your fingertips. After reading the chapters, check out these resources for further help:

- **Review Cards,** found in the back of your book, include all learning outcomes, key terms and definitions, and visual summaries for each chapter.

- **Online Printable Flash Cards** give you additional ways to check your comprehension of key Management Information Systems concepts.

- Other great tools to help you review include **interactive games, videos, and online tutorial quizzes.**

Go to CourseMate for MIS2 to find plenty of resources to help you *Review!* Access at login.cengagebrain.com.

COMPUTERS: THE MACHINES BEHIND COMPUTING

In this chapter, you will learn about the major components of a computer and what factors distinguish computing power. We will review a brief history of computer hardware and software and provide an overview of computer operations. We will also go into more detail on specific computer components: input, output, and memory devices. You will learn how computers are classified, based on size, speed, and sophistication, and about the two major types of software—system software and application software—and the five generations of computer languages.

learning outcomes

After studying this chapter you should be able to:

LO1 Define a computer system and describe its components.

LO2 Discuss the history of computer hardware and software.

LO3 Explain the factors distinguishing the computing power of computers.

LO4 Summarize the binary system and data representation.

LO5 Discuss the types of input, output, and memory devices.

LO6 Explain how computers are classified.

LO7 Describe the two major types of software.

LO8 List the generations of computer languages.

If airplanes had developed as computers have developed, today you could go around the globe in less than 20 minutes for just 50 cents.

1 Defining a Computer

If airplanes had developed as computers have developed, today you could go around the globe in less than 20 minutes for just 50 cents. Computers have gone through drastic changes in a short time. For example, a computer that weighed more than 18 tons 60 years ago has been replaced by one that now weighs less than two pounds. Today's computer is 100 times more powerful and costs less than one percent of the 60-year-old computer.

As you learned in Chapter 1, you use computers every day for a multitude of purposes. You even use them indirectly when you use appliances with embedded computers, such as TVs and microwaves. Computers have become so ubiquitous, in fact, that a cashless and checkless society is likely just around the corner. Similarly, computers might eliminate the need for business travel. Even now, executives seldom need to leave their offices for meetings in other locations because of technologies such as computer conferencing and telepresence systems.

Computers are used in a wide variety of tasks, including report distribution in businesses, rocket guidance control in the NASA space program, and DNA analysis in medical research. This book could not have been published without the use of computers. The text was typed

A **computer** is a machine that accepts data as input, processes data without human intervention by using stored instructions, and outputs information.

and revised with word-processing software, and composition software was used to typeset the pages. Printing, warehousing, inventory control, and shipping were accomplished with the help of computers.

© Yellowj/Shutterstock.com

So what is a computer? Many definitions are possible, but in this book, a **computer** is defined as a machine that accepts data as input, processes data without human intervention by using stored instructions, and outputs information. The instructions, also called a "program," are step-by-step directions for performing a specific task, written in a language the computer can understand. Remember that a computer only processes data (raw facts); it can't change or correct the data that's entered. If data is erroneous, the information the computer provides is also erroneous. This rule is sometimes called GIGO: garbage in, garbage out.

To write a computer program, first you must know what needs to be done, and then you must plan a method to achieve this goal, including selecting the right language for the task. Many computer languages are available—the language you select depends on the problem being solved and the type of computer you're using. Regardless of the language, a program is also referred to as the "source code." This source code must be translated into object code—consisting of binary 0s and 1s. Binary code is a set of instructions used to control the computer, and

uses 0s and 1s, which the computer understands as on or off signals. You will learn more about the binary system and computer languages later in this chapter.

1.1 Components of a Computer System

A computer system consists of hardware and software. Hardware components are physical devices, such as keyboards, monitors, and processing units. The software component consists of programs written in computer languages.

Exhibit 2.1 shows the building blocks of a computer. Input devices, such as keyboards, are used to send data and information to the computer. Output devices,

Exhibit 2.1 *The building blocks of a computer*

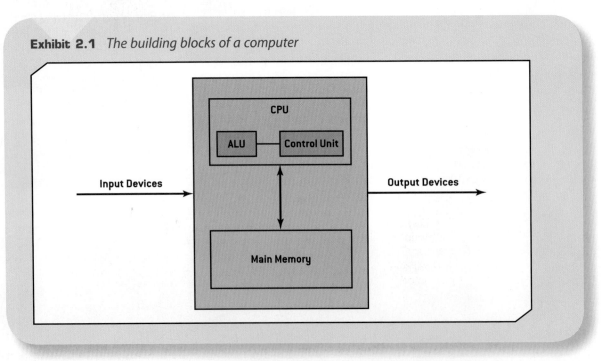

such as monitors and printers, display the output a computer generates.

Main (primary) memory is where computers store data and instructions, similar to a human brain. The **central processing unit (CPU)** is the heart of a computer. It's divided into two components: the **arithmetic logic unit (ALU)** and the **control unit**. The ALU performs arithmetic operations (+, −, *, /) and comparison or relational operations (<, >, =), which are used to compare numbers. The control unit tells the computer what to do, such as instructing the computer which device to read or send output to.

Some computers have a single processor; other computers, called "multiprocessors," contain multiple processors. Multiprocessing is the use of two or more CPUs in a single computer system. Generally, a multiprocessor computer has better performance than a single-processor computer in the same way that a team would have better performance than an individual on a large, time-consuming project. Some computers use a dual-core processor, which is essentially two processors in one, to improve processing power. Dual-core processors are common in new PCs and Apple computers.

Another component that affects computer performance is a **bus**, which is the link between devices connected to the computer. A bus can be parallel or serial, internal (local) or external. An internal bus enables communication between internal components, such as a video card and memory, and an external bus is capable of communicating with external components, such as a USB device.

Other factors that affect computer performance include the processor size and the operating system (OS). In recent years, 32-bit and 64-bit processors and OSs have created a lot of interest. A 32-bit processor can use 2^{32} bytes (4 GB) of RAM; and, in theory, a 64-bit processor can use 2^{64} bytes (16 EB, or exabytes) of RAM. So a computer with a 64-bit processor can perform calculations with larger numbers and be more efficient with smaller numbers; and overall, it has better performance than a 32-bit system. However, to take advantage of this higher performance, you must also have a 64-bit OS.

Exhibit 2.2 shows additional components of a computer system.

> The **central processing unit (CPU)** is the heart of a computer. It's divided into two components: the arithmetic logic unit (ALU) and the control unit.
>
> The **arithmetic logic unit (ALU)** performs arithmetic operations (+, −, *, /), and comparison or relational operations (<, >, =) are used to compare numbers.
>
> The **control unit** tells the computer what to do, such as instructing the computer which device to read or send output to.
>
> A **bus** is a link between devices connected to the computer. It can be parallel or serial, internal (local) or external.
>
> A **disk drive** is a peripheral devise for recording, storing, and retrieving information.
>
> A **CPU case** is also known as a computer chassis or tower. It is the enclosure containing the computer's main components.
>
> A **motherboard** is the main circuit board containing connectors for attaching additional boards. It usually contains the CPU, Basic Input/Output System (BIOS), memory, storage interfaces, serial and parallel ports, expansion slots, and all the controllers for standard peripheral devices, such as the display monitor, disk drive, and keyboard.

A **disk drive** is a peripheral device for recording, storing, and retrieving information. A **CPU case** (also known as a computer chassis or tower) is the enclosure containing the computer's main components. A **motherboard** is the main circuit board containing connectors for attaching additional boards. In addition, it usually contains the CPU, Basic Input/Output System (BIOS), memory, storage, interfaces, serial and parallel ports, expansion slots, and all the controllers for standard peripheral devices, such as the display monitor, disk drive, and keyboard.

Exhibit 2.2 *Components of a computer system*

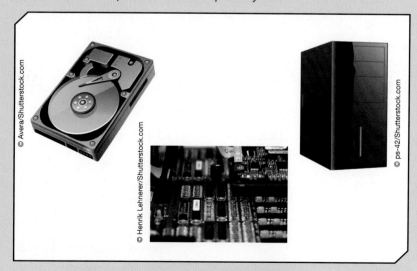

© Avera/Shutterstock.com
© Henrik Lehnerer/Shutterstock.com
© ps-42/Shutterstock.com

A serial port is a communication interface through which information is transferred one bit at a time; a parallel port is an interface between a computer and a printer, and the computer transfers multiple bits of information to the printer simultaneously.

2 The History of Computer Hardware and Software

ajor developments in hardware have taken place over the past 60 years. To make these developments more clear, computers are often categorized into "generations" to mark technological breakthroughs. Beginning in the 1940s, first-generation computers used vacuum tube technology. They were bulky and unreliable, generated excessive heat, and were difficult to program. Second-generation computers used transistors and were faster, more reliable, and easier to program and maintain. Third-generation computers operated on integrated circuits, which enabled computers to be even smaller, faster, more reliable, and more sophisticated. Remote data entry and telecommunications were introduced during this generation. Fourth-generation computers continued several trends that further improved speed and ease of use: miniaturization, very large scale integration (VLSI) circuits, widespread use of personal computers, and optical discs (discs written or encoded and read using a laser optical device). The current fifth-generation computers include parallel-processing (computers containing hundreds or thousands of CPUs for rapid data processing), gallium arsenide chips that run at higher speeds and consume less power than silicon chips, and optical technologies. Table 2.1 summarizes these hardware generations.

Because silicon can't emit light and has speed limitations, computer designers have concentrated on technology using gallium arsenide, in which electrons move almost five times faster than in silicon. Devices made with this synthetic compound can emit light, withstand higher temperatures, and survive much higher doses of radiation than silicon devices. The major problems with gallium arsenide are difficulties in mass production. This material is softer and more fragile than silicon, so it breaks more easily during slicing and polishing. Because of the high costs and difficulty of production, the military is currently the major user of this technology. However, research continues to eliminate some shortcomings of this technology.

The field of optical technologies involves the applications and properties of light, including its interactions with lasers, fiber optics, telescopes, and so forth. These technologies offer faster processing speed, parallelism (several thousand light beams can pass through an ordinary device), and interconnection—much denser arrays of interconnections are possible because light rays don't affect each other. Optical computing is in its infancy, and more research is needed to produce a full-featured optical computer. Nevertheless, storage devices using this technology are revolutionizing the computer field by enabling massive amounts of data to be stored in very small spaces.

Computer languages and software have also developed through five generations. They are discussed in more detail in the "Computer Languages" section, but Table 2.2 summarizes these generations.

Table 2.1

Hardware generations

Generation	Date	Major technologies	Example
First	1946–1956	Vacuum tube	ENIAC
Second	1957–1963	Transistors	IBM 7094, 1401
Third	1964–1970	Integrated circuits, remote data entry, telecommunications	IBM 360, 370
Fourth	1971–1992	Miniaturization, VLSI, personal computers, optical discs	Cray XMP, Cray II
Fifth	1993–present	Parallel processing, gallium arsenide chips, optical technologies	IBM System z10

Table 2.2

Computer language trends

Generation	Major attribute
First	Machine language
Second	Assembly language
Third	High-level language
Fourth	Fourth-generation language
Fifth	Natural language processing (NLP)

3 The Power of Computers

omputers draw their power from three factors that far exceed human capacities: speed, accuracy, and storage and retrieval capabilities; all three factors are discussed in the following sections.

3.1 Speed

Computers process data with amazing speed. They are capable of responding to requests faster than humans can, which improves efficiency. Today's high-speed computers make it possible for knowledge workers to perform tasks much faster than with the slower computers of the past. Typically, computer speed is measured as the number of instructions performed during the following fractions of a second:

- Millisecond: 1/1,000 of a second
- Microsecond: 1/1,000,000 of a second
- Nanosecond: 1/1,000,000,000 of a second
- Picosecond: 1/1,000,000,000,000 of a second

3.2 Accuracy

Unlike humans, computers don't make mistakes. To understand computer accuracy more clearly, take a look at these two numbers:

4.000000000000000000000001
4.000000000000000000000002

To humans, these two numbers are so close that they are usually considered equal. To a computer, however, these two numbers are completely different. This degree of accuracy is critical in many computer applications. On a space mission, for example, computers are essential for calculating reentry times and locations for space shuttles. A small degree of inaccuracy could lead the space shuttle to land in Canada instead of the United States.

3.3 Storage and Retrieval

Storage means saving data in computer memory, and retrieval is accessing data from memory. Computers can store vast quantities of data and locate a specific item quickly, which makes knowledge workers more efficient in performing their jobs.

In computers, data is stored in bits. A bit is a single value of *0* or *1*, and 8 bits equal 1 byte. A byte is the size of a character. For example, the word *computer* consists of 8 characters or 8 bytes (64 bits). Table 2.3 shows storage measurements.

Table 2.3

Storage measurements (approximations)

1 bit	A single value of 0 or 1
8 bits	1 byte or character
2^{10} bytes	1000 bytes, or 1 kilobyte (KB)
2^{20} bytes	1,000,000 bytes, or 1 megabyte (MB)
2^{30} bytes	1,000,000,000 bytes, or 1 gigabyte (GB)
2^{40} bytes	1,000,000,000,000 bytes, or 1 terabyte (TB)
2^{50} bytes	1,000,000,000,000,000 bytes, or 1 petabyte (PB)
2^{60} bytes	1,000,000,000,000,000,000 bytes, or 1 exabyte (EB)

Computers draw their power from three factors that far exceed human capacities: speed, accuracy, and storage and retrieval capabilities.

> You can store the text of more than one million books in a memory device about the size of your fist.

Every character, number, or symbol on the keyboard is represented as a binary number in computer memory. A binary system consists of *0*s and *1*s, with a *1* representing "on" and a *0* representing "off," similar to a light switch.

Computers and communication systems use data codes to represent and transfer data between computers and network systems. The most common data code for text files, PC applications, and the Internet is American Standard Code for Information Interchange (ASCII), developed by the American National Standards Institute. In an ASCII file, each alphabetic, numeric, or special character is represented with a 7-bit binary number (a string of *0*s or *1*s). Up to 128 (2^7) characters can be defined. There are two additional data codes used by many operating systems: Unicode and Extended ASCII. Unicode is capable of representing 256 (2^8) characters, and Extended ASCII is an 8-bit code that also allows representing 256 characters.

Before the ASCII format, IBM's Extended Binary Coded Decimal Interchange Code (EBCDIC) was popular. In an EBCDIC file, each alphabetic, numeric, or special character is represented with an 8-bit binary number.

4 Computer Operations

Computers can perform three basic tasks: arithmetic operations, logical operations, and storage and retrieval operations. All other tasks are carried out by one or a combination of these operations. For example, playing a computer game could be a combination of all three operations. During a game, your computer may perform calculations to reach to a point, it may compare two numbers, and it may perform storage and retrieval functions for going forward with the process.

Computers can add, subtract, multiply, divide, and raise numbers to a power (exponentiation), as shown in these examples:

> **Input devices** send data and information to the computer. Examples include keyboards and mouses.

Spotlight on Computer Storage

IBM has developed a new storage technology called Millipede, which allows the storing of one trillion bits of data per square inch, about 20 times more than magnetic disks.[1] In other words, you could store 600,000 digital photos, 127,000 MP3 files, or 1,500,000 books on a device the size of a postage stamp! You could buy a storage device of 1 TB for less than $400. This process uses thousands of very fine silicon tips to punch holes into a thin film of plastic. The tiny holes represent bits of data. The technology is considered nanotechnology (discussed in Chapter 14) because it is at an atomic level.

A + B (addition)	5 + 7 = 12
A − B (subtraction)	5 − 2 = 3
A * B (multiplication)	5 * 2 = 10
A / B (division)	5 / 2 = 2.5
A ^ B (exponentiation)	5 ^ 2 = 25

Computers can perform comparison operations by comparing two numbers. For example, a computer can compare A and B to determine which number is larger.

Computers can store massive amounts of data in very small spaces and locate a particular item quickly. For example, you can store the text of more than one million books in a memory device about the size of your fist. Later in this chapter, you will learn about different storage media, such as magnetic disks and tape.

5 Input, Output, and Memory Devices

5.1 Input Devices

Input devices send data and information to the computer. The following input devices are constantly being improved to make data input easier.

- *Keyboards* are the most widely used input devices. Originally, they were designed to resemble typewriters, but several modifications have been made to improve their ease of use. For example, most keyboards include control keys, arrow keys, function keys, and other special keys. In addition, some keyboards, such as the split keyboard, have been developed for better ergo-

nomics. You can perform most computer input tasks with keyboards, but for some tasks, a scanner or mouse is faster and more accurate.

- *A mouse* is a pointing device for moving the cursor on the screen and allows fast, precise cursor positioning. With programs that use graphical interfaces, such as Microsoft Windows or Mac OS, the mouse has become the input device of choice.

- *Touch screens*, which usually work with menus, are actually a combination of input devices. Some touch screens rely on light detection to determine which menu item has been selected, and others are pressure sensitive. Touch screens are often easier to use than keyboards, but they might not be as accurate because selections can be misread. You probably saw touch screens used extensively during the 2008 presidential election to show electoral maps and analyze election data in different ways quickly.

- *Light pens* are connected to the monitor with a cable. When the pen is placed on an on-screen location, the data in this spot is sent to the computer. The data can be characters, lines, or blocks. A light pen is easy to use, inexpensive, and accurate, and it is particularly useful for engineers and graphic designers who need to make modifications to technical drawings.

- *A trackball* is kept in a stationary location, but it can be rolled on its axis to control the on-screen cursor. Track-balls occupy less space than a mouse, so they are ideal for notebook computers. However, positioning with a trackball is sometimes less precise than with a mouse.

- *A data tablet* consists of a small pad and a pen. Menus are displayed on the tablet, and you make selections with the pen. Data tablets are used most widely in computer-aided design and manufacturing applications.

- *Barcode readers* are optical scanners that use lasers to read codes in bar form. These devices are fast and accurate and have many applications in inventory, data entry, and tracking systems. They are used mostly with UPC systems in retail stores.

- *Optical character readers (OCRs)* work on the same principle as barcode readers but read text. They must recognize many special characters and distinguish between uppercase and lowercase letters, so using an OCR is more difficult than using a barcode reader. Nevertheless, OCRs have been used successfully in many applications and are improving steadily. The United States Postal Service uses OCRs to sort mail.

- *Magnetic ink character recognition (MICR) systems* read characters printed with magnetic ink and are used primarily by banks for reading the information at the bottom of checks.

- *Optical mark recognition (OMR) systems* are sometimes called "mark sensing" systems because the machine reads marks on paper. OMRs are often used to grade multiple-choice and true/false tests.

5.2 Output Devices

Many **output devices** are available for both mainframes and personal computers. Output displayed on a screen is called "soft copy." The most common output devices for soft copy are cathode ray tube (CRT), plasma display, and liquid crystal display (LCD).

The other type of output is "hard copy," for which the most common output device is a printer. Inkjet and laser printers are standard printers used today. Inkjet printers produce characters by projecting electrically charged droplets of ink onto paper that create an image. High-quality inkjet printers use multicolor ink cartridges for near-

> An **output device** is capable of representing information from a computer. The form of this output might be visual, audio, or digital and examples include printers, display monitors, and plotters.

photo quality output, and are often used to print digital photographs. Inkjet printers are suitable for home users who have limited text and photo printing needs. When selecting a printer, consider cost (initial and maintenance), quality, speed, space, and networking facilities.

Laser printers use laser-based technology that creates electrical charges on a rotating drum to attract toner. The toner is fused to paper using a heat process that creates high-quality output. Laser printers are better suited to larger office environments with high-volume and high-quality printing requirements. Other output devices include plotters for converting computer output to graphics, and voice synthesizers for converting computer output to voice. Voice synthesis has become common. Cash registers at grocery stores use it to repeat item prices. When you call directory assistance, the number you hear is probably computer generated. Voice output is also being used for marketing. A computer can dial a long list of phone numbers and deliver a message. If a number is busy, the computer makes a note and dials it later. Although the value of this method has been questioned, it is positively used in some political campaigns to deliver messages about voting.

 ## 5.3 Memory Devices

Two types of memory are common to any computer: main memory and secondary memory. **Main memory** stores data and information and is usually volatile, meaning its contents are lost when electrical power is turned off. **Secondary memory**, which is nonvolatile, holds data when the computer is off or during the course of a program's operation. It also serves as archival storage. Main memory plays a major role in a computer's performance; to some extent, the more memory a computer has, the faster and more efficient its input/output (I/O) operations are. Graphics cards, also called video adapters, enhance computer performance, too. High-end graphics cards are important for graphic designers, who need fast rendering

---

Main memory stores data and information and is usually volatile; its contents are lost when electrical power is turned off. It plays a major role in a computer's performance.

Secondary memory, which is nonvolatile, holds data when the computer is off or during the course of a program's operation. It also serves as archival storage.

Random access memory (RAM) is volatile memory, in which data can be read from and written to; it is also called read-write memory.

Cache RAM resides on the processor. Because memory access from main RAM storage takes several clock cycles (a few nanoseconds), cache RAM stores recently accessed memory so that the processor isn't waiting for the memory transfer.

Read-only memory (ROM) is nonvolatile; data can't be written to ROM.

---

of 3D images. Many video games also require high-end graphics cards for the best display.

5.3.1 Main Memory Devices

The most common type of main memory is semiconductor memory chips made of silicon. A semiconductor memory device can be volatile or nonvolatile. Volatile memory is called **random access memory (RAM)**, although you could think of it as "read-write memory." In other words, data can be read from and written to RAM. Some examples of the type of information stored in RAM include open files, the Clipboard's contents, running programs, and so forth.

A special type of RAM, called **cache RAM**, resides on the processor. Because memory access from main RAM storage generally takes several clock cycles (a few nanoseconds), cache RAM stores recently accessed memory so that the processor isn't waiting for the memory transfer.

Nonvolatile memory is called **read-only memory (ROM)**: data can not be written to ROM. The type of data usually stored in ROM includes BIOS information and the computer system's clock. There are two other types of ROM. Programmable read-only memory (PROM) is a type of ROM chip that can be programmed with a special device. However, after it has been programmed, the contents can't be erased. Erasable programmable read-only memory (EPROM) is similar to PROM, but its contents can be erased and reprogrammed.

5.3.2 Secondary Memory Devices

Secondary memory devices are nonvolatile and used for storing large volumes of data for long periods. As mentioned earlier, they can also hold data when the computer is off or during the course of a program's operation. There are three main types: magnetic disks, magnetic tape, and optical discs. Large enterprises also use storage area networks and network-attached storage (discussed in the next section) for storing massive amounts of data in a network environment.

A **magnetic disk**, made of Mylar or metal, is used for random-access processing. In other words, data can be accessed in any order, regardless of its order on the surface. Magnetic disks are much faster but more expensive than tape devices. Exhibit 2.3 shows both types of magnetic memory devices.

Magnetic tape, made of a plastic material, resembles a cassette tape and stores data sequentially. Records can be stored in a block or separately, with a gap between each record or block called the inter-record gap (IRG). Magnetic tape is sometimes used for storing backups, although other media are more common now. Exhibit 2.3 shows both types of magnetic memory devices.

Optical discs use laser beams to access and store data. Optical technology can store vast amounts of data and is durable. Three common types of optical storage are CD-ROMs, WORM discs, and DVDs.

Compact disc read-only memory (CD-ROM), as the name implies, is a read-only medium. CD-ROMs are easy to duplicate and distribute and are widely used in large permanent databases, such as for libraries, real estate firms, and financial institutions. They are sometimes used for multimedia applications and to distribute software products. However, because of their larger capacity—a minimum of 4.7 GB—digital versatile disc read-only memory (DVD-ROMs) are used more often now, particularly for software distribution.

A write once, read many (WORM) disc is also a permanent device. Information can be recorded once and can't be altered. A major drawback is that a WORM disc can't be duplicated. It is used mainly to store information that must be kept permanently but not altered—for example, annual reports and information for nuclear power plants, airports, and railroads. SanDisk recently announced a Secure Digital (SD) card, a type of WORM disc, which can store data for 100 years but can be written on only once.[2]

Unlike with CD-ROMs and WORM discs, information stored on an erasable optical disc can be erased and altered repeatedly. These discs are used when high-volume storage and updating are essential.

Other secondary memory devices include hard disks, USB flash drives, and memory cards (see Exhibit 2.4). Hard disks come in a variety of sizes and can be internal or external, and their costs have been decreasing steadily. Memory sticks have become popular because of their small size, high storage

A **magnetic disk**, made of Mylar or metal, is used for random-access processing. In other words, data can be accessed in any order, regardless of its order on the surface.

Magnetic tape, made of a plastic material, resembles a cassette tape and stores data sequentially.

Optical discs use laser beams to access and store data. Examples include CD-ROMs, WORM discs, and DVDs.

Exhibit 2.3 *Magnetic memory devices*

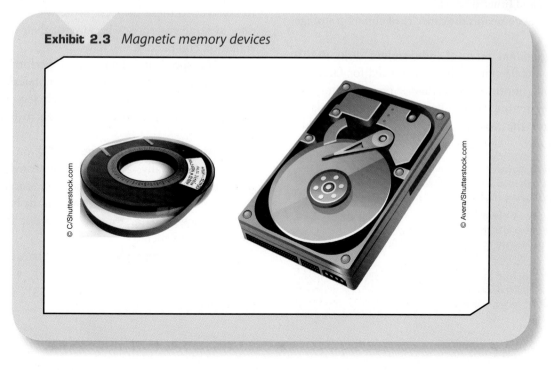

© C/Shutterstock.com

© Avera/Shutterstock.com

Exhibit 2.4 *Examples of memory devices*

© Timurpix/Shutterstock.com © Martin Petransky/Shutterstock.com © Juba Sompinmaki/Shutterstock.com

Table 2.4

Capacity of secondary memory devices

Device	Storage capacity
Memory stick	16 GB
Hard disk	2 TB
CD-ROM, CD-R, CD-RW	800 MB
DVD-ROM, DVD-R, DVD-RW	4.7 GB or more
Blu-Ray (next generation optical disc)	Up to 25 GB on a single-layer disc and 50 GB on a dual-layer disc

capacity, and decreasing cost. Flash memory is nonvolatile memory that can be electronically erased and reprogrammed. It is used mostly in memory cards and USB flash drives for storing and transferring data between computers and other devices.

Table 2.4 compares the capacity of common storage devices.

A **redundant array of independent disks (RAID) system** is a collection of disk drives used for fault tolerance and improved performance, typically in large network systems. Data can be stored in multiple places to improve the system's reliability. In other words,

---

A **redundant array of independent disks (RAID) system** is a collection of disk drives used for fault tolerance and improved performance, typically in large network systems.

A **storage area network (SAN)** is a dedicated high-speed network consisting of both hardware and software used to connect and manage shared storage devices, such as disk arrays, tape libraries, and optical storage devices.

Network-attached storage (NAS) is essentially a network-connected computer dedicated to providing file-based data storage services to other network devices.

---

if one disk in the array fails, data isn't lost. In some RAID configurations, sequences of data can be read from multiple disks simultaneously, which improves performance.

5.3.3 Storage Area Networks and Network-Attached Storage

A **storage area network (SAN)** is a dedicated high-speed network consisting of both hardware and software used to connect and manage shared storage devices, such as disk arrays, tape libraries, and optical storage devices. A SAN network makes storage devices available to all servers on a local area network (LAN) or wide area network (WAN). (LANs and WANs will be discussed in Chapter 6.) Because a SAN is a dedicated network, servers can access storage devices more quickly and don't have to use their processing power to connect to these devices. Typically, SANs are used only in large enterprises because of their cost and installation complexity.

SANs speed up data access performance, and despite their cost, they're more economical than having storage devices attached to each server. A SAN's capacity can be extended easily, even to hundreds of terabytes.

Network-attached storage (NAS), on the other hand, is essentially a network-connected computer dedicated to providing file-based data storage services to other network devices. Software on the NAS handles features such as data storage, file access, and file and storage management.

When choosing a SAN or NAS system, consider the following factors:

- Are hybrid solutions (combining a SAN and a NAS) available?

> A notebook computer today has more power than a mainframe of the 1970s, and all indications suggest that this trend will continue.

- A SAN offers only storage; a NAS system offers both storage and file services.
- NAS is popular for Web servers and e-mail servers because it lowers management costs and helps make these servers more fault tolerant. It's also becoming a useful solution for providing large amounts of heterogeneous data (text, documents, voice, images, movies, and so forth) for consumer applications.
- The biggest issue with NAS is that, as the number of users increases, its performance deteriorates. However, it can be expanded easily by adding more servers or upgrading the CPU.

 ## 6 Classes of Computers

sually, computers are classified based on cost, amount of memory, speed, and sophistication. Using these criteria, computers are classified as subnotebooks, notebooks, personal computers, minicomputers, mainframes, or supercomputers. Supercomputers are the most powerful; they also have the highest storage capabilities and the highest price.

Ubiquitous Computing

Many experts describe the current state of computing as ubiquitous computing or pervasive computing. It's also called the third wave. The first wave was identified by mainframe computers, the second wave by personal computers, and the third wave by small computers embedded into many devices used daily, such as cell phones, cameras, watches, and so forth. Because people usually carry these devices around, the term "wearable computers" has been coined to describe them. Wearable computers are also used in medical monitoring systems and can be useful when people need to use computers—to enter text, for example—while standing or walking around.

Mainframe computers are usually compatible with the IBM System/360 line introduced in 1965. As mentioned in the Industry Connection later in this chapter, IBM System z10 is the latest example in this class. Systems that aren't based on System/360 are referred to as "servers" (discussed in the next section) or supercomputers. Supercomputers are more expensive, much bigger in size, faster, and have more memory than personal computers, minicomputers, and mainframes.

Applications for computers include anything from doing homework (subnotebook, notebook, and personal computer) to launching space shuttles (supercomputer). Because all computers are increasing steadily in speed and sophistication, delineating different classes of computers is more difficult now. For example, a notebook computer today has more power than a mainframe of the 1970s, and all indications suggest that this trend will continue. The information boxes highlight some of the applications available for the iPad and describe ubiquitous computing.

Popular iPad Business Applications

The iPad is a tablet computer designed and developed by Apple. iPad users can browse the Web, read and send e-mail, share photos, watch HD videos, listen to music, play games, read eBooks, and much more by using a multitouch user interface. Currently, iPads are used by many business professionals, including the following[3,4]:

Health care workers—Access medical applications and for bedside care

Sales agents and service workers—Perform on-the-road sales presentations and to display product information

Insurance agents—Quote displays

Real estate agents—Provide remote, interactive, visual home tours for interested home buyers

Legal professionals—Access legal documents and conduct a NexisLexis search from a car, office, or courtroom

Teachers and students—Access Windows applications and resources

Financial professionals—Access Windows trading applications, dashboards, documents, real-time quotes, Bloomberg Anywhere, and portfolio analysis tools

Corporate campus workers—Access corporate data while employees move from office to office for collaborations with other colleagues

Remote and mobile workers—Access Windows business applications, desktops, and data while on the road

6.1 Server Platforms: An Overview

A **server** is a computer and all the software for managing network resources and offering services to a network. Many different server platforms are available for performing specific tasks, including the following:

- Application servers store computer software, which users can access from their workstations.
- Database servers store and manage vast amounts of data for access from users' computers.
- Disk servers contain large-capacity hard drives and enable users to store files and applications for later retrieval.
- Fax servers contain software and hardware components that enable users to send and receive faxes.
- File servers contain large-capacity hard drives for storing and retrieving data files.
- Mail servers are configured for sending, receiving, and storing e-mails.
- Print servers enable users to send print jobs to network printers.

A **server** is a computer and all the software for managing network resources and offering services to a network.

An **operating system (OS)** is a set of programs for controlling and managing computer hardware and software. An OS provides an interface between a computer and the user and increases computer efficiency by helping users share computer resources and by performing repetitive tasks for users.

- Remote access servers (RAS) allow off-site users to connect to network resources, such as network file storage, printers, and databases.
- Web servers store Web pages for access over the Internet.

7 What Is Software?

Software is all the programs that run a computer system. It can be classified broadly as system software and application software. For example, Microsoft Windows is the OS for most PCs and belongs to the system software group. This type of software works in the background and takes care of housekeeping tasks, such as deleting files that are no longer needed. Application software is used to perform specialized tasks. Microsoft Excel, for example, is used for spreadsheet analyses and number-crunching tasks.

7.1 Operating System Software

An **operating system (OS)** is a set of programs for controlling and managing computer hardware and software. An OS provides an interface between a computer and the user and increases computer efficiency by helping users share computer resources and by performing repetitive tasks for users. A typical OS consists of control programs and supervisor programs.

Control programs manage computer hardware and resources by performing the following functions:

- *Job management*—Control and prioritize tasks performed by the CPU.
- *Resource allocation*—Manage computer resources, such as storage and memory. In a network, control programs are also used for tasks, such as assigning a print job to a particular printer.
- *Data management*—Control data integrity by generating checksums to verify that data hasn't been corrupted or changed. Today's OSs use 256-bit checksums that guarantee integrity to almost 100 percent. Briefly, when the OS writes data to storage, it generates a value (the checksum) along with the data. The next time this data is retrieved, the checksum is recalculated and compared with the original

Google Docs: Applications and Challenges

Google Docs is a free Web-based application for creating word-processor documents, spreadsheets, presentations, and forms. You can use it to create and edit documents online while collaborating in real time with other users, and you can send files you create to others via the Internet or e-mail. You can save files in a variety of formats: .doc, .xls, .rtf, .csv, .ppt, and so forth. By default, files are saved to Google's servers, using the Google cloud computing platform. Cloud computing is covered in more detail in Chapter 14, but, briefly put, it makes data and applications more portable so that you can work with a program from anywhere. Another popular feature of Google Docs is collaboration. Multiple users can share and edit files at the same time. With spreadsheets, users can even be notified of changes by e-mail.[5]

Cloud computing and Google Docs do present some challenges and security risks, however. On March 10, 2009, Google Docs revealed some private documents to unauthorized users because of a security flaw. According to Google, the flaw has been corrected.[6]

checksum. If they match, the integrity is intact. If they don't, the data has been corrupted somehow. In addition, the OS can correct some corrupt data (but not all), back up data automatically to prevent data loss, and control access to data for improved security.

- *Communication*—Control the transfer of data among parts of a computer system, such as communication between the CPU and I/O devices.

The supervisor program, also known as the kernel, is responsible for controlling all other programs in the OS, such as compilers, interpreters, assemblers, and utilities for performing special tasks.

In addition to single-tasking and multitasking OSs, time-shared OSs allow several users to use computer resources simultaneously. OSs are also available in a variety of platforms for both mainframes and personal computers. Microsoft Windows, Mac OS, and Linux are examples of personal computer OSs, and mainframe OSs include UNIX and OpenVMS, as well as some versions of Linux.

7.2 Application Software

A personal computer can perform a variety of tasks by using **application software**, which can be commercial software or software developed in house. In-house software is usually more expensive than commercial software but is more customized and often fits the users' needs better. For almost any task you can imagine, a software package is available. The following sections give you an overview of common categories of commercial application software for personal computers. In addition to these, many other categories of software are available, such as information management software, Web authoring software, and photo and graphics software.

7.2.1 Word-Processing Software

You're probably most familiar with word-processing software used to generate documents. Typically, this includes editing features, such as deleting, inserting, and copying text, and advanced word-processing software often includes sophisticated graphics and data management features. Word-processing software saves time, particularly for repetitive tasks, such as sending the same letter to hundreds of customers. Most word-processing software offers spell checkers and grammar checkers. Some popular word-processing programs are Microsoft Word, Corel WordPerfect, and OpenOffice.

7.2.2 Spreadsheet Software

A spreadsheet is a table of rows and columns, and spreadsheet software is capable of performing numerous tasks with the information in a spreadsheet. You can even prepare a budget and perform a "what-if" analysis on the data. For example, you could calculate the effect on other budget items of reducing your income by 10%, or you might want to see the effect of a 2% reduction in your mortgage interest rate. Common spreadsheet software includes Microsoft Excel as well as IBM's Lotus 1-2-3, and Corel Quattro Pro.

7.2.3 Database Software

Database software is designed to perform operations such as creating, deleting, modifying, searching, sorting, and joining data. A database is essentially a collection of tables consisting of rows and columns. Database software makes accessing and working with data faster and more efficient. For example, manually searching a database containing thousands of records would be almost impossible. With database software, users can search information quickly and even tailor searches to meet specific criteria, such as finding all accounting students younger than 20 who have GPAs higher than 3.6. You will learn more about databases in Chapter 3. Popular database software for personal computers includes Microsoft Access, FileMaker Pro, and Alpha Software's Alpha Five. High-end database software used in large enterprises includes Oracle, IBM DB2, and Microsoft SQL Server.

7.2.4 Presentation Software

Presentation software is used to create and deliver slide shows. Microsoft PowerPoint is the most commonly used presentation software; other examples include Adobe Persuasion and Corel Presentations. You can include many types of content in slide shows, such as bulleted and numbered lists, charts, and graphs. You can also embed graphics as well as sound and movie clips.

Presentation software also offers several options for running slide shows, such as altering the time interval between slides. In addition, you can usually convert presentations into other formats, including Web pages and photo albums with music and narration. Another option in some presentation software is

Application software can be commercial software or software developed in house and is used to perform a variety of tasks on a personal computer.

capturing what's on the computer screen, and then combining several screen captures into a video for demonstrating a process, which can be useful in educational settings or employee training seminars, for example.

7.2.5 Graphics Software

Graphics software is designed to present data in a graphical format, such as line graphs and pie charts. These formats are useful for illustrating trends and patterns in data and for showing correlations. Graphics are created with integrated packages, such as Excel, Lotus 1-2-3, and Quattro Pro, or dedicated graphics packages, such as Adobe Illustrator and IBM Freelance. Exhibit 2.5 shows the types of graphs you can create in Microsoft Excel.

7.2.6 Desktop Publishing Software

Desktop publishing software is used to produce professional-quality documents without expensive hardware and software. This software works on a "what-you-see-is-what-you-get" (WYSIWYG, pronounced "wizzy-wig") concept, so the high-quality screen display gives you a good idea of what you'll see in the printed output. Desktop publishing software is used for creating newsletters, brochures, training manuals, transparencies, posters, and even books. Many desktop publishing packages are available; three popular ones are Adobe InDesign, QuarkXPress, and Microsoft Office Publisher.

7.2.7 Financial Planning and Accounting Software

Financial planning software, which is more powerful than spreadsheet software, is capable of performing many types of analysis on large amounts of data. These analyses include present value, future value, rate of return, cash flow, depreciation, retirement planning, and budgeting. A widely used financial planning package is Intuit Quicken. Using this package, you can plan and analyze all kinds of financial scenarios. For example, you can calculate how much your $2,000 IRA will be worth at 5% interest in 30 years or determine how to save $150,000 in 18 years toward your child's college education, using a fixed interest rate.

In addition to spreadsheet software, dedicated accounting software is available for performing many sophisticated accounting tasks, such as general ledgers, accounts receivable, accounts payable, payroll, balance sheets, and income statements. Some popular accounting software packages include QuickBooks, a small-business accounting software, and Sage Software's Peachtree.

7.2.8 Project Management Software

A project, such as designing a Web site or setting up an order entry system, consists of a set of related tasks. The goal of project management software is to help project managers keep time and budget under control by solving scheduling problems, planning and setting goals, and highlighting potential bottlenecks. You can use such software to study the cost, time, and resource impact of schedule changes. There are several project management software packages on the market, including Microsoft Project and Micro Planning International's Micro Planner.

7.2.9 Computer-Aided Design Software

Computer-aided design (CAD) software is used for drafting and design and has replaced traditional tools, such as T-squares, triangles, paper, and pencils. It's used extensively in architecture and engineering firms, but because of major price reductions and increases in PC power, small companies and home users can now afford this software. Some widely used CAD software includes: Autodesk AutoCAD, Cadkey, and VersaCAD.

Exhibit 2.5 *Types of graphs in Microsoft Excel*

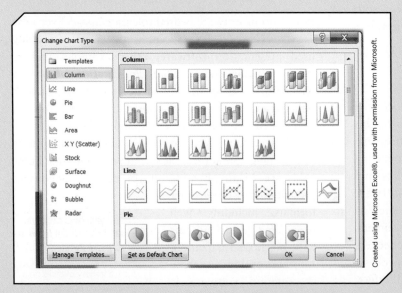

Created using Microsoft Excel®, used with permission from Microsoft.

8 Computer Languages

as mentioned, computer languages have developed through four generations, and the fifth generation is currently being developed. The first generation of computer languages, **machine language**, consists of a series of *0*s and *1*s representing data or instructions. Machine language depends on the machine, so code written for one type of computer doesn't work on another type of computer. Writing a machine-language program is time consuming and painstaking.

Assembly language, the second generation of computer languages, is a higher-level language than machine language but is also machine dependent. It uses a series of short codes, or mnemonics, to represent data or instructions. For example, ADD and SUBTRACT are typical commands in assembly language. Writing programs in assembly language is easier than in machine language.

Third-generation computer languages are machine independent and are called **high-level languages**. Three of the most widely used languages are C++, Java, and VB.NET. These languages are used mostly for Web development and Internet applications. High-level languages are more like English, so they're easier to learn and code. In addition, they're self-documenting, meaning that you can usually understand the programs without needing additional documentation.

Fourth-generation languages (4GLs) are the easiest computer languages to use. The commands are powerful and easy to learn, even for people with little computer training. Sometimes, 4GLs are called nonprocedural languages, which means you don't need to follow a rigorous command syntax to use them. Instead, 4GLs use macro codes that can take the place of several lines of programming. For example, in a 4GL you might issue the PLOT command, a macro code that takes the place of 100 or more lines of high-level programming code. One simple command does the job for you. SQL (structured query language), which will be discussed in Chapter 3, is an example of a 4GL.

Fifth-generation languages (5GLs) use some of the artificial intelligence technologies, which will be discussed in Chapter 13, such as knowledge-based systems, natural language processing (NLP), visual programming, and a graphical approach to programming. Codes are automatically generated and designed to make the computer solve

MachinVoguage, the first generation of computer languages, consists of a series of *0*s and *1*s representing data or instructions. It is dependent on the machine, so code written for one type of computer does not work on another type of computer.

Assembly language, the second generation of computer languages, is a higher-level language than machine language but is also machine dependent. It uses a series of short codes, or mnemonics, to represent data or instructions.

High-level languages are machine independent and part of the third-generation computer languages. Many languages are available, and each is designed for a specific purpose.

Fourth-generation languages (4GLs) use macro codes that can take the place of several lines of programming. The commands are powerful and easy to learn, even for people with little computer training.

Fifth-generation languages (5GLs) use some of the artificial intelligence technologies, such as knowledge-based systems, natural language processing (NLP), visual programming, and a graphical approach to programming. These languages are designed to facilitate natural conversations between you and the computer.

a given problem without a programmer or with minimum programming effort. These languages are designed to facilitate natural conversations between you and the computer. Imagine that you could ask your computer, "What product generated the most sales last year?" Your computer, equipped with a voice synthesizer, could respond, "Product X." Dragon NaturallySpeaking Solutions is an example of NLP. Research continues in this field because of the promising results so far. Some of the programming languages used for Internet programming and Web development include ActiveX, C++, Java, JavaScript, Perl, Visual Basic, and Extensible Stylesheet Language (XSL). The most important Web development languages are Hypertext Markup Language (HTML) and Extensible Markup Language (XML). Both languages are markup languages, not full-featured programming languages.

The Industry Connection highlights IBM and its most popular product areas.

9 Chapter Summary

In this chapter, you've learned about components and distinguishing factors of computers and reviewed a brief history of computer hardware and software. You've also learned about a variety of input, output, and memory devices as well as the criteria for classifying computers. Finally, you've learned about different types of software—system software and application software—and reviewed the five generations of computer languages.

INTERNATIONAL BUSINESS MACHINES (IBM)*

IBM, the largest computer company in the world, is active in almost every aspect of computing, including hardware, software, services, and collaboration tools such as groupware and e-collaboration. IBM has also been a leader in developing mainframe computers; its latest mainframe is the IBM System z10. IBM's most popular product areas include the following:

- *Software*—IBM offers software suites for all types of computers. Lotus, for example, includes features for e-mail, calendaring, and collaborative applications as well as business productivity software, much like Microsoft Office does. Tivoli, another software suite, has many features for asset management, security management, backup and restore procedures, and optimization of storage systems and data management.

- *Storage*—IBM's storage devices include disk and tape systems, storage area networks, network-attached storage, hard disks, and microdrives. A microdrive is a 1-inch hard disk designed to fit in a CompactFlash Type II slot.

- *Servers*—IBM offers a variety of servers, including UNIX and Linux servers, Intel-based servers, AMD-based servers, and more.

IBM is also active in e-commerce software, hardware, and security services, such as encryption technologies, firewalls, antivirus solutions, and more.

*This information has been gathered from the IBM Web site and other promotional materials. For more detailed information, visit *www.ibm.com*.

Key Terms

application software (33)

arithmetic logic unit (ALU) (23)

assembly language (35)

bus (23)

cache RAM (28)

central processing unit (CPU) (23)

computer (22)

control unit (23)

CPU case (23)

disk drive (23)

fifth-generation languages (5GLs) (35)

fourth-generation languages (4GLs) (35)

high-level languages (35)

input devices (26)

machine language (35)

magnetic disk (29)

magnetic tape (29)

main memory (28)

motherboard (23)

network attached storage (NAS) (30)

operating system (OS) (32)

optical discs (29)

output devices (27)

random access memory (RAM) (28)

read-only memory (ROM) (28)

redundant array of independent disks (RAID) system (30)

secondary memory (28)

server (32)

storage area network (SAN) (30)

Problems, Activities, and Discussions

1. How is computer speed measured?
2. What are the unique technologies of each generation of computer hardware?

3. What kinds of decisions can be improved by spreadsheet software?
4. What kinds of decisions can be improved by project management software?

5. How long will the most current SanDisk WORM disc last?

6. What are some of the popular iPad business applications?

7. After reading the information at the following links and from other sources, compare and contrast the Android and iPhone OS. Write a one-page paper that highlights the distinguishing features of each OS.

   www.infoworld.com/d/mobilize/7-ways-the-new-android-22-os-beats-the-iphone-915?source=IFWNLE_nlt_daily_2010-06-01

   wwww.infoworld.com/d/mobilize/apple-ios-40-vs-android-os-22-business-use-599?source=IFWNLE_nlt_daily_2010-06-11

   www.infoworld.com/d/mobilize/apples-iphone-still-ahead-android-globally-says-

   gartner-343?source=IFWNLE_nlt_wrapup_2010-05-19

8. After reading the information at the following link and other sources, write a one-page paper that highlights the top-five Office 2010 features for business.

   www.infoworld.com/d/applications/top-10-office-2010-features-business-068?source=IFWNLE_nlt_daily_2010-05-26

9. Which of the following is a distinguishing factor in computer power? (Choose all that apply.)

   a. Speed
   b. Accuracy
   c. Storage and retrieval capabilities
   d. BIOS

10. What-if analysis is a key feature of spreadsheet software. True or false?

casestudy

LINUX, AN OPERATING SYSTEM ON THE RISE

Linux (*www.linux.org/info/* and *http://tldp.org/LPD/intro-linux/html/chap_01.html*) is a full-featured, multiuser, multitasking OS, and its source code, developed under the GNU General Public License, is free. Linux can run on almost any type of computer, from very small to very large. Many versions are available for all major microprocessor and server platforms, and most are free or inexpensive. Although Linux was originally developed with a command-line interface (a mechanism for interacting with a computer operating system or software by typing commands to perform specific tasks), you can find several graphical user interface

© Asia Images Group/Getty Images

versions now. Because of its many features, low cost, adaptability, and ease of use, this OS has become a popular alternative to proprietary OSs, including Windows and UNIX.

Answer the following questions:

1. What are some advantages of Linux compared to other operating systems?

2. With all its advantages, why do you think Linux isn't used more widely?

3. What are some security features of Linux?

4. What does the GNU General Public License require software developers to do when modifying Linux versions?

DATABASE SYSTEMS, DATA WAREHOUSES, AND DATA MARTS

this chapter gives you an overview of databases and database management systems and their importance in information systems. You learn about the types of data in a database, methods for accessing files, and physical and logical views of information. You also review the most common data model, the relational model, and the major components of a database management system. This chapter also discusses recent trends in database use, including data-driven Web sites, distributed databases, client/server databases, and object-oriented databases. Finally, you get an overview of data warehouses and data marts and their role in generating business intelligence.

learning outcomes

After studying this chapter, you should be able to:

LO1 Define a database and a database management system.

LO2 Explain logical database design and the relational database model.

LO3 Define the components of a database management system.

LO4 Summarize recent trends in database design and use.

LO5 Explain the components and functions of a data warehouse.

LO6 Describe the functions of a data mart.

In a database system, all files are integrated, meaning information can be linked.

1 Databases

a **database** is a collection of related data that can be stored in a central location or in multiple locations. You can think of it as being similar to a filing cabinet, where data is organized and stored in drawers and file folders. As you can imagine, however, retrieving data from a database is much faster.

Although a database can consist of only a single file, it's usually a group of files. A college database, for example, might have files for students, staff, faculty, and courses. In a database, a file is a group of related records, and a record is a group of related fields. This structure is called **data hierarchy**, as shown in Exhibit 3.1. In this university database example, fields consist of Social Security number, student name, and address. All the fields storing information for Mary Smith, for

A **database** is a collection of related data that can be stored in a central location or in multiple locations.

Data hierarchy is the structure and organization of data, which involves fields, records, and files.

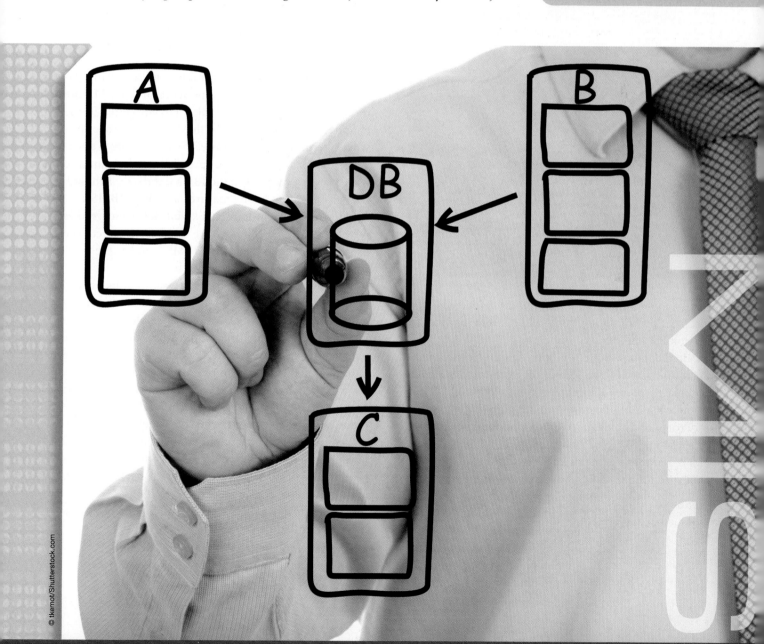

MIS

Exhibit 3.1 *Data hierarchy*

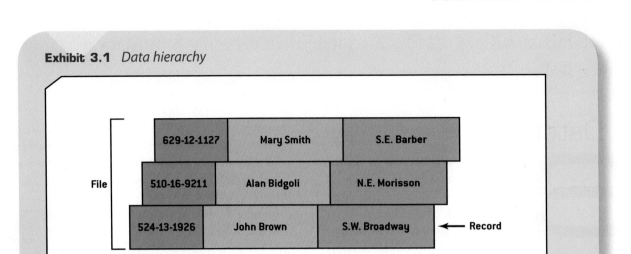

File

| 629-12-1127 | Mary Smith | S.E. Barber |
| 510-16-9211 | Alan Bidgoli | N.E. Morisson |
| 524-13-1926 | John Brown | S.W. Broadway | ← Record

Field

© Rana Faure/Getty Images

instance, constitute a record, and all three records in Exhibit 3.1 make up the student file.

In a database system, all files are integrated, meaning information can be linked. For example, you should be able to retrieve all students enrolled in Professor Thomas's MIS 480 course from the courses file, or look up Professor Thomas's record to find out other courses he's teaching in a particular semester.

A database is a critical component of information systems because any type of analysis that's done is based on data available in the database. To make using databases more efficient, a **database management system (DBMS)** is used, which is software

for creating, storing, maintaining, and accessing database files. The major components of a DBMS are discussed in "Components of a DBMS" later in the chapter. If you're familiar with Microsoft Office software, you know that you use Word to create a document and Excel to create a spreadsheet. You can also use Access to create and modify a database, although it doesn't have as many features as other DBMSs.

For now, take a look at Exhibit 3.2, which shows how users, the DBMS, and the database interact. The user issues a request, and the DBMS searches the database and returns the information to the user.

In the past, data was stored in a series of files called "flat files," because they weren't arranged in a hierarchy and there was no relation among these files. The problem with this flat file organization was that the same data could be stored in more than one file, creating data redundancy. For example, in a customer database, a customer's name might be stored in more than one table. This duplication takes up unnecessary storage space and can make retrieving data inefficient. In addition, in a flat file system, data might not be updated in all files consistently, resulting in conflicting reports generated from these files. Updating a flat file system can also be time consuming.

In summary, a database has the following advantages over a flat file system:

- More information can be generated from the same data.

- Complex requests can be handled more easily.

A **database management system (DBMS)** is software for creating, storing, maintaining, and accessing database files. A DBMS makes using databases more efficient.

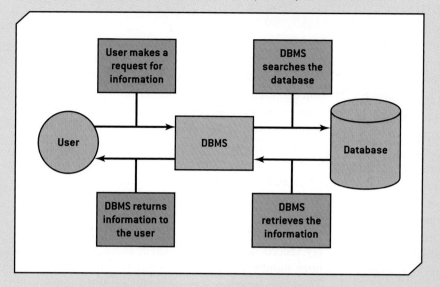

Exhibit 3.2 *Interaction between the user, DBMS, and database*

User makes a request for information

DBMS searches the database

User

DBMS

Database

DBMS returns information to the user

DBMS retrieves the information

Collecting and using external data can be more challenging. External data comes from a variety of sources and is often stored in a data warehouse (discussed later in the chapter). The following are examples of sources for external data:

- Competitors, customers, and suppliers
- Distribution networks
- Economic indicators (for example, the consumer price index)
- Government regulations
- Labor and population statistics
- Tax records

In the section called "Data Warehouses," found later in this chapter, you learn how information from external data sources is used to conduct analyses and generate reports for BI.

The following information box discusses an example of how BI is used in other fields—in this case, law enforcement.

- Data redundancy is eliminated or minimized.
- Programs and data are independent, so more than one program can use the same data.
- Data management is improved.
- A variety of relationships among data can be maintained easily.
- More sophisticated security measures can be used.
- Storage space is reduced.

1.1 Types of Data in a Database

As you learned in Chapter 1, to generate business intelligence (BI), the database component of an information system needs access to two types of data: internal and external. Internal data is collected from within an organization and can be transaction records, sales records, personnel records, and so forth. An organization might use internal data on customers' past purchases to generate BI about future buying patterns, for example. Internal data is usually stored in the organization's internal databases and can be used by functional information systems.

© Phase4Photography/Shutterstock.com

BI in Action: Law Enforcement

Business intelligence (BI) is used in law enforcement as well as in the business world. In Richmond, Virginia, data entered into the information system includes crime reports from the past 5 years, records of 911 phone calls, details about weather patterns, and information about special events. The system generates BI reports that help pinpoint crime patterns and are useful for personnel deployment, among other purposes. The system has increased public safety, reduced 911 calls, and helped management make better use of Richmond's 750 officers.

Recently, the department refined its reports by separating violent crimes into robberies, rapes, and homicides to help them discover patterns for certain types of crime. For example, the department discovered that Hispanic workers were often robbed on paydays. By entering workers' paydays into the system and looking at robbery patterns, law enforcement officers were able to identify days and locations where these incidents were likely to occur. Moving additional officers into those areas on paydays has reduced the number of robberies.[1]

1.2 Methods for Accessing Files

In a database, files are accessed by using a sequential, random, or indexed sequential method. In a **sequential access file structure**, records in files are organized and processed in numerical or sequential order, typically the order in which they were entered. Records are organized based on what is known as a "primary key," discussed later in this chapter, such as Social Security numbers or account numbers. For example, to access record number 10, records 1 through 9 must be read first. This type of access method is effective when a large number of records are processed less frequently, perhaps on a quarterly or yearly basis. Because access speed usually isn't critical, these records are typically stored on magnetic tape. A sequential file structure is usually used for backup and archive files, because they need updating only rarely.

In a **random access file structure**, records can be accessed in any order, regardless of their physical location in storage media. This method of access is fast and very effective when a small number of records needs to be processed daily or weekly. To achieve this speed, these records are often stored on magnetic disks. Disks are random access devices, whereas tapes are sequential access devices (consider how much quicker it is to skip a song on a CD as opposed to a tape cassette).

With the **indexed sequential access method (ISAM)**, records can be accessed sequentially or randomly, depending on the number being accessed. For a small number, random access is used, and for a large number, sequential access is used. This file structure is similar to a book index that lists page numbers where you can find certain topics. The advantage of this method is that both types of access can be used, depending on the situation and the user's needs.[2]

Indexed sequential access, as the name suggests, uses an index structure and has two parts: the indexed value and a pointer to the disk location of the record matching the indexed value. Retrieving a record requires at least two disk accesses, once for the index structure and once for the actual record. Because every record needs to be indexed, if the file contains a huge number of records, the index is also quite large. Therefore, an index is more useful when the number of records is small. Access speed with this method is fast, so it's recommended when records must be accessed frequently. This advice was more applicable when processors were slow and memory and storage were expensive, however. Given the speed and low storage cost of today's computers, the number of records might not be as important, meaning more records could be accessed and processed with this method.

2 Logical Database Design

before designing a database, you need to know the two ways information is viewed in a database. The **physical view** involves how data is stored on and retrieved from storage media, such as hard disks, magnetic tapes, or CDs. For each database, there's only one physical view of data. The **logical view** involves how information appears to users and how it can be organized and retrieved. There can be more than one logical view of data, depending on the user. For example, marketing executives might want to see data organized by

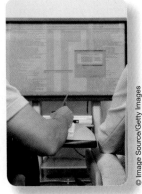

© Image Source/Getty Images

---

top-selling products in a specific region; the finance officer might need to see data organized by cost of raw materials for each product.

The first step in database design is defining a **data model**, which determines how data is created, represented, organized, and maintained. A data model usually includes these three components:

- *Data structure*—Describes how data is organized and the relationship among records.

- *Operations*—Describes methods, calculations, and so forth that can be performed on data, such as updating and querying data.

- *Integrity rules*—Defines the boundaries of a database, such as maximum and minimum values allowed for a field, constraints (limits on what type of data can be stored in a field), and access methods.

Many data models are used. The most common, the relational model, is described in the next section, and you learn about the object-oriented model later in the chapter, in "Object-Oriented Databases." Two other common data models are hierarchical and network, although they aren't used as much now.

In a **hierarchical model**, shown in Exhibit 3.3, the relationships among records form a treelike structure (hierarchy). Records are called nodes, and relationships among records are called branches. The node at the top is called the root, and every other node (called a child) has a parent. Nodes with the same parents are called twins or siblings. In Exhibit 3.3, the root node is a supplier, which provides three product lines, all considered siblings. Each product line has categories of products, and each category has specific products. (The product lines and categories are also siblings.) For example, Supplier A supplies three product lines: soap, shampoo, and toothpaste. The toothpaste product line has two

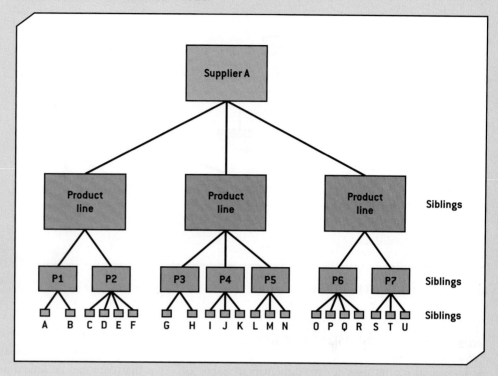

Exhibit 3.3 *A hierarchical model*

categories (P6 and P7): whitening and cavity-fighting toothpaste. The P7 category has three specific cavity-fighting products: S, T, and U.

The **network model** is similar to the hierarchical model, but records are organized differently, as shown in Exhibit 3.4. This model links invoice number, customer number, and method of payment. For example, invoice #111 belongs to customer #2000, who paid with cash. Unlike the hierarchical model, each record in the network model can have multiple parent and child records. For example, in Exhibit 3.4, a customer can have several invoices, and an invoice can be paid by more than one method.

A **data model** determines how data is created, represented, organized, and maintained. It usually contains data structure, operations, and integrity rules.

In a **hierarchical model**, the relationships between records form a treelike structure (hierarchy). Records are called nodes, and relationships between records are called branches. The node at the top is called the root, and every other node (called a child) has a parent. Nodes with the same parents are called twins or siblings.

The **network model** is similar to the hierarchical model, but records are organized differently. Unlike the hierarchical model, each record in the network model can have multiple parent and child records.

Exhibit 3.4 *A network model*

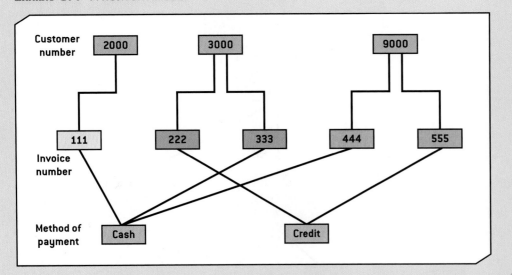

2.1 The Relational Model

A **relational model** uses a two-dimensional table of rows and columns of data. Rows are records (also called *tuples*), and columns are fields (also referred to as *attributes*). To begin designing a relational database, you must define the logical structure by defining each table and the fields in it. For example, the Students table has fields for StudentID, StudentFirstName, StudentLastName, and so forth. The collection of these definitions is stored in the **data dictionary**. The data dictionary can also store other definitions, such as data types for fields, default values for fields, and validation rules for data in each field, as described in the following list:

- *Field name*—Student name, admission date, age, and major.

- *Field data type*—Character (text), date, and number.

- *Default value*—The value entered if none is available; for example, if no major is declared, the value is "undecided."

- *Validation rule*—A rule determining whether a value is valid; for example, a student's age can't be a negative number.

In a relational database, every record must be uniquely identified by a **primary key**. Student ID numbers, Social Security numbers, account numbers, and invoice numbers are examples of primary keys. To establish relationships among tables so that data can be linked and retrieved more efficiently, a primary key for one table can appear in other tables. In this case, it's called a **foreign key**. For example, a student ID number is the primary key for the Students table, and the code for a student's major (such as MKT, MIS, or FIN) is the primary key for the Majors table. Each student can have one major, and a number of students can be enrolled in each major. The primary key of the Majors table is a foreign key in the Students table.

To improve database efficiency, a process called **normalization** is used, which eliminates redundant data (storing customer names in only one table, for example) and ensures that only related data is stored in a table. Normalization can go through several stages, from first normal form (1NF) to fifth normal form (5NF). Typically, however, only stages 1NF through 3NF are used. For example, the following tasks are performed in the 1NF stage:

- Eliminate duplicated fields from the same table.
- Create separate tables for each group of related data.
- Identify each record with a unique field (the primary key).

Data stored in a relational model is retrieved from tables by using operations that pick and combine data from one or more tables. There are several operations, such as select, project, join, intersect, union, and difference. The first three are the most commonly used and are explained in the following paragraphs.

A **relational model** uses a two-dimensional table of rows and columns of data. Rows are records (also called *tuples*), and columns are fields (also referred to as *attributes*).

The **data dictionary** stores definitions, such as data types for fields, default values, and validation rules for data in each field.

A **primary key** uniquely identifies every record in a relational database. Examples include: Student ID numbers, account numbers, Social Security numbers, and invoice numbers.

A **foreign key** is a field in a relational table that matches the primary key column of another table. It can be used to cross-reference tables.

Normalization improves database efficiency by eliminating redundant data and ensuring that only related data is stored in a table.

Table 3.1

Data in the Students table

Student ID#	Name	Major	Age	GPA
111	Mary	MIS	25	4.00
222	Sue	CS	21	3.60
333	Debra	MGT	26	3.50
444	Bob	MKT	22	3.40
555	George	MIS	28	3.70

Table 3.2

Results of the select operation

Student ID#	Name	Major	Age	GPA
111	Mary	MIS	25	4.00
555	George	MIS	28	3.70

Table 3.3

Results of the project operation

Student ID#	Name	Major	GPA
111	Mary	MIS	4.00
222	Sue	CS	3.60
333	Debra	MGT	3.50
444	Bob	MKT	3.40
555	George	MIS	3.70

Table 3.4

The Customers table

Customer#	Name	Address
2000	ABC	Broadway
3000	XYZ	Jefferson
9000	TRY	Madison

Table 3.5

The Invoices table

Invoice#	Customer#	Amount	Payment
1110	2000	$2000.00	Cash
2220	3000	$4000.00	Credit
3330	3000	$1500.00	Cash
4440	9000	$6400.00	Cash
5550	9000	$7000.00	Credit

A *select operation* searches data in a table and retrieves records based on certain criteria (also called *conditions*). Table 3.1 shows data stored in the Students table. Using the select operation "Major=MIS," you can generate a list of only the students majoring in MIS, as Table 3.2 shows.

A *project operation* pares down a table by eliminating columns (fields) according to certain criteria. For example, you need a list of students but don't want to include their ages. Using the project operation "PROJECT Student ID#, Name, Major, GPA (Table 3.1)," you can retrieve the data shown in Table 3-3. The "(Table 3.1)" in this statement means to use the data in Table 3.1.

A *join operation* combines two tables based on a common field (for example, the primary key in the first table and the foreign key in the second table). Table 3.4 shows data in the Customers table, and Table 3.5 shows data in the Invoices table. The Customer# is the primary key for the Customers table and is a foreign key in the Invoices table; the Invoice# is the primary key for the Invoices table. Table 3.6 shows the table resulting from joining these two tables.

Table 3.6

Joining the Invoices and Customers tables

Invoice#	Customer#	Amount	Payment	Name	Address
1110	2000	$2000.00	Cash	ABC	Broadway
2220	3000	$4000.00	Credit	XYZ	Jefferson
3330	3000	$1500.00	Cash	XYZ	Jefferson
4440	9000	$6400.00	Cash	TRY	Madison
5550	9000	$7000.00	Credit	TRY	Madison

Now that you've learned about the components of a database and a common data model, you can examine the software used to manage databases.

3 Components of a DBMS

BMS software includes these components, discussed in the following sections:

- Database engine
- Data definition
- Data manipulation
- Application generation
- Data administration

3.1 Database Engine

A database engine, the heart of DBMS software, is responsible for data storage, manipulation, and retrieval. It converts logical requests from users into their physical equivalents (reports, for example) by interacting with other components of the DBMS (usually the data manipulation component). For example, a marketing manager wants to see a list of the top three salespeople in the Southeast region (a logical request). The database engine interacts with the data manipulation component to find where these three names are stored and displays them on screen or in a printout (the physical equivalent). Because more than one logical view of data is possible, the database engine can retrieve and return data to users in many different ways.

Structured Query Language (SQL) is a standard fourth-generation query language used by many DBMS packages, such as Oracle 11*g* and Microsoft SQL Server. SQL consists of several keywords specifying actions to take.

With **query by example (QBE)**, you request data from a database by constructing a statement made up of query forms. With current graphical databases, you simply click to select query forms instead of having to remember keywords, as you do with SQL. You can add AND, OR, and NOT operators to the QBE form to fine-tune the query.

Create, read, update, and delete (CRUD) refers to the range of functions that data administrators determine who has permission to perform certain functions.

3.2 Data Definition

The data definition component is used to create and maintain the data dictionary and define the structure of files in a database. Any changes to a database's structure, such as adding fields, deleting fields, changing a field's size, and changing the data type stored in a field, are made with this component.

3.3 Data Manipulation

The data manipulation component is used to add, delete, modify, and retrieve records from a database. Typically, a query language is used for this component. Many query languages are available, but Structured Query Language (SQL) and Query By Example (QBE) are two of the most widely used.

Structured Query Language (SQL) is a standard fourth-generation query language used by many DBMS packages, such as Oracle 11*g* and Microsoft SQL Server. SQL consists of several keywords specifying actions to take. The basic format of an SQL query is as follows:

SELECT *field* FROM *table or file* WHERE *conditions*

After the SELECT keyword, you list the fields you want to retrieve. After FROM, you list the tables or files the data is retrieved from, and after WHERE, you list conditions (criteria) for retrieving the data. The following example retrieves the name, Social Security number, title, gender, and salary from the Employee and Payroll tables for all employees with the job title "engineer":

SELECT NAME, SSN, TITLE, GENDER, SALARY
FROM EMPLOYEE, PAYROLL
WHERE   EMPLOYEE.SSN=PAYROLL.SSN   AND
    TITLE="ENGINEER"

This query means that data in the NAME, SSN, TITLE, GENDER, and SALARY fields from the two tables EMPLOYEE and PAYROLL should be retrieved. Line 3 indicates on which field the EMPLOYEE and PAYROLL tables are linked (the SSN field) and specifies a condition for displaying data: only employees who are engineers. You can add many other conditions to SQL statements by using AND, OR, and NOT operators (discussed next).

With **query by example (QBE)**, you request data from a database by constructing a statement made up of query forms. With current graphical databases, you simply click to select query forms instead of having to remember keywords, as you do with SQL. You can add AND, OR, and NOT operators to the QBE form to fine-tune the query:

- *AND*—Means that all conditions must be met. For example, "Major=MIS AND GPA > 3.8" means a student must be majoring in MIS and have a GPA higher than 3.8 to be retrieved.
- *OR*—Means only one of the conditions must be met. For example, "Major=MIS OR GPA > 3.8" means the DBMS could retrieve a student with a GPA higher than 3.8 but majoring in another field, such as accounting.
- *NOT*—Searches for records that don't meet the condition. For example, "Major NOT ACC" retrieves all students except accounting majors.

3.4 Application Generation

The application generation component is used to design elements of an application using a database, such as data entry screens, interactive menus, and interfaces with other programming languages. These applications might be used to create a form or generate a report, for example. If you're designing an order entry application for users, you could use the application generation component to create a menu system that makes the application easier to use. Typically, IT professionals and database administrators use this component.

3.5 Data Administration

The data administration component, also used by IT professionals and database administrators, is used for tasks such as backup and recovery, security, and change management. In addition, this component is used to determine who has permission to perform certain func-

© zhu difeng/Shutterstock.com

tions, often summarized as **create, read, update, and delete (CRUD)**.

In large organizations, database design and management is handled by the **database administrator (DBA)**, although with complex databases, this task is sometimes handled by an entire department. The DBA's responsibilities include the following:

- Designing and setting up a database
- Establishing security measures to determine users' access rights
- Developing recovery procedures in case data is lost or corrupted
- Evaluating database performance
- Adding and fine-tuning database functions

In the following sections, you learn about recently developed database types other than relational databases.

4 Recent Trends in Database Design and Use

Recent trends in database design and use include data-driven Web sites, natural language processing, distributed databases, client/server databases, and object-oriented databases. In addition to these trends, advances in artificial intelligence and natural language processing will have an impact on database design and use, such as improving user interfaces.[3] Chapter 13 covers natural language processing, and the other trends are discussed in the following sections.

4.1 Data-Driven Web Sites

With the popularity of e-commerce applications, data-driven Web sites are used more widely to provide dynamic content. A **data-driven Web site** acts as an

> **Database administrators (DBA)**, found in large organizations, design and set up databases, establish security measures, develop recovery procedures, evaluate database performance, and add and fine-tune database functions.
>
> A **data-driven Web site** acts as an interface to a database, retrieving data for users and allowing users to enter data in the database.

interface to a database, retrieving data for users and allowing users to enter data in the database. Without this feature, Web site designers must edit the HTML code every time a Web site's data contents change. This type of site is called a "static" Web site. A data-driven Web site, on the other hand, changes automatically because it retrieves content from external dynamic data sources, such as MySQL, Microsoft SQL Server, Microsoft Access, Oracle, IBM DB2, and other databases.

A data-driven Web site improves access to information so users' experiences are more interactive, and it reduces the support and overhead needed to maintain static Web sites. A well-designed data-driven Web site is easier to maintain because most content changes require no change to the HTML code. Instead, changes are made to the data source, and the Web site adjusts automatically to reflect these changes. With a data-driven Web site, users can get more current information from a variety of data sources. Data-driven Web sites are useful for the following applications, among others:

- E-commerce sites that need frequent updates
- News sites that need regular updating of content
- Forums and discussion groups
- Subscription services, such as newsletters

4.2 Distributed Databases

The database types discussed so far use a central database for all users of an information system. However, in some situations, a **distributed database**, in which data is stored on multiple servers placed throughout an organization, is preferable. Organizations might choose a distributed database for the following reasons:[4]

- The design reflects the organization's structure better. For example, an organization with several branch offices might find a distributed database more suitable because it allows faster local queries and can reduce network traffic.

- Local storage of data decreases response time but increases communication costs.
- Distributing data among multiple sites minimizes the effects of computer failures. If one database server goes down, it doesn't affect the entire organization.
- The number of users of an information system aren't limited by one computer's capacity or processing power.
- Several small integrated systems might cost less than one large server.
- Accessing one central database server could increase communication costs for remote users. Storing some data at remote sites can help reduce these costs.
- Distributed processing, which includes database design, is used more widely now and is often more responsive to users' needs than centralized processing.
- Most importantly, a distributed database isn't limited by data's physical location.

There are three approaches to setting up a distributed DBMS (DDBMS), although these approaches can be combined:[5]

- *Fragmentation*—The **fragmentation** approach addresses how tables are divided among multiple locations. Horizontal fragmentation breaks a table into rows, storing all fields (columns) in different locations. Vertical fragmentation stores a subset of columns in different locations. Mixed fragmentation, which combines vertical and horizontal fragmentation, stores only site-specific data in each location. If data from other sites is needed, the DDBMS retrieves it.

- *Replication*—With the **replication** approach, each site stores a copy of data of the organization's database. Although this method can increase costs, it also increases availability of data, and each site's copy can be used as a backup for other sites.

- *Allocation*—The **allocation** approach combines fragmentation and replication. Generally, each site stores the data it uses most often. This method improves response time for local users (those in the same location as the database storage facilities).

Security issues are more challenging in a distributed database because of multiple access points from both inside and outside the organization. Security policies, scope of user access, and user privileges must be clearly defined, and authorized users must be identified. Distributed database designers should also keep in mind that distributed processing isn't suitable for every situation, such as a company with all its departments in one location.

4.3 Client/Server Databases

With the increased power and reduced costs of computers, client/server database processing is widely used now. In a **client/server database**, users' workstations (clients) are linked in a local area network (LAN) to share the services of a single server. In contrast to the older file server method, in which the entire file is sent to the client to be processed, clients send requests to the server, which processes the data and returns only the records meeting the request.[6]

4.4 Object-Oriented Databases

The relational model discussed previously is designed to handle homogenous data organized in a field-and-record format. Including different types of data, such as multimedia files, can be difficult, however. In addition, a relational database has a simple structure: Relationships between tables are based on a common value (the key). Representing more complex data relationships sometimes isn't possible with a relational database.[7]

To address these problems, **object-oriented databases** were developed. Like object-oriented programming, this data model represents real-world entities with database objects. An object consists of attributes (characteristics describing an entity) and methods (operations or calculations) that can be performed on the object's data. For example, as shown in Exhibit 3.5, a real-world car can be represented by an object in the Vehicle class. Objects in this class have attributes of year, make, model, and license number, for example. You can then use methods to work with data in a Vehicle object, such as the AddVehicle method to add a car to the database. Thinking of classes as categories or types of objects can be helpful.

Grouping objects along with their attributes and methods into a class is called **encapsulation**, which essentially means grouping related items into a single unit. Encapsulation helps handle more complex types of data, such as images and graphs. Object-oriented databases can also use **inheritance**, which means new objects can be created faster and more easily by entering new data in attributes. For example, you can add a BMW as a Vehicle object by having it inherit attributes and methods of the Vehicle class. You don't have to redefine an object—in other words, specifying all its attributes and methods—every time you want to add a new one.

This data model expands on the relational model by supporting more complex data management, so modeling real-world problems is easier. In addition, object-oriented databases can handle storing and manipulating all types of multimedia as well as numbers and characters. Being able to handle many file types is useful in many fields. In the medical field, for example, doctors need to access X-ray images and graphs of vital signs in addition to patient histories consisting of text and numbers.

In contrast to the query languages used to interact with a relational database, interaction with an object-oriented database takes places via methods, which are called by sending a message to an object. Messages are usually generated by an event of some kind, such as pressing Enter or clicking the mouse button. Typically, a high-level language, such as C++, is used to create methods. Some examples of object-oriented DBMSs are Progress ObjectStore and Objectivity/DB.

In a **client/server database**, users' workstations (clients) are linked in a local area network (LAN) to share the services of a single server.

In **object-oriented databases**, both data and their relationships are contained in a single object. An object consists of attributes and methods that can be performed on the object's data.

Encapsulation refers to the grouping into a class of various objects along with their attributes and methods—i.e., grouping related items into a single unit. This helps handle more complex types of data, such as images and graphs.

Inheritance refers to new objects being created faster and more easily by entering new data in attributes.

A **data warehouse** is a collection of data from a variety of sources used to support decision-making applications and generate business intelligence.

5 Data Warehouses

a **data warehouse** is a collection of data from a variety of sources used to support decision-making applications and generate business intelligence.[8] Data warehouses store multidimensional data, so they're sometimes called "hypercubes." Typically, data in a data warehouse is described as having the following characteristics in contrast to data in a database:

- *Subject oriented*—Focused on a specific area, such as the home-improvement business or a university, whereas data in a database is transaction/function oriented.

- *Integrated*—Comes from a variety of sources, whereas data in a database usually doesn't.
- *Time variant*—Categorized based on time, such as historical information, whereas data in a database only keeps recent activity in memory.
- *Type of data*—Captures aggregated data, whereas data in a database captures raw transaction data.
- *Purpose*—Used for analytical purposes, whereas data in a database is used for capturing and managing transactions.

> **Extraction, transformation, and loading (ETL)** refers to the processes used in a data warehouse. It includes extracting data from outside sources, transforming it to fit operational needs, and loading it into the end target (database or data warehouse).

Designing and implementing a data warehouse is a complex task, but specific software is available to help. Oracle, IBM, Microsoft, Teradata, SAS, and Hewlett-Packard are market leaders in data-warehousing platforms. The following information box discusses how a data warehouse was used at InterContinental Hotels Group (IHG).

Exhibit 3.6 shows a data warehouse configuration with four major components: input; extraction, transformation, and loading (ETL); storage; and output. These components are explained in the following sections.

5.1 Input

Data can come from a variety of sources, including external data sources, databases, transaction files, enterprise resource planning (ERP) systems, and customer relationship management (CRM) systems. ERP systems collect, integrate, and process data that can be used by all functional areas in an organization. CRM systems collect and process customer data to provide information for improving customer service. (ERP and CRM systems are discussed in Chapter 11.) Together, these data sources provide the input a data warehouse needs to perform analyses and generate reports.

5.2 ETL

Extraction, transformation, and loading (ETL) refers to the processes used in a data warehouse. Extraction

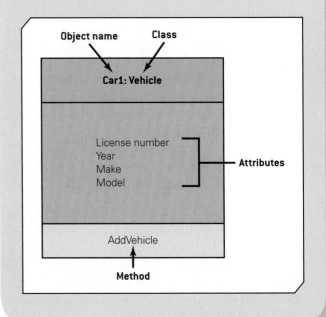

Exhibit 3.5 *Objects, classes, attributes, and methods*

means collecting data from a variety of sources and converting it into a format that can be used in transformation processing. The extraction process can also parse (divide into pieces) data to make sure it meets the data warehouse's structural needs. For example, parsing can be used to separate the street number, street name, city, and state in an address if you want to find out how many customers live in a particular region of a city.

Transformation processing is done to make sure data meets the data warehouse's needs. Its tasks include the following:

- Selecting only certain columns or rows to load
- Translating coded values, such as replacing *Yes* with *1* and *No* with *2*
- Performing select, project, and join operations on data

> Data warehouses store multidimensional data, so they're sometimes called "hypercubes."

- Sorting and filtering data
- Aggregating and summarizing data before loading it in the data warehouse

Loading is the process of transferring data to the data warehouse. Depending on the organization's needs and the data warehouse's storage capacity, loading might overwrite existing data or add collected data to existing data.

5.3 Storage

Collected information is organized in a data warehouse as raw data, summary data, or metadata. Raw data is information in its original form. Summary data gives users subtotals of various categories, which can be useful. For example, sales data for a company's southern regions can be added and represented by one summary number. However, maintaining both raw data (disaggregated data) and summary data (aggregated data) is a good idea for decision-making purposes, as you learned in Chapter 1. Metadata is information about data—its content, quality, condition, origin, and other characteristics. Metadata tells users how, when, and by whom data was collected and how data has been formatted and converted into its present form. For example, metadata in a financial database could be used to generate a report for shareholders explaining how revenue, expenses, and profits from sales transactions are calculated and stored in the data warehouse.

5.4 Output

As Exhibit 3.6 shows, a data warehouse supports different types of analysis and generates reports for decision making. The databases discussed so far support **online transaction processing (OLTP)** to generate reports such as the following:[10]

- Which product generated the highest sales last month?
- Which region generated the lowest sales last month?
- Which salespersons increased sales by more than 30% last quarter?

Data Warehouse Applications at InterContinental Hotels Group (IHG)[9]

KIHG, which operates more than 4,400 hotels throughout the world, needed to improve its data management system. The old system wasn't accommodating the company's BI requirements, and there were some performance issues. There was also some inconsistency in the reports that were being generated by the old system, as well as problems with reporting speed and the handling of large volumes of data. IHG decided to migrate from an entry-level data mart to an enterprise data warehouse (EDW), and it chose Teradata Data Warehouse Appliance to help make this transformation. The implementation has been a success, repositioning the company to meet its short- and long-term goals in the areas of growth, profit, and organizational management. The new system has increased the company's query response time from hours to minutes, and it has generated valuable BI on both its customers and the competition. Future plans include the migration of financial data, which will enable IHG to perform side-by-side analyses of operations, marketing, sales, and financial data. IHG also plans to build a master data management discipline for its sales group.

Data warehouses, however, use online analytical processing and data-mining analysis to generate reports. These are discussed in the following sections.

5.4.1 Online Analytical Processing

Online analytical processing (OLAP), unlike OLTP, is an approach to quickly answer multidimensional analytical queries. It generates business intelligence. It uses multiple sources of information and provides multidimensional analysis, such as viewing data based on time,

> **Online transaction processing (OLTP)** systems are used to facilitate and manage transaction-oriented applications, such as point of sale, data entry, and retrieval transaction processing. They usually utilize internal data and respond in real time.
>
> **Online analytical processing (OLAP)** generates business intelligence. It uses multiple sources of information and provides multidimensional analysis, such as viewing data based on time, product, and location.

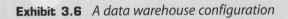

Exhibit 3.6 *A data warehouse configuration*

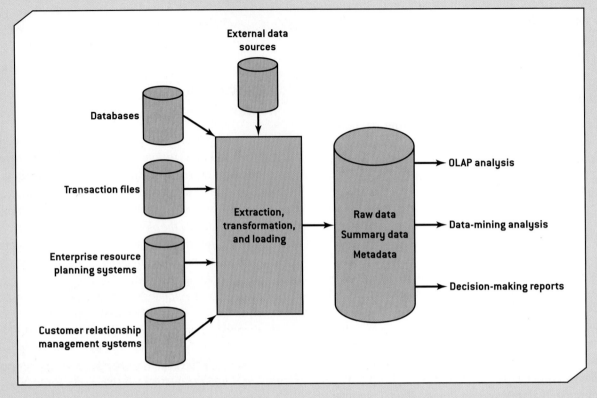

product, and location. For example, it can be used to find out how Product *X* performed last quarter in the Northwest region. Sometimes this analysis is called slicing and dicing data. Using the hypercube in Exhibit 3.7, you can slice it in different directions. You can think of this hypercube as a multidimensional spreadsheet, with each side representing a dimension, such as region ("Geography" in the exhibit). The advantage of a hypercube is that it enables fast manipulations and calculations. In the hypercube in Exhibit 3.7, each smaller cube in a dimension represents a subdivision of data. Therefore, data in one of these smaller cubes could pertain to the sale of canned goods in the Northeast region in 2004. Each smaller cube can be subdivided further; for example, 2004 could be divided into financial quarters: Q1, Q2, Q3, and Q4. The number of cubes is determined by the "granularity" (specificity) of each dimension.

OLAP allows you to analyze information that has been summarized in multidimensional views. OLAP tools are used to perform trend analysis and sift through massive amounts of statistics to find specific information. These tools usually have a "drill-down" feature for ac-

cessing multilayer information. For example, an OLAP tool might access the first layer of information to generate a report on sales performance in a company's eight regions. If a marketing executive is interested in more information on the Northwest region, the OLAP tool can access the next layer of information for a more detailed analysis. OLAP tools are also capable of "drilling up," proceeding from the smallest unit of data to higher levels. For example, an OLAP tool might examine sales data for each region, and then drill up to generate sales performance reports for the entire company.

5.4.2 Data-Mining Analysis

Data-mining analysis is used to discover patterns and relationships. For example, data-mining tools can be used to examine point-of-sale data to generate reports on customers' purchase histories. Based on this information, a company could better target marketing promotions to certain customers. Similarly, a company could mine demographic data from comment or warranty cards and use it to develop products that appeal to a certain customer group, such as teenagers or women over 30. When Netflix.com recommends movies to you based on your rental history, the information is generated by using data-mining tools. Netflix awarded a $1 million prize in September 2009 to the team that devised the best algorithm for substantially improving the accuracy of movie recommendations (*www.netflixprize.com*). American Express conducts the same type of analysis to suggest products and services to cardholders based on their monthly expenditures—patterns discovered by using data-mining tools. The following are typical questions you can answer by using data-mining tools:

- Which customers are likely to respond to a new product?

- Which customers are likely to respond to a new ad campaign?

- What product should be recommended to this customer based on his or her past buying patterns?

Vendors of data-mining software include SAP Business Objects (*www.sap.com*), SAS (*www.sas.com*), Cognos (*http://cognos.com*), and Informatica (*www. informatica.com*).

5.4.3 Decision-Making Reports

A data warehouse can generate all types of information as well as reports used for decision making. The following are examples of what a data warehouse can allow you to do:

- Cross-reference segments of an organization's operations for comparison purposes. For example, you can compare personnel data with data from the finance department, even if they have been stored in different databases with different formats.

- Generate complex queries and reports faster and easier with data warehouses than with databases.

- Generate reports efficiently using data from a variety of sources in different formats and

stored in different locations throughout an organization.

- Find patterns and trends that can't be found with databases.

- Analyze large amounts of historical data quickly.

- Assist management in making well-informed business decisions.

- Manage a high demand for information from many users with different needs and decision-making styles.

Data-mining analysis is used to discover patterns and relationships.

A **data mart** is usually a smaller version of a data warehouse, used by a single department or function.

6 Data Marts

a **data mart** is usually a smaller version of a data warehouse, used by a single department or function. Data marts focus on business functions for a specific user group in an organization, such as a data mart

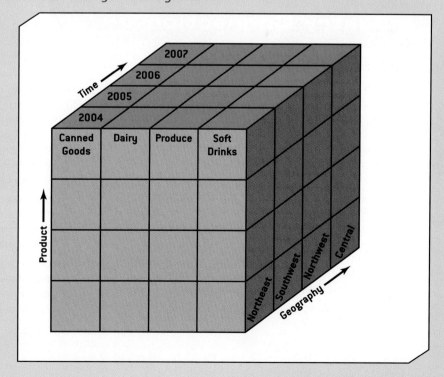

Exhibit 3.7 *Slicing and dicing data*

> When Netflix.com recommends movies to you based on your rental history, the information was generated by using data-mining tools.

for the Marketing Department. Despite being smaller, data marts can usually perform the same types of analysis as a data warehouse. Data marts have the following advantages over data warehouses:

- Access to data is often faster because of their smaller size.
- Response time for users is improved.
- They're easier to create because they're smaller and often less complex.
- They're less expensive.
- Users are targeted better, because a data mart is designed for a specific department or division; identifying their requirements and the functions they need is easier. A data warehouse is designed for an entire organization's use.

Data marts, however, usually have more limited scope than data warehouses, and consolidating information from different departments or functional areas

(such as sales and production) is more difficult.

The Industry Connection summarizes Oracle Corporation's database products and services.

7 Chapter Summary

This chapter has given you an overview of databases and DBMSs and their importance in information systems. You've learned about methods for accessing files, database design principles, and components of a DBMS. You have also reviewed recent trends in database design and use: data-driven Web sites, distributed databases, client/server databases, and object-oriented databases. Finally, you have learned about data warehouses and data marts and their uses.

Industry Connection

ORACLE CORPORATION*

Oracle offers database software and services. It's a major vendor of software for enterprise-level information management and was the first software company to offer Web-based products. In addition, Oracle offers "single-user versions" of some of its database products. The following list describes some of Oracle's database software and services:

- *Oracle Database 11g*—A relational DBMS that runs on Windows, Linux, and UNIX platforms and includes a variety of features for managing transaction processing, business intelligence, and content management applications.

- *Oracle OLAP*—An option in Oracle Database 11g Enterprise Edition that is a calculation engine for performing analyses such as planning, budgeting, forecasting, sales, and marketing reports. It improves the performance of complex queries with multidimensional data.

- *Oracle PeopleSoft*—An enterprise-level product for customer/supplier relationship management and human capital management (HCM).

- *Oracle Fusion Middleware*—A suite of products for service-oriented architecture (SOA), business process management, business intelligence, content management, identity management, and Web 2.0.

- Other Oracle products and services include Oracle E-Business Suite as well as Siebel customer-relationship management applications.

*This information has been gathered from the company Web site and other promotional materials. For detailed information and updates, visit *www.oracle.com*.

allocation (48)

client/server database (49)

create, read, update, and delete (CRUD) (46)

data dictionary (44)

data hierarchy (39)

data mart (53)

data model (43)

data warehouse (49)

database (39)

database administrator (DBA) (47)

database management system (DBMS) (40)

data-driven Web site (47)

data-mining analysis (53)

distributed database (48)

encapsulation (49)

extraction, transformation, and loading (ETL) (50)

foreign key (44)

fragmentation (48)

hierarchical model (43)

inheritance (49)

indexed sequential access method (ISAM) (42)

logical view (42)

network model (43)

normalization (44)

object-oriented databases (49)

online analytical processing (OLAP) (51)

online transaction processing (OLTP) (51)

physical view (42)

primary key (44)

query by example (QBE) (46)

random access file structure (42)

relational model (44)

replication (48)

sequential file structure (42)

Structured Query Language (SQL) (46)

Problems, Activities, and Discussions

1. Explain the difference between a logical view and a physical view of data in a database.

2. Explain the difference between primary keys and foreign keys.

3. What are the components of ETL?

4. Explain whether SQL or QBE is easier to learn and why.

5. Write a two-page report on how OLAP analysis is used in retail businesses. What decisions could be improved by OLAP analysis?

6. Go to *www.eweek.com/c/a/Enterprise-Applications/At-WalMart-Worlds-Largest-Retail-Data-Warehouse-Gets-Even-Larger/* and *www.informationweek.com/news/storage/showArticle.jhtml?articleID=201203024* and read about Wal-Mart's use of data warehouses. Write a two-page report on how data warehousing has improved retail operations at Wal-Mart.

7. Read the information at the following link and from other sources, and write a one-page paper that summarizes how to generate better business intelligence:

www.cio.com/article/170201/Four_Tips_for_Better_Business_Intelligence_in_2008?source=nlt_cioinsider

8. Read the information at the following link and from other sources, and write a one-page paper that outlines the limitations of data-mining tools.

www.infoworld.com/d/data-management/ny-bomb-plot-highlights-limitations-data-mining-762?source=IFWNLE__2010-05-10

9. Which of the following is not a DBMS component?

    a Database engine

    b Data description

    c Data manipulation

    d Application generation

10. A foreign key must not be a primary key in another table. True or False?

casestudy

BUSINESS INTELLIGENCE AND DATA WAREHOUSING AT HARRAH'S ENTERTAINMENT, INC.

With over 80,000 employees, Harrah's Entertainment, Inc., is the world's largest provider of casino entertainment, and the information it has gathered on its more than 40 million customers is stored in an NCR/Teradata data warehouse. The data has been collected over the years through the company's Total Rewards Card program, which records its customers' activities at the gaming tables and elsewhere. But there was a need to analyze that data to gain insight into customer preferences, gaming patterns, and so on. For this reason, Harrah's chose IBM Cognos business intelligence software to analyze the data stored in its data warehouse. With the system now in place, the company is able to classify customers into various profiles and use this classification to develop campaigns targeted to specific customer groups. For example, the company may target customers who haven't visited in more than 6 months and offer incentives to get

© Kevin Tietz/Shutterstock.com

them to return. The system is also able to measure the effectiveness of various campaigns. The implementation has significantly increased revenue, reduced operational and personnel costs, and enhanced customer service.[11]

Answer the following questions:

1. Summarize Harrah's Entertainment, Inc., information needs.

2. What main goals did the new information system achieve?

3. What were some additional benefits of using the new system?

LEARN YOUR WAY!

We know that no two students are alike. You come from different walks of life and with many different preferences. You need to study just about anytime and anywhere. **MIS2** was developed to help you learn Management Information Systems in a way that works for you.

Not only is the format fresh and contemporary, it's also concise and focused. And, **MIS2** is loaded with a variety of study tools, like in-text review cards, printable flash cards, and more.

Go to CourseMate for MIS2 to find plenty of resources to help you study—no matter what learning style you like best! Access at login.cengagebrain.com.

PERSONAL, LEGAL, ETHICAL, AND ORGANIZATIONAL ISSUES OF INFORMATION SYSTEMS

t his chapter begins by discussing the impact of information technology tools on privacy and how these tools can be used to commit computer crimes. It then examines privacy issues, such as censorship and data collection on the Internet, as well as intellectual property and copyright laws. Finally, it reviews some broader issues of information technologies, including the digital divide, electronic publishing, and effects on the workplace and green computing.

learning outcomes

After studying this chapter, you should be able to:

LO1 Describe information technologies that could be used in computer crimes.

LO2 Review privacy issues and methods for improving privacy of information.

LO3 Explain the effects of e-mail, data collection, and censorship on privacy.

LO4 Discuss ethical issues of information technology.

LO5 Describe intellectual property principles and infringement issues.

LO6 Explain information system issues affecting organizations, including the digital divide, electronic publishing, and effects on the workplace and employees' health.

LO7 Discuss green computing and ways it could help improve the quality of the environment.

Information technologies can be misused to invade users' privacy and commit computer crimes.

 1 Risks Associated with Information Technologies

nformation technologies can be misused to invade users' privacy and commit computer crimes. The following sections describe some of these misuses and discuss related privacy issues. Keep in mind, however, that you can minimize or prevent many of these risks by installing operating system updates regularly, using antivirus and antispyware software, and using e-mail security features.

1.1 Cookies

Cookies are small text files with unique ID tags that are embedded in a Web browser and saved on the user's hard drive. Sometimes, cookies are useful or innocuous, such as those used by a Web page that welcomes you or those used by a Web site that remembers your personal information for online ordering. Typically, users rely on Web sites to keep this information from being compromised. Cookies also make it possible for Web sites to customize pages for users, such as Amazon. com recommending books based on your past purchases.

> **Cookies** are small text files with unique ID tags that are embedded in a Web browser and saved on the user's hard drive.

Other times, cookies can be considered an invasion of privacy, and some people believe their information should be collected only with their consent. Cookies also provide information about the user's location and computer equipment, and this information could be used for unauthorized purposes, such as corporate espionage.

For these reasons, many users disable cookies by installing a cookie manager, which can eliminate

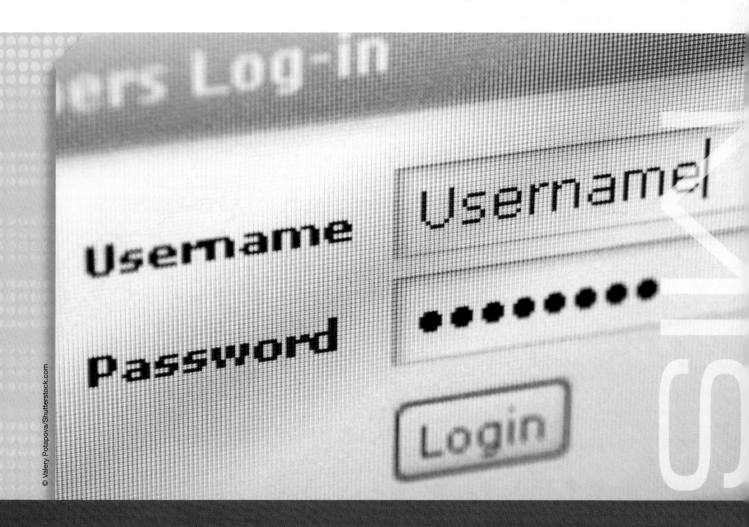

existing cookies and prevent additional cookies from being saved to a user's hard drive. Popular Web browsers such as Internet Explorer and Firefox provide a range of options for accepting and restricting cookies.

1.2 Spyware and Adware

Spyware is software that secretly gathers information about users while they browse the Web. This information could be used for malicious purposes. Spyware can also interfere with users' control of their computers by installing additional software and redirecting Web browsers, for example. Some spyware changes computer settings, resulting in slow Internet connections, changes to users' default home pages, and loss of functions in other programs. To protect against spyware, you should install antivirus software that also checks for spyware or install antispyware software, such as Spy Sweeper, CounterSpy, STOPzilla, and Spyware Doctor.

Adware is a form of spyware that collects information about the user (without the user's consent) to display advertisements in the Web browser, based on information it collects from the user's browsing patterns. In addition to antivirus software, installing an ad-blocking feature in Web browsers is recommended to protect against adware.

1.3 Phishing

Phishing is sending fraudulent e-mails that seem to come from legitimate sources, such as a bank or university. They usually direct e-mail recipients to false Web sites that look like the real thing for the purpose of capturing private information, such as bank account numbers or Social Security numbers.

1.4 Keyloggers

Keyloggers monitor and record keystrokes and can be software or hardware devices. Sometimes, companies use these devices to track employees' use of e-mail and the Internet, and this use is legal. However, keyloggers can be used for malicious purposes, too, such as collecting the credit card numbers that users enter while shopping online. Some antivirus and antispyware programs guard against software keyloggers, and utilities are available to install as additional protection.

1.5 Sniffing and Spoofing

Sniffing is capturing and recording network traffic. Although it can be done for legitimate reasons, such as monitoring network performance, hackers often use it to intercept information. **Spoofing** is an attempt to gain access to a network by posing as an authorized user to find sensitive information, such as passwords and credit card information. Spoofing can also be illegitimate programs posing as legitimate ones.

1.6 Computer Crime and Fraud

Computer fraud is the unauthorized use of computer data for personal gain, such as transferring money from another's account or charging purchases to someone else's account. Many of the technologies discussed previously can be used for committing computer crimes. In addition, social networking sites, such as Facebook and MySpace, have been used for committing computer crimes. The following information box discusses the cost of Internet fraud.

© Michael D Brown/Shutterstock.com

Spyware is software that secretly gathers information about users while they browse the Web.

Adware is a form of spyware that collects information about the user (without the user's consent) to display advertisements in the Web browser, based on information it collects from the user's browsing patterns.

Phishing is sending fraudulent e-mails that seem to come from legitimate sources, such as a bank or university, for the purpose of capturing private information, such as bank account numbers or Social Security numbers.

Keyloggers monitor and record keystrokes and can be software or hardware devices.

Sniffing is capturing and recording network traffic.

Spoofing is an attempt to gain access to a network by posing as authorized user to find sensitive information, such as passwords and credit card information.

Computer fraud is the unauthorized use of computer data for personal gain.

In addition to phishing, which was introduced earlier in the chapter, computer crimes include the following:

- Denial-of-service attacks, which inundate a Web site or network with e-mails and other network traffic so that it becomes overloaded and can't handle legitimate traffic.
- Identity theft, such as stealing Social Security numbers for unauthorized use.
- Software piracy and other infringements of intellectual property (discussed later in the chapter).
- Distributing child pornography.
- E-mail spamming.
- Writing or spreading viruses, worms, Trojan programs, and other malicious code.
- Stealing files for industrial espionage.
- Changing computer records illegally.
- Virus hoaxes, in which individuals intentionally spread false statements or information through the Internet in such a way that readers believe they are true.

Another computer crime is sabotage, which involves destroying or disrupting computer services. Computer criminals change, delete, hide, or use computer files for personal gain. Usually called "hackers," many of them break into computer systems for personal satisfaction, but others seek financial gain. Surprisingly, most computer crimes are committed by company insiders, which makes protecting information resources even more difficult.

2 Privacy Issues

Information technologies have brought many benefits, but they've also created concerns about privacy in the workplace. For example, some employers search social networking sites, such as Facebook or MySpace, to find background information on applicants, and this information can influence their hiring decisions. Is this use of social networking sites legal or ethical? What about users' privacy? Because information posted on these sites is often considered public domain, you should be careful about what you post, in case it comes back to haunt you. See the information box called "Social Networking Sites and Privacy Issues" for an example of what could happen.

With employee monitoring systems, managers can also supervise employees' performance—the number of errors they make, their work speed, and their time away from the desk. Naturally, this monitoring has made some workers concerned about their privacy.

© Monkey Business Images/Shutterstock.com

Social Networking Sites and Privacy Issues

Stacy Snyder, a former student at Millersville University of Pennsylvania, posted on MySpace a photo of herself wearing a pirate's hat while drinking. The photo was captioned "Drunken Pirate." Although Snyder was of legal drinking age at the time, Millersville administrators considered the image unprofessional and refused to grant her a degree in Education and a teaching certificate. Instead, she was given a degree in English. Did the university violate Stacy's privacy?[2]

> ...with employee monitoring systems, managers can supervise employees' performance—the number of errors they make, their work speed, and their time away from the desk.

Health care organizations, financial institutions, legal firms, and even online-ordering firms gather a great deal of personal data and enter it in databases. Misuse and abuse of this information can have serious consequences. For this reason, organizations should establish comprehensive security systems (discussed in Chapter 5) to protect their employees' or clients' privacy.

Some "information paranoia" is valid, because information about almost every aspect of people's lives is now stored on various databases, and misuse of extremely sensitive information (such as medical records) could prevent someone from getting employment, health insurance, or housing. Laws are in place to prevent these problems, but taking legal action is often costly, and by that point, the damage has often already been done.

You can probably give examples of things you expect to be private, such as your personal mail, your bank account balances, and your phone conversations. Defining privacy is difficult, however. In terms of electronic information, most people believe they should be able to keep their personal affairs to themselves and should be told how information about them is used. Based on this definition, many practices of government agencies, credit agencies, and marketing companies using databases would represent an invasion of privacy. Unfortunately, information technologies have increased ease of access to information for hackers as well as for legitimate organizations.

The number of databases is increasing rapidly. In the United States, for example, the top three credit-rating companies—Experian, Equifax, and TransUnion—have records on nearly every person living in the United States. Although these organizations and agencies are reputable and supply information only to people using it for its intended purpose, many small companies buy information from credit-rating companies and use it in ways that were never intended. This action is clearly illegal, but enforcement of federal laws has been lax. You may have noticed the effects of this problem if you recently joined an organization and then began getting mail from other organizations that you have not given your address to.

Advances in computer technology have made it easy to do what was once difficult or impossible. Information in databases can now be cross-matched to create profiles of people and predict their behavior, based on their transactions with educational, financial, government, and other institutions. This information is often used for direct marketing and for credit checks on potential borrowers or renters.

The most common way to index and link databases is by using Social Security numbers (typically obtained from credit bureaus), although names are sometimes used to track transactions that don't require Social Security numbers, such as credit card purchases, charitable contributions, and movie rentals. Direct marketing companies are a major user of this information. You may think that the worst result of this information sharing is an increase in junk mail (postal mail or e-mail), but there are more serious privacy issues than that. Should information you give to a bank to help establish a credit record be repackaged (that is, linked with other databases) and used for other purposes?

In 1977, the U.S. government began linking large databases to find information. The Department of Health, Education, and Welfare decided to look for people collecting welfare who were also working for the government. (Collecting welfare while being employed is illegal.) By comparing records of welfare payments and government payroll, the department was able to identify these workers. In this case, people abusing the system were discovered, so this use of databases was useful.

The Housing and Urban Development Department keeps records showing whether mortgage borrowers are in default on federal loans and made this information available to large banking institutions, such as Citibank, to add to their credit files. This action led Congress to pass the first of several laws intended to protect people's rights of privacy with regard to their credit records.

Several federal laws now regulate the collecting and using of information on people and corporations, but the laws are narrow in scope and contain loopholes. For example, the 1970 Fair Credit Reporting Act prohibits credit agencies from sharing information with anyone but "authorized customers." An authorized customer, however, is simply defined as anyone with a "legitimate need," and the act doesn't specify what a legitimate need is.

There are three important concepts regarding Internet and network privacy: **acceptable use policies**, accountability, and nonrepudiation. To guard against possible legal ramifications and the consequences of using the Internet and networks, organizations usually establish an acceptable use policy, which is a set of rules specifying the legal and ethical use of a system and the consequences of noncompliance. Having a clear, specific policy can help prevent users from taking legal action against an organization, as in cases of termination. Most organizations have new employees sign an acceptable use policy before they can access the network. Accountability refers to issues involving both the user's and the organization's responsibility and liability. Nonrepudiation is basically a method for binding all involved parties to a contract; it is covered in more detail in Chapter 5.

Because of concerns about privacy, hardware or software controls should be used to determine what personal information is provided on the Web. Chapter 5 explains these controls in more detail, but users and organizations should adhere to the following guidelines to eliminate or minimize the invasion of privacy:[3]

- Conduct business only with Web sites whose privacy policies are easy to find, read, and understand.

- Organizations must limit access to personal information to people who have authorization.

- Any organization creating, maintaining, using, or disseminating records of personal data must ensure the data's reliability and take precautions to prevent misuse of the data.

- Any data collection must have a stated purpose. Organizations should keep collected information only as long as it's needed for the stated purpose.

- There must be a way for people to prevent personal information that was gathered about them for one purpose from being used for other purposes or being disclosed to others without their consent.

- Organizations should monitor data collection and entry, and use verification procedures to ensure data accuracy; they should also collect only the data that's necessary.

- Records kept on an individual should be accurate and up to date. Organizations must correct or delete incorrect data and delete data when it's no longer needed for the stated purpose.

- Users should be able to review their records and correct any inaccuracies.

- The existence of record-keeping systems storing personal data shouldn't be kept secret. In addition, there must be a way for people to find out what information about them has been stored and how it's used.

- Organizations must take all necessary measures to prevent unauthorized access to data and misuse of data.

> An **acceptable use policies** is a set of rules specifying the legal and ethical use of a system and the consequences of noncompliance.
>
> **Spam** is unsolicited e-mail sent for advertising purposes.

Privacy-protection software can take many forms. For example, to guard against cookies that record your navigations around the Web, you can use cookie control features within your browser. One company attempting to address this problem is Anonymizer, Inc., which is discussed in the Industry Connection box at the end of this chapter. Using privacy-protection software has some drawbacks, however. For example, eBay often contends with sellers who, by using different user accounts, bid on their own item to inflate the price. Currently, eBay can trace these sellers' user accounts, but privacy-protection software would make this tracking impossible.

2.1 E-mail

Although e-mail is widely used, it presents some serious privacy issues. One issue is junk e-mail, or **spam**—unsolicited e-mail sent for advertising purposes. Because sending these e-mails is so inexpensive,

even a small response—a fraction of a percent, for example—is a worthwhile return on the investment. Usually, spam is sent in bulk using automated mailing software, and many spammers sell their address lists. For these reasons, the volume of spam can rise to an unmanageable level quickly, clogging users' in-boxes and preventing access to legitimate e-mails.

Another privacy concern is ease of access. Whether an e-mail is distributed through the Internet or through a company network, people should assume that others have access to their messages. In addition, many organizations have policies stating that any e-mails sent on company-owned computers are the property of the organization, and that the organization has the right to access them. In other words, employees often have no right to privacy, although there's a lot of controversy over this point and several lawsuits have resulted.

Spamming has also created decency concerns, because these e-mails often contain explicit language or nudity and can be opened by children. The following list provides some 2009 statistics for e-mail and spam[4]:

- **90 trillion**—Number of e-mails sent on the Internet
- **247 billion**—Average number of e-mail messages per day
- **1.4 billion**—Number of e-mail users worldwide
- **100 million**—New e-mail users from the year before
- **81 percent**—Percentage of e-mails that were spam
- **92 percent**—Peak spam levels late in the year
- **24 percent**—Increase in spam over previous year

2.2 Data Collection on the Internet

The number of people shopping online is increasing rapidly because of convenience, the array of choices, and lower prices. Many customers, however, are re-luctant to make online purchases because of concerns about hackers getting access to their credit card numbers and charging merchandise to their accounts. To lessen consumers' concerns, many credit card companies reimburse fraudulent charges. In addition, other electronic payment systems are being developed, such as e-wallets and smart cards, that reduce the risks of exposing consumers' information on the Web (discussed in Chapter 8).

Some Web sites require you to enter your name, address, and employment information before you're allowed to use the site. Privacy issues include the concern that this personal information will be sold to telemarketing firms, and consumers don't want to be bombarded with spam. Also, some consumers are concerned about their computers' contents being searched while they're connected to the Internet, and personal information could be used without their consent for solicitation and other purposes.

Information that users provide on the Web can also be combined with other information and technologies to produce new information. For example, by collecting a person's employment information, a financial profile could be created and used for other purposes. Two commonly used technologies for data collection are cookies (discussed previously) and log files. **Log files**, which are generated by Web server software, record a user's actions on a Web site.

Sometimes, users give incorrect information on purpose—on chatting or dating sites, for example, or when opening e-mail accounts. If the information collected isn't accurate, the result could be identity misrepresentation. For example, if someone claims to be younger on an online dating site, any demographic data collected would be flawed. Similarly, if a TV network collects data on viewing trends through online surveys and people supply answers that aren't truthful, any analyses the network attempts to conduct wouldn't be accurate. Therefore, data collected on the Internet must be used and interpreted with caution.

Log files, which are generated by Web server software, record a user's actions on a Web site.

> The distinction between what's legal and what's illegal is usually clear, but drawing a line between what's ethical and what's unethical is more difficult.

3 Ethical Issues of Information Technologies

Companies such as Enron, Arthur Andersen, WorldCom, and Tyco, to mention a few, have highlighted the ethics issues that corporations face in the 21st century. In essence, ethics means doing the right thing, and its meaning can vary in different cultures and even from person to person.[5]

The distinction between what's legal and what's illegal is usually clear, but drawing a line between what's ethical and what's unethical is more difficult. Exhibit 4.1 shows a grid that can be used for assessing whether an action is legal and/or ethical.

Review the following situations and try to determine where they might fall in Exhibit 4.1's grid:

1. You make two copies of a software package you just bought and sell one to a friend.

2. You make two copies of a software package you just bought for personal use, in case the original software fails and you need a backup.

3. A banker uses the information a client enters in a loan application to sell other financial products to this client.

4. A credit card company sells its customers' mailing addresses to other competitors.

5. A supervisor fires a programmer who has intentionally spread viruses to the organization's network.

Exhibit 4.1 *Ethical versus legal grid*

	Legal	Illegal
Ethical	I	II
Unethical	III	IV

Number 1 is clearly illegal and unethical (quadrant IV). Number 2 is ethical because you made the copy for your own use, but some software vendors who prohibit making copies might consider it illegal (quadrant II). Numbers 3 and 4 are legal but not ethical (quadrant III). In number 5, the supervisor's behavior is both legal and ethical. The supervisor has a clear legal reason for firing the programmer, and allowing the programmer to continue working there wouldn't be ethical. As a future knowledge worker, watch your own actions, and make sure you behave both legally and ethically. Be careful about decisions you make affecting coworkers so you can help maintain an ethical working environment.

Some information systems professionals believe that information technology offers many opportunities for unethical behavior, particularly because of the ease of collecting and disseminating information. Cybercrime, cyberfraud, identity theft, and intellectual property theft (discussed later in this chapter) are on the rise. For example, the incidence of identity theft increased by 11 percent from 2008 to 2009, affecting 11 million Americans.[7]

Many experts believe management can reduce employees' unethical behavior by developing and enforcing codes of ethics. Many associations promote ethically responsible use of information systems and technologies and have developed codes of ethics for their members. The Association for Computing Machinery (ACM), for example, has a code of ethics and professional conduct to help guide the actions of IT professionals. The ACM's code of ethics (*www.acm.org/about/code-of-ethics*) includes the following moral guidelines:

- 1.1: Contribute to society and human well-being.
- 1.2: Avoid harm to others.
- 1.3: Be honest and trustworthy.
- 1.4: Be fair and take action not to discriminate.
- 1.5: Honor property rights, including copyrights and patents.
- 1.6: Give proper credit for intellectual property.
- 1.7: Respect the privacy of others.
- 1.8: Honor confidentiality.

As a knowledge worker, you should consider the following questions and statements before making a decision:

- Does this decision comply with my organization's values?
- How will I feel about myself after making this decision?
- If I know this decision is wrong, I must not make it.
- If I'm not sure about this decision, I must ask my supervisor before making it.
- Is the decision right?
- Is the decision fair? How will I feel if somebody else makes this decision on my behalf?
- Is the decision legal?
- Would I want everyone to know about this decision after I make it?

3.1 Censorship

No organization controls the whole Internet, so who decides what content should be on it? Two types of information are available on the Web: public and private. Public information, posted by an organization or public agency, can be censored for public policy reasons—such as not allowing military secrets to be published, lest the information fall into enemy hands. Public information can also be censored if the content is deemed offensive to a political, religious, or cultural group. However, private information—what's posted by a person—isn't subject to censorship because of our constitutional freedom of expression. Of course, whether or not something can be censored depends in part on who is doing the censoring. For example, if you agree to abide by an organization's (e.g., a company's or an Internet service provider's) terms of service or policies and then post something

that violates that, you might be censored or denied access.

Another type of censorship is restricting access to the Internet. Some countries, such as Burma, China, and Singapore, restrict or forbid their citizens' access or try to censor the information posted on the Internet. These governments believe that the racist, pornographic, and politically extreme content of some Web sites could affect national security. In other countries, only employees of multinational corporations have direct access to the Internet.

Although U.S. citizens don't want the government controlling Internet access, many parents are concerned about what their children are exposed to while using the Web, such as pornography, violence, and adult language.

Another concern is children searching for information on the Web. If a search includes keywords such as *toys, pets, boys,* or *girls,* for example, the results could list pornography sites. Guidelines for Web use have been published to inform parents of the benefits and hazards of the Internet, and parents can use these to teach their children to use good judgment while on the Internet. For example, Microsoft posts a guideline called "Help protect kids online: 4 things you can do" (*www.microsoft.com/protect/family/guidelines/basics.mspx*).

© Rob Marmion/Shutterstock.com

In addition, many parents use programs such as CyberPatrol, CyberSitter, Net Nanny, and SafeSurf to prevent their children's access to certain Web sites. Web browser software has also been developed to improve children's security. For example, a Web browser may accept e-mail only from an address that uses the same Web browser software. This helps ensure that children receive e-mail only from other children. Another possibility is creating different levels of user access, similar to movie ratings, to prevent children from accessing controversial or pornographic information. This system could use techniques such as requiring passwords or using biometrics, including fingerprints or retinal scans (discussed in Chapter 5).

3.2 Intellectual Property

Intellectual property is a legal umbrella covering protections that involve copyrights, trademarks, trade secrets, and patents for "creations of the mind" developed by people or businesses.[8] Intellectual property can be divided into two categories: industrial property (inventions, trademarks, logos, industrial designs, and so on) and copyrighted material, which covers literary and artistic works.

Generally, copyright laws protect tangible material, such as books, drawings, and so forth. However, they also cover online materials, including Web pages, HTML code, and computer graphics, as long as the content can be printed or saved on a storage device. Copyright laws give only the creator exclusive rights, meaning no one else can reproduce, distribute, or perform the work without permission.[9]

> **Intellectual property** is a legal umbrella covering protections that involve copyrights, trademarks, trade secrets, and patents for "creations of the mind" developed by people or businesses.

Copyright laws do have some exceptions, however, usually under the Fair Use Doctrine. This exception means you can use copyrighted material for certain purposes, such as quoting passages of a book in literary reviews. There are limits on the length of material you can use. In addition, some copyrighted material can be used to create new work, particularly for educational purposes. Checking copyright laws carefully before using this material is strongly recommended. The United States Copyright Office (*www.copyright.gov*) offers detailed information on copyright issues. Exhibit 4.2 shows the home page.

Other intellectual property protections include trademarks and patents. A trademark protects product names and identifying marks (logos, for instance). A patent protects new processes. (Note that laws governing trademarks, patents, and copyrights in the United States might not apply in other countries.) The length of a copyright varies based on the type of work, but, in general, copyrights last for the author's lifetime plus 70 years and do not need to be renewed, and patents last 20 years (14 years for design patents).

An organization can benefit from a patent in at least three ways:[10]

- It can generate revenue by licensing its patent to others.
- It can use the patent to attract funding for further research and development.
- It can use the patent to keep competitors from entering certain market segments.

Another copyright concern is software piracy, but the laws covering it are very straightforward. The 1980 revisions to the Copyright Act of 1976 include computer programs, so both people and organizations can be held liable for unauthorized duplication and use of copyrighted programs. Sometimes, contracts are used to supplement copyrights and give the software originator additional protection. For example, a software vendor might have a university sign a contract specifying how many people can use the software. Companies

Exhibit 4.2 *The United States Copyright Office home page*

Courtesy of the US Copyright Office

Cybersquatting is registering, selling, or using a domain name to profit from someone else's trademark.

Information technology and the Internet have created a **digital divide**. Computers still aren't affordable for many people. The digital divide has implications for education.

also make use of laws on trade secrets, which cover ideas, information, and innovations, as extra protection.

Most legal issues related to information technologies in the United States are covered by the Telecommunications Act of 1996, the Communications Decency Act (CDA), and laws against spamming. The CDA was partially overturned in the 1997 Reno v. ACLU case, in which the U.S. Supreme Court unanimously voted to strike down the CDA's anti-indecency provisions, finding they violated the freedom of speech provisions of the First Amendment. To avoid the legal risks listed here,[11] organizations should have an Internet use policy.

- *Risk 1*—If employees download pornographic materials to their office computers over the corporate network, the organization could be liable for harassment charges as well as infringement of privacy and even copyright laws.

- *Risk 2*—Indecent e-mail exchanges among employees can leave the corporation open to discrimination and sexual harassment charges.

- *Risk 3*—Employees using the corporate network to download and distribute unlicensed software could leave the corporation open to serious charges of copyright infringement and other legal issues.

Verizon's Cybersquatting Suit

In June 2008, Verizon sued OnlineNic, accusing it of trademark infringement and illegal cybersquatting. According to Verizon, OnlineNic registered domain names containing Verizon trademarks. The registered names included myverizonwireless.com, iphoneverizonplans.com, and verizon-cellular.com, among others, and Verizon was concerned about the names misleading consumers. Verizon won this suit and was awarded a $33 million judgment.[12]

One aspect of intellectual property that has attracted attention recently is **cybersquatting**, which is registering, selling, or using a domain name to profit from someone else's trademark. Often, it involves buying domains containing the names of existing businesses and then selling the names later for a profit. The information box called "Verizon's Cybersquatting Suit" describes such a case.

3.3 Social Divisions and the Digital Divide

Some believe that information technology and the Internet have created a **digital divide** between the information rich and the information poor. Although prices have been decreasing steadily, computers still aren't affordable for many people. In addition, a type of economic "red-lining" can occur when companies installing coaxial and fiber-optic cables for Internet connections focus on higher-income communities, where more residents are expected to use the Internet.[13]

Children, in particular, are often victims of the digital divide. Those without computers or Internet access at home, as well as students who can't afford computer equipment, are at a disadvantage and often become further behind in their education. Students without access to the wide array of resources on the Web have more difficulty writing papers and learning about topics that interest them. Interactive and virtual reality educational games available on the Internet can widen the gap more, when some children have access and others don't. Increasing funding for computer equipment at schools and adding more computers in public places, such as libraries, can help offset this divide. Some schools have even started loaner programs so that students can borrow a portable computer for use after school hours.

4 The Impact of Information Technology in the Workplace

lthough information technology has eliminated some clerical jobs, it has created many new jobs (described in Chapter 1) for programmers, systems analysts, database and network administrators, network engineers, Webmasters, Web page developers, e-commerce specialists, chief information officers (CIOs), and technicians. In e-commerce, jobs for Web designers, Java programmers, and Web troubleshooters have been created, too. Some argue that the jobs eliminated have been clerical and the jobs created have been mostly technical, requiring extensive training. Others believe that information technologies have reduced production costs and, therefore, improved and increased consumers' purchasing power, resulting in a stronger economy.

Information technologies have a direct effect on the nature of jobs. Telecommuting or virtual work, for example, has enabled some people to perform their jobs from home. With telecommunications technology, a

Sakala/Shutterstock.com

worker can send and receive data to and from the main office, and organizations can use the best and most cost-effective human resources in a large geographical region. Table 4.1 lists some benefits and drawbacks of telecommuting.

By handling repetitive and boring tasks, information technologies have made our jobs more interesting, resulting in more worker satisfaction. Information technologies has also led to "job deskilling." This occurs when skilled labor is eliminated by high technology or when a job is downgraded from a skilled to a semiskilled or unskilled position. It usually takes place when a job is

Table 4.1

The Benefits and Potential Drawbacks of Telecommuting

Benefit	Potential drawback
Can care for small children or elderly parents and spend more time with family	Can become a workaholic (no hard boundaries between "at work" and "at home")
Have fewer restrictions on clothing for work, thereby saving the expense of work wear	No regulated work routine
No commute, so distance and time factors are reduced as well as the effects of car emissions on air quality	Less interaction with coworkers
Able to work in more pleasant surroundings	No separation between work and home life
Increased productivity	Potential legal issues about workers' injuries
Decreased neighborhood crime because of more people being home during the day	Family interruptions and household distractions
Easier work environment for employees with disabilities	Lack of necessary supplies or equipment
Reduced costs for office space and utilities	Could create a two-tiered workforce—telecommuters and on-site workers—that affects promotions and raises
Reduced employee turnover and absenteeism	
Able to find and hire people with special skills, regardless of where they're located	
Fewer interruptions from coworkers	

automated or when a complex job is fragmented into a sequence of easily performed tasks. An example is when a computer-aided design (CAD) program performs the technical tasks that used to be performed by a designer. On the other hand, information technologies have created "job upgrading," as when clerical workers use computers for word-processing tasks. This upgrading makes it possible to add new tasks to employees' responsibilities, too; for example, clerical workers could be responsible for updating the company's Web site. Job upgrading has some limitations, however. Even with information technologies, training clerical workers to write programs for the company Web site would be difficult, for instance.

With information technologies, one skilled worker might be capable of doing the job of several workers. For example, with mail-merge programs, an office worker can generate thousands of letters, eliminating the need for additional workers. Information technologies can also make workers more efficient—being able to send a message throughout an entire organization by using e-mail instead of interoffice memos, for example. Similarly, mass-marketing efforts for new product announcements have been streamlined, reducing the expense and personnel needed to reach millions of customers.

Another impact of information technology is the creation of **virtual organizations**, which are networks of independent companies, suppliers, customers, and manufacturers connected via information technologies so that they can share skills and costs and have access to each other's markets.[14] A virtual organization doesn't need central offices or an organizational hierarchy for participants to contribute their expertise. Advantages of virtual organizations include the following:[15]

- Each participating company can focus on what it does best, thus improving the ability to meet customers' needs.
- Because skills are shared among participating companies, the cost of hiring additional employees is reduced.
- Companies can respond to customers faster and more efficiently.
- The time needed to develop new products is reduced.
- Products can be customized more to respond to customers' needs.

In 2001, Dell, Microsoft, and Unisys Corporation created a partnership to design a voting system for several U.S. states. Microsoft offered software, Dell offered hardware, and Unisys served as the systems integrator. This example illustrates the principle of virtual organizations—the idea that several organizations working together can do what one organization can't.

4.1 Information Technology and Health Issues

Although there have been reports of health problems caused by video display terminals (VDTs), no conclusive study indicates that VDTs are the cause, despite all the complaints. Work habits can cause some physical problems, however, and so can the work environment in which computers are used—static electricity, inadequate ventilation, poor lighting, dry air, unsuitable furniture, and too few rest breaks.

Other reports of health problems related to computer equipment include vision problems, such as fatigue, itching, and blurred vision; musculoskeletal problems (back strain and wrist pain); skin problems, such as rashes; reproductive problems, such as miscarriage; and stress-related problems (headaches and depression). Ergonomics experts believe that using better-designed furniture as well as flexible or wireless keyboards, correct lighting, special monitors for workers with vision problems, and so forth can solve many of these problems.

Another recent health issue is the amount of time some people spend on the Web playing games, participating in chat rooms, and other activities. Although the

Health and Social Issues of Online Gaming

The online games World of Warcraft and EverQuest, both MMORPG (massively multiplayer online role-playing games) have been blamed for a host of problems, including poor academic performance, divorces, suicide, and the death of a child because of parental neglect. Mental health professionals believe the fantasy worlds in online games can be addicting and affect marriages and careers. According to Dr. Timothy Miller, a clinical psychologist, it's a growing problem with teenage and young adult men.[16]

Internet can provide valuable educational resources, too much time on the Web can create psychological, social, and health problems, especially for young people. The information box called "Health and Social Issues of Online Gaming" mentions some of these problems.

● 5 Green Computing

green computing is computing that promotes a sustainable environment and consumes the least amount of energy. Information and communications technology (ICT) generates approximately 2% of the world's carbon dioxide emissions, roughly the same amount as the aviation industry.[17] Although ICT is a part of the problem, however, it could also be part of the solution. Many IT applications and tools can help reduce carbon dioxide emissions. Green computing not only helps an organization save on energy costs, it improves the quality of the environment that we live and work in.

Green computing involves the design, manufacture, use, and disposal of computers, servers, and computing devices (such as monitors, printers, storage devices, and networking and communications equipment) in such a way that there is minimal impact on the environment.[18] It is one of the methods for combating global warming. In some states, certain computer manufacturers collect a fee from their customers, called an advance recovery fee, in order to dispose of the computer after its useful life. A successful green computing strategy cannot be fully implemented without the cooperation of both the private and the public sector. Furthermore, both employees and top management must be involved.

Industry Connection

ANONYMIZER, INC. *

Anonymizer, Inc. provides online privacy services so that users can browse the Web anonymously and securely. Its features include hiding users' IP addresses and removing cookies, spyware, and adware. Although using Anonymizer can slow down connection and surfing speeds, many consumers, businesses, and government agencies take advantage of its products and services, which include the following:

- *Digital Shredder Lite*—Erases traces of your Internet activity and Windows use, including Internet history, cache, and cookies as well as recently opened files, the Temp folder, and the Recycle Bin.

- *Nyms*—Enable you to create and destroy alias e-mail addresses to protect your real e-mail address from spamming and phishing attempts.

- *Antispyware*—Removes adware and spyware that have accumulated on your computer and prevents new spyware and adware from being installed on your computer.

- *Enterprise Web harvesting tools*—Web harvesting is collecting data from Web sites, usually for competitive intelligence. Anonymizer helps businesses protect their corporate information from Web harvesting tools.

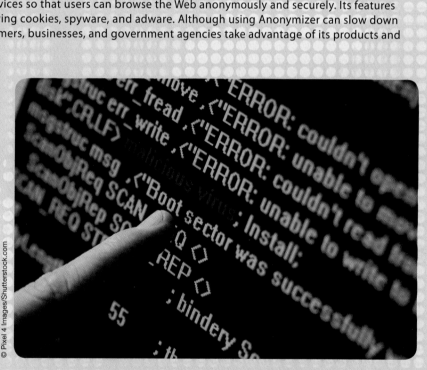

© Pixel 4 Images/Shutterstock.com

*This information has been gathered from the company Web site (*http://anonymizer.com*) and other promotional materials. For more information and updates, visit the Web site.

There are several ways to pursue a green computing strategy. Some can be easily done with no cost to the organization. Others are more challenging and require an initial investment. Here are some examples:

- Designing products that last longer and are modular in design, so that certain parts can be upgraded without replacing the entire system.
- Designing search engines and other computing routines that are faster and consume less energy.
- Replacing several underutilized smaller servers with one large server using a virtualization technique. In this case, multiple operating systems are hosted on a single hardware platform and can share this hardware platform. IBM's Project Big Green is an example of virtualization, with energy savings of approximately 42 percent for an average data center.[19]
- Using computing devices that consume less energy and are biodegradable.
- Allowing certain employees to work from their homes, resulting in fewer cars on the roads (discussed earlier in this chapter).
- Replacing actual face-to-face meetings with meetings over computer networks (discussed in Chapter 6).
- Using video conferencing, electronic meeting systems, and groupware (discussed in Chapter 12). These technologies can also reduce business travel.
- Using a virtual world (discussed in Chapter 14). This technology can also reduce face-to-face meetings, resulting in less travel.

- Using cloud computing as promoted by companies such as Amazon.com (discussed in Chapter 14). This platform can also reduce energy consumption.
- Turning off idle PCs, recycling computer-related materials, and encouraging carpool and nonmotorized transportation for employees.

The Industry Connection highlights Anonymizer Inc.'s online privacy services.

6 Chapter Summary

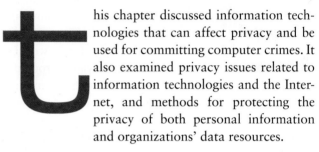

this chapter discussed information technologies that can affect privacy and be used for committing computer crimes. It also examined privacy issues related to information technologies and the Internet, and methods for protecting the privacy of both personal information and organizations' data resources.

The chapter reviewed ethical issues related to information systems including censorship, laws related to intellectual property, and the digital divide. It covered the effect of information systems on the workplace and reviewed possible heath issues. Finally green computing and its impacts on the environment were highlighted.

Key Terms

acceptable use policy (63)

adware (60)

computer fraud (60)

cookies (59)

cybersquatting (68)

digital divide (68)

intellectual property (67)

keyloggers (60)

log files (64)

phishing (60)

sniffing (60)

spam (63)

spoofing (60)

spyware (60)

virtual organizations (70)

Problems, Activities, and Discussions

1. What are some information technologies that could affect your privacy?

2. Give two examples of combined illegal and unethical behavior.

3. What are some ways to minimize the digital divide for children?

4. Visit *www.youtube.com/watch?v=SdxPgjfQ9yk*, consult other sources, and write a one-page paper on green computing that describes three ways to promote green IT.

5. Visit *www.youtube.com/watch?v=p11lJOnALS4*, consult other sources, and write a one-page paper on server virtualization that describes how it can help organizations reduce their energy consumption. Also, name two computer companies that are active in virtualization techniques.

6. Visit the following Web page, consult other sources, and write a one-page paper that summarizes why executives are easy targets for cybercriminals: *www.infoworld.com/d/security-central/4-reasons-why-execs-are-the-easiest-social-engineering-targets-264*

7. Visit the following Web page, consult other sources, and write a one-page paper that summarizes how spammers get your e-mail address: *www.cloudmark.com/media/spamnetnews/01/*

8. Visit the following Web page, consult other sources, and write a one-page paper that summarizes how telecommuters can handle more work: *www.infoworld.com/d/green-it/research-telecommuters-flex-time-can-handle-50-percent-more-work-866*

9. Intellectual property laws don't cover which of the following?
   a. Copyrights
   b. Trademarks
   c. Vendor contracts
   d. Patents

10. Indecent e-mail exchanges among employees can leave corporations open to charges of discrimination and sexual harassment. True or False?

casestudy

PRIVACY AND SECURITY BREACHES AT ACXIOM

Acxiom Corporation, based in Little Rock, Arkansas, provides companies with consumer information that they can use to develop and build relationships with customers (*www.acxiom.com/overview*). The companies include credit card companies, retail companies, automakers, and insurers. Acxiom is the world's largest processor of consumer data, collecting and analyzing more than a billion records a day. Its access to such a vast database has been useful for other purposes, too. After September 11, 2001, when the FBI released the names of the 19 hijackers, Acxiom was able to locate 11 of them in its database.

In 2003, Acxiom encountered a major security breach of its consumer data. Daniel Baas, a computer systems administrator at a data-marketing company, had stolen the personal data of millions of people from Acxiom's

© Harry B. Lamb/Shutterstock.com

servers over a period of 2 years. Although the information doesn't appear to have been used to defraud consumers, this event raises security concerns about the vulnerability of databases such as Acxiom's.[20]

Answer the following questions:

1. What could Acxiom do to increase the security and privacy of its customers' data?

2. Based on the article cited in this case study and your own research, do you think the government should have access to private-sector data?

3. In addition to government, what organizations might need access to private data for security reasons?

PROTECTING INFORMATION RESOURCES

t his chapter discusses security issues and measures related to computer and network environments. A comprehensive security system can protect an organization's information resources, which are its most important asset after human resources. The chapter then discusses major types of security threats and a variety of measures designed to protect against those threats. Some organizations even use the resources of a Computer Emergency Response Team (CERT) to handle threats and their effects. Finally, we cover the principles behind devising a comprehensive security system and the use of business continuity planning to help an organization recover from a disaster.

 learning outcomes

After studying this chapter, you should be able to:

LO1 **Describe basic safeguards in computer and network security.**

LO2 **Explain the major security threats.**

LO3 **Describe security and enforcement measures.**

LO4 **Summarize the guidelines for a comprehensive security system, including business continuity planning.**

Is Facebook a Friend or Fiend?

It wasn't the price, it was the volume. In 2010, researchers at VeriSign's iDefense group announced that they had discovered a hacker named Kirllos who was peddling 1.5 million stolen Facebook accounts for as little as 2.5 cents per account. If true, that would mean that one out of every 300 Facebook users were, unbeknownst to them, on the market. Their accounts had been hacked into and were now up for sale. Stolen social-networking accounts were nothing new, but Kirllos could get them for you wholesale—and the prices, which varied according to the number of contacts a particular account had, were literally a steal. Cybercriminals use such stolen accounts to spam, scam, and otherwise profit from unwary Facebook users, who are likely to respond to a familiar face or name without realizing that the friend is a fiend.[1]

1 Computer and Network Security: Basic Safeguards

omputer and network security has become critical for most organizations, especially in recent years with hackers, or computer criminals, becoming more numerous and more adept at stealing and altering private information. The various types of hackers are described in an information box. Hackers use a variety of tools to break into computers and networks, such as sniffers, password crackers, rootkits, and many others, which can be found free on the Web. Also, journals such as *Phrack* and *2600: The Hacker Quarterly* offer hackers informative tips. The Facebook information box describes a crime committed by a hacker against some Facebook users.

A comprehensive security system protects an organization's resources, including information and computer and network equipment. The information an organization needs to protect can take many forms: e-mails, invoices transferred via electronic data interchange (EDI), new product designs, marketing campaigns, and financial statements. Security threats involve more than stealing data; they include such actions as sharing passwords with co-workers, leaving a computer unattended while logged on to the network,

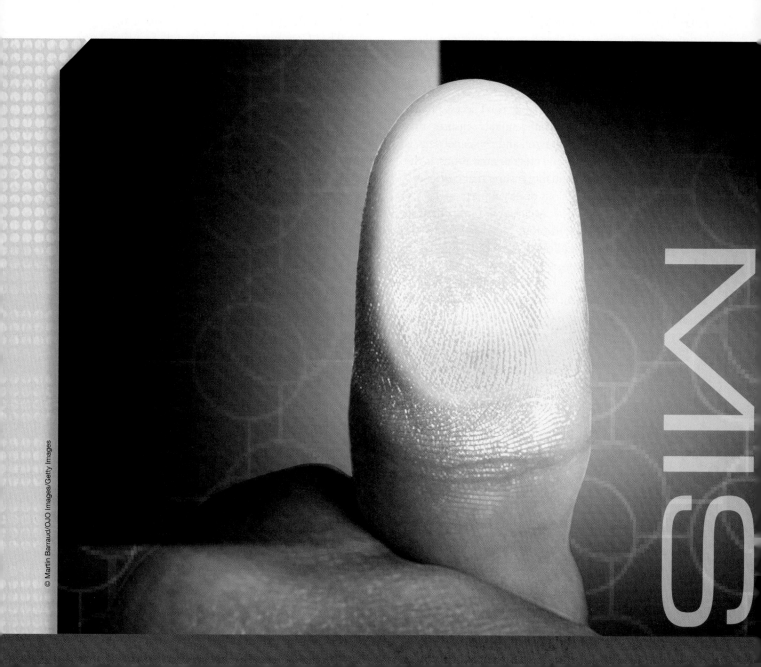

MIS

Types of Hackers

Script kiddie: An inexperienced, usually young hacker who uses programs that others have developed to attack computer and network systems and deface Web sites.

Black hat: Hackers who specialize in unauthorized penetration of information systems. They attack systems for profit, fun, or political motivation or as part of a social cause. These penetration attacks often involve modifying and destroying data.

White hat, also known as "ethical hackers:" Computer security experts who specialize in penetration testing and other testing methods to ensure that a company's information systems are secure.

or even spilling coffee on a keyboard. A comprehensive security system includes hardware, software, procedures, and personnel that collectively protect information resources and keep intruders and hackers at bay. There are three important aspects of computer and network security: confidentiality, integrity, and availability, collectively referred to as "the CIA triangle."[2]

Confidentiality means that a system must not allow the disclosing of information by anyone who isn't authorized to access it. In highly secure government agencies, such as the Department of Defense, the CIA, and the IRS, confidentiality ensures that the public can't access private information. In businesses, confidentiality ensures that private information, such as payroll and personnel data, is protected from competitors and other organizations. In the e-commerce world, confidentiality ensures that customers' data can't be used for malicious or illegal purposes.

Integrity refers to the accuracy of information resources within an organization. In other words, the security system must not allow data to be corrupted or allow unauthorized changes to a corporate database. In financial transactions, integrity is probably the most important aspect of a security system, because incorrect or corrupted data can have a huge impact. For example, imagine a hacker breaking into a financial network and changing a customer's balance from $10,000 to $1000—a small change, but one with a serious consequence. Database administrators and Webmasters are essential in this aspect of security. In addition, part of ensuring integrity is identifying authorized users and granting them access privileges.

Availability means that computers and networks are operating, and authorized users can access the information they need. It also means a quick recovery in the event of a system failure or disaster. In many cases, availability is the most important aspect for authorized users. If a system isn't accessible to users, the confidentiality and integrity aspects can't be assessed.

The Committee on National Security Systems (CNSS) has proposed another model, called the "McCumber cube." John McCumber created this framework for evaluating information security. Represented as a three-dimensional cube (see Exhibit 5.1), it defines nine characteristics of information security.[3] The McCumber cube is more specific than the CIA triangle and helps designers of security systems consider many crucial issues for improving the effectiveness of security measures. Note that this model includes the different states in which information can exist in a system: transaction, storage, and processing.

Confidentiality means that a system must prevent disclosing information to anyone who isn't authorized to access it.

Integrity refers to the accuracy of information resources within an organization.

Availability means that computers and networks are operating, and authorized users can access the information they need. It also means a quick recovery in the event of a system failure or disaster.

Exhibit 5.1 *The McCumber cube*

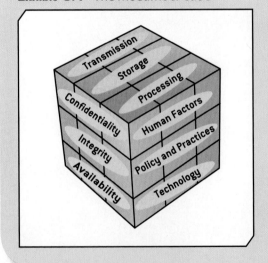

In addition, a comprehensive security system must provide three levels of security:

- *Level 1*—Front-end servers, those available to both internal and external users, must be protected against unauthorized access. Typically, these systems are e-mail and Web servers.

- *Level 2*—Back-end systems (such as users' work-stations and internal database servers) must be protected to ensure confidentiality, accuracy, and integrity of data.

- *Level 3*—The corporate network must be protected against intrusion, denial-of-service attacks, and unauthorized access.

When planning a comprehensive security system, the first step is designing **fault-tolerant systems**, which use a combination of hardware and software for improving reliability—a way of ensuring availability in case of a system failure. Some commonly used methods include the following:

- *Uninterruptible power supply (UPS)*—This backup power unit continues to provide electrical power in the event of blackouts and other power interruptions and is most often used to protect servers. It performs two crucial tasks: It serves as a power source to continue running the server (usually for a short period), and it safely shuts down the server. More sophisticated UPS units can prevent users from accessing the server and send an alert to the network administrator.

- *Redundant array of independent disks (RAID)*—As you learned in Chapter 2, a RAID system is a collection of disk drives used to store data in multiple places. RAID systems also store a value called a "checksum," used to verify that data has been stored

or transmitted without error. If a drive in the RAID system fails, data stored on it can be reconstructed from data stored on the remaining drives. RAID systems vary in cost, performance, and reliability.

- *Mirror disks*—This method uses two disks containing the same data so that, if one fails, the other is available, allowing operations to continue. Mirror disks are usually a less expensive, level-1 RAID system and can be a suitable solution for small organizations.

Fault-tolerant systems ensure availability in the event of a system failure by using a combination of hardware and software.

2 Security Threats: An Overview

Computer and network security are important to prevent loss of, or unauthorized access to, important information resources. Some threats can be controlled completely or partially, but some can't be controlled. For example, you can control power fluctuations and blackouts to some degree by using surge suppressors and UPSs, but you can't control whether natural disasters strike. You can, however, minimize the effects of a natural disaster by making sure fire suppression systems are up to code or by making structural changes to your organization's facility for earthquake protection—such as bolting the foundation.

Threats can also be categorized by whether they're unintentional (such as natural disasters, a user's accidental deletion of data, and structural failures) or intentional. Intentional threats include hacker attacks and attacks by disgruntled employees—spreading a virus on the company network, for instance. The following sections describe the most common intentional threats.

2.1 Intentional Threats

Discussed in the following sections, intentional computer and network threats, include:

- Viruses
- Worms
- Trojan programs
- Logic bombs
- Backdoors
- Blended threats (e.g., worm launched by Trojan)
- Rootkits
- Denial-of-service attacks
- Social engineering

2.1.1 Viruses

The most well-known computer and network threat are viruses, which are a type of malware. (Malware, or malicious software, is any program or file that's harmful to computers or networks.) According to Symantec Corporation, the number of computer viruses exceeded 1 million in 2008, an increase of 486% from 2007.[4] According to Panda Labs, over 25 million unique malware programs were identified in 2009, more than all the malware programs created up to that time.[5]

Estimating the dollar cost of damage that viruses cause, however, can be difficult; many organizations are reluctant to report their losses, because they don't want to publicize how vulnerable they are.

Viruses are usually given names; you've probably heard of the I Love You and Michelangelo viruses, for example. A **virus** consists of self-propagating program code that's triggered by a specified time or event. When the program or operating system containing the virus is used, the virus attaches itself to other files, and the cycle continues. The seriousness of viruses varies,

ranging from playing a prank, such as displaying a funny (but usually annoying) image on the user's screen, to destroying programs and data.

Viruses can be transmitted through a network or through e-mail attachments. Some of the most dangerous ones come through bulletin boards or message boards, because they can infect any system using the board. Experts believe that viruses infecting large servers, such as those used by air traffic control systems, pose the most risk to national security.

Sometimes, virus hoaxes are spread, too. These reports about viruses that turn out not to exist can cause panic and even prompt organizations to shut down their networks. In some ways, hoaxes can cause as much damage as real viruses.

The following list describes some of the indications that a computer might be infected by a virus:

- Some programs have suddenly increased in size.
- Files have been corrupted, or you can't open some files.
- Hard disk free space is reduced drastically.
- The keyboard locks up, or the screen freezes.
- Available memory dips down more than usual.
- Disk access is slow.

A **virus** consists of self-propagating program code that's triggered by a specified time or event. When the program or operating system containing the virus is used, the virus attaches itself to other files, and the cycle continues.

- Your computer takes longer than normal to start.
- There is unexpected disk activity, such as the disk light flashing even though you're not trying to save or open a file.
- There are unfamiliar messages on screen.

Installing and updating an antivirus program is the best measure against viruses. Widely used antivirus programs include McAfee Virus Scan (*www.mcafee.com/us/*), Norton Antivirus (*www.norton.com*), and Trend Micro (*www.trendmicro.com*). You can even download free or low-cost programs on the Internet. Most computers now have antivirus software already installed, but you should check for the most current version of the antivirus software. New viruses are released constantly, so use automatic updating to make sure your computer's protection is current.

2.1.2 Worms

A **worm** travels from computer to computer in a network, but it doesn't usually erase data. Unlike a virus, it is an independent program that can spread itself without having to be attached to a host program. It might corrupt data, but it usually replicates itself into a full-blown version that eats up computing resources, eventually bringing a computer or network to a halt. Well-known worms include Code Red, Melissa, and Sasser. Conficker, a recent worm, has infected millions of Windows computers.

2.1.3 Trojan Programs

A **Trojan program** (named after the Trojan horse that the Greeks used to enter Troy during the Trojan Wars) contains code intended to disrupt a computer, network, or Web site, and it is usually hidden inside a popular program. Users run the popular program, unaware that the malicious program is also running in the background. Disgruntled programmers trying to get even with an organization have created many Trojan programs. These programs can erase data and wreak havoc on computers and networks, but they don't replicate themselves, as viruses and worms do.

2.1.4 Logic Bombs

A **logic bomb** is a type of Trojan program used to release a virus, worm, or other destructive code. Logic bombs are triggered at a certain time (sometimes the birthday of a famous person) or by a specific event, such as a user pressing Enter or running a certain program.

2.1.5 Backdoors

A **backdoor** (also called a "trapdoor") is a programming routine built into a system by its designer or programmer. This routine enables the designer or programmer to bypass system security and sneak back into the system later to access programs or files. Usually, system users aren't aware a backdoor has been activated, although a user logon or combination of keystrokes can be used to activate backdoors.

2.1.6 Blended Threats

A **blended threat** is a security threat that combines the characteristics of computer viruses, worms, and other malicious codes with vulnerabilities found on public and private networks. Blended threats search

A **worm** travels from computer to computer in a network, but it doesn't usually erase data. Unlike viruses, worms are independent programs that can spread themselves without having to be attached to a host program.

A **Trojan program** contains code intended to disrupt a computer, network, or Web site, and it is usually hidden inside a popular program. Users run the popular program, unaware that the malicious program is also running in the background.

A **logic bomb** is a type of Trojan program used to release a virus, worm, or other destructive code. Logic bombs are triggered at a certain time (sometimes the birthday of a famous person) or by a specific event, such as a user pressing Enter or running a certain program.

A **backdoor** (also called a "trapdoor") is a programming routine built into a system by its designer or programmer. It enables the designer or programmer to bypass system security and sneak back into the system later to access programs or files.

A **blended threat** is a security threat that combines the characteristics of computer viruses, worms, and other malicious codes with vulnerabilities found on public and private networks.

for vulnerabilities in computer networks and then take advantage of these vulnerabilities by embedding malicious codes in the server's HTML files or by sending unauthorized e-mails from compromised servers with a worm attachment. They may launch a worm through a Trojan horse or launch a denial-of-service (DoS) attack at a targeted IP address. Their goal is not just to start and transmit an attack—but to spread it. A multilayer security system, as discussed in this chapter, can guard against blended threats.

2.1.7 Denial-of-Service Attacks

A **denial-of-service (DoS) attack** floods a network or server with service requests to prevent legitimate users' access to the system. Think of it as 5000 people surrounding a store and blocking customers who want to enter; the store is open, but it can't provide service to legitimate customers. Typically, DoS attackers target Internet servers (usually Web, FTP, or mail servers), although any system connected to the Internet running TCP services is subject to attack.

In February 2000, hackers launched DoS attacks against Web sites such as eBay, Yahoo!, Amazon.com, CNN.com, and E*TRADE and slowed down their services drastically. This assault was a distributed denial-of-service (DDoS) attack, in which hundreds or thousands of computers worked together to bombard a Web site with thousands of requests for information in a short period, causing it to grind to a halt. Because these attacks come from multiple computers, they're difficult to trace.

> A **denial-of-service (DoS) attack** floods a network or server with service requests to prevent legitimate users' access to the system.

> In the context of security, **social engineering** means using people skills—such as being a good listener and assuming a friendly, unthreatening air—to trick others into revealing private information. This attack takes advantage of the human element of security systems.

2.1.8 Social Engineering

In the context of security, **social engineering** means using "people skills"—such as being a good listener and assuming a friendly, unthreatening air—to trick others into revealing private information. This attack takes advantage of the human element of security systems. Social engineers use the private information they've gathered to break into servers and networks and steal data, thus compromising the integrity of information resources. Social engineers use a variety of tools and techniques to gather private information, including publicly available sources of information: Google Maps, company Web sites, newsgroups, and blogs, for example.

In addition, two commonly used social-engineering techniques are called "dumpster diving" and "shoulder surfing." Social engineers often search through dumpsters or trash cans looking for discarded material—such as phone lists and bank statements—that they can use to help break into a network. For example, a social engineer might look up the phone number of a receptionist he or she can call and pretend to be someone else in the organization. Shoulder surfing—in other words, looking over someone's shoulder—is the easiest form of collecting information. Social engineers use this technique to observe an employee entering a password or a person entering a PIN at the cash register, for example.

In addition to these intentional threats, loss or theft of equipment and computer media is a serious problem, particularly when a computer or flash drive contains confidential data. The data theft and data loss information box discusses this problem and offers some protective measures.

Protecting Against Data Theft and Data Loss

Memory sticks, PDAs, CDs, USB flash drives, smartphones, and other portable storage media pose a serious security threat to organizations' data resources. Theft or loss of these devices is a risk, of course, but disgruntled employees can also use these devices to steal company data. The following guidelines are recommended to protect against these potential risks:[6]

- *Do a risk analysis to determine the effects of confidential data being lost or stolen.*
- *Ban portable media devices and remove or block USB ports, floppy drives, and CD/DVD-ROM drives, particularly in organizations that require tight security. This measure might not be practical in some companies, however.*
- *Make sure employees have access only to data they need to do their jobs, and set up rigorous access controls.*
- *Store data in databases instead of in spreadsheet files for better access control.*
- *Have clear, detailed policies about what employees can do with confidential data, including whether data can be removed from the organization.*
- *Encrypt data downloaded from the corporate network.*

3 Security Measures and Enforcement: An Overview

In addition to backing up data and storing it securely, organizations can take many other steps to guard against threats. A comprehensive security system should include the following:

- Biometric security measures
- Nonbiometric security measures
- Physical security measures
- Access controls
- Virtual private networks
- Data encryption
- E-commerce transaction security measures
- Computer Emergency Response Team

3.1 Biometric Security Measures

Biometric security measures use a physiological element to enhance security measures. This element is unique to a person and can't be stolen, lost, copied, or passed on to others. The following list describes some biometric devices and measures, and some of these can be seen in Exhibit 5.2:

- *Facial recognition*—Identify users by analyzing the unique shape, pattern, and positioning of facial features.
- *Fingerprints*—Scan users' fingerprints and verify them against prints stored in a file.
- *Hand geometry*—Compare the length of each finger, the translucence of fingertips, and the webbing between fingers against stored data to verify users' identities.
- *Iris analysis*—Use a video camera to capture an image of the user's iris, then use software to compare the data against stored templates.
- *Palm prints*—The palm's unique characteristics are used to identify users. A palm reader uses near

infrared light to capture a user's vein pattern, which is unique to each individual. This is then compared to a database that contains existing patterns. This method is often used by law enforcement agencies.

- *Retinal scanning*—Scan the retina using a binocular eye camera, then check against data stored in a file.
- *Signature analysis*—Check the user's signature as well as deviations in pen pressure, speed, and length of time used to sign the name.
- *Vein analysis*—Analyze the pattern of veins in the wrist and back of the hand without making any direct contact with the veins.
- *Voice recognition*—Translate words into digital patterns, which are recorded and examined for tone and pitch. Using voice to verify user identity has one advantage most other biometric measures don't: It can work over long distances via ordinary telephones. A well-designed voice-recognition security system can improve the security of financial transactions conducted over the phone.

> **Biometric security measures** use a physiological element to enhance security measures. These elements are unique to a person and can't be stolen, lost, copied, or passed on to others.

Although biometric techniques are effective security measures, they might not be right for all organizations. Some drawbacks of biometrics are high costs, users' reluctance, and complex installation. However, with improvements being made to address these drawbacks, biometrics can be a viable alternative to traditional security measures. The Phoebe Putney Memorial Hospital information box presents a real-life application of biometrics.

3.2 Nonbiometric Security Measures

The three main nonbiometric security measures are callback modems, firewalls, and intrusion detection systems.

Biometrics at Phoebe Putney Memorial Hospital

Phoebe Putney Memorial Hospital, a 443-bed community hospital in Albany, Georgia, needed to improve its electronic health system record (EHR) system. Doctors and nurses were complaining about the number of passwords they had to provide to access clinical records, so the hospital switched to fingerprint scanners, which, along with a single sign-on application, made the EHR system both easier to use and more secure. With the scanners, it's possible to audit usage, thereby ensuring that only authorized users have access to sensitive information. Another advantage of fingerprint scanners: They don't tend to get lost like smart cards.[7]

Exhibit 5.2 *Examples of biometric devices*

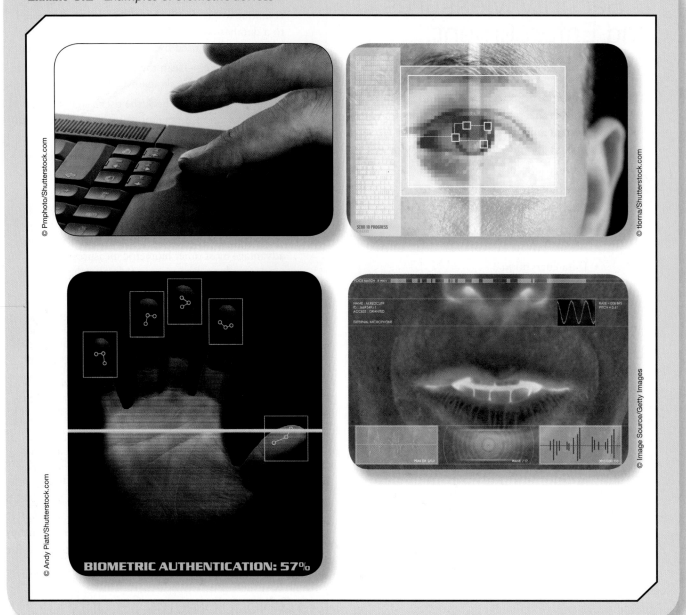

BIOMETRIC AUTHENTICATION: 57%

3.2.1 Callback Modems

A **callback modem** verifies whether a user's access is valid by logging the user off (after he or she attempts to connect to the network) and then calling the user back at a predetermined number. This method is useful in organizations with many employees who work off-site and who need to connect to the network from remote locations.

3.2.2 Firewalls

A **firewall** is a combination of hardware and software that acts as a filter or barrier between a private network and external computers or networks, including the Internet. A network administrator defines rules for access, and all other data transmissions are blocked. An effective firewall should protect outgoing data from the network as well as incoming data into the network. Exhibit 5.3 shows a basic firewall configuration.

A **callback modem** verifies whether a user's access is valid by logging the user off (after he or she attempts to connect to the network) and then calling the user back at a predetermined number.

A **firewall** is a combination of hardware and software that acts as a filter or barrier between a private network and external computers or networks, including the Internet. A network administrator defines rules for access, and all other data transmissions are blocked.

Exhibit 5.3 *A basic firewall configuration*

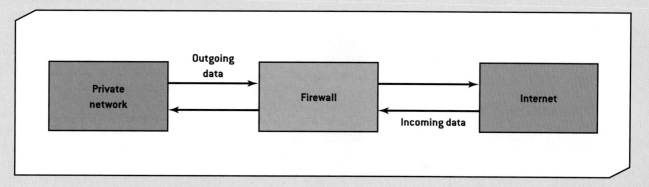

A firewall can examine data passing into or out of a private network and decide whether to allow the transmission based on users' IDs, the transmission's origin and destination, and the transmission's contents. Information being transmitted is stored in what's called a "packet," and after examining a packet, a firewall can take one of the following actions:

- Reject the incoming packet.
- Send a warning to the network administrator.
- Send a message to the packet's sender that the attempt failed.
- Allow the packet to enter (or leave) the private network.

The main types of firewalls are packet-filtering firewalls, application-filtering firewalls, and proxy servers. Packet-filtering firewalls control data traffic by configuring a router to examine packets passing into and out of the network. The router examines the following information in a packet: source IP address and port, destination IP address and port, and protocol used. Based on this information, rules called "packet filters" determine whether a packet is accepted, rejected, or dropped. For example, a packet filter can be set up to deny packets coming from specific IP addresses. A packet-filtering firewall informs senders if packets are rejected but does nothing if packets are dropped; senders have to wait until their requests time out to learn that the packets they sent weren't received.

In addition, these firewalls record all incoming connections, and packets that are rejected might be a warning sign of an unauthorized attempt. Packet-filtering firewalls are somewhat inefficient, however, because they have to examine packets one by one, and they might be difficult to install. In addition, they can't usually record every action taking place at the firewall,

so network administrators could have trouble finding out whether and how intruders are trying to break into the network.

Application-filtering firewalls are generally more secure and flexible than packet-filtering firewalls, but they are also more expensive. Typically, they are software that is installed on a host computer (a dedicated workstation or server) to control use of network applications, such as e-mail, Telnet, and FTP. In addition to checking which applications are requested, these firewalls monitor the time at which application requests take place. This information can be useful, because many unauthorized attempts take place after normal work hours. Application-filtering firewalls also filter viruses and log actions more effectively than packet-filtering firewalls, which helps network administrators spot potential security breaches. Because of all the application-filtering that these firewalls do, however, they're often slower than other types of firewalls, which can affect network performance.

A proxy server, shown in Exhibit 5.4, is software that acts as an intermediary between two systems—between network users and the Internet, for example. It's often used to help protect the network against unauthorized access from outside the network by hiding the network addresses of internal systems. A proxy server can also be used as a firewall that scans for malware and viruses, speeds up network traffic, or takes some load off internal servers (which firewalls can't do). It can also block requests from certain servers.

Although firewalls can do a lot to protect networks and computers, they don't offer complete security. Sophisticated hackers and computer criminals can circumvent almost any security measure. For example, some hackers use a technique called "IP spoofing" to trick firewalls into treating packets as coming from

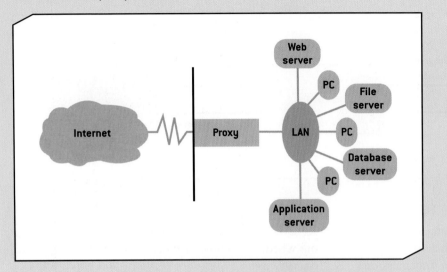

Exhibit 5.4 *A proxy server*

legitimate IP addresses. This technique is the equivalent of forgery. To provide comprehensive security for data resources, firewalls should be used along with other security measures. Other guidelines for improving a firewall's capabilities include the following:

- Identify what data must be secured, and conduct a risk analysis to assess the costs and benefits of a firewall.

- Compare a firewall's features with the organization's security needs. For example, if your organization uses e-mail and FTP frequently, make sure the application-filtering firewall you're considering can handle these network applications.

- Compare features of packet-filtering firewalls, application-filtering firewalls, and proxy servers to determine which of these types addresses your network's security needs the best.

- Examine the costs of firewalls, and remember that the most expensive firewall isn't necessarily the best. Some inexpensive firewalls might be capable of handling everything your organization needs.

- Compare the firewall's security with its ease of use. Some firewalls emphasize accuracy and security

An **intrusion detection system (IDS)** can protect against both external and internal access. It's usually placed in front of a firewall and can identify attack signatures, trace patterns, generate alarms for the network administrator, and cause routers to terminate connections with suspicious sources.

Physical security measures primarily control access to computers and networks, and include devices for securing computers and peripherals from theft.

rather than ease of use and functionality. Determine what's most important to your organization when considering the trade-offs.

- Check the vendor's reputation, technical support, and update policies before making a final decision. As the demand for firewalls has increased, so has the number of vendors, and not all vendors are equal. Keep in mind that you might have to pay more for a product from a vendor with a good reputation that offers comprehensive technical support.

Another alternative is to build a firewall instead of purchase one. This option might be more expensive (and requires having an employee with the necessary skills), but the customized features and flexibility offered by a firewall developed in-house could outweigh the cost.

3.2.3 Intrusion Detection Systems

Firewalls protect against external access, but they leave networks unprotected from internal intrusions. An **intrusion detection system (IDS)** can protect against both external and internal access. They're usually placed in front of a firewall and can identify attack signatures, trace patterns, generate alarms for the network administrator, and cause routers to terminate connections with suspicious sources. These systems can also prevent DoS attacks. An IDS monitors network traffic and uses the "prevent, detect, and react" approach to security. Although it improves security, it requires a lot of processing power and can affect network performance. In addition, it might need additional configuration to prevent it from generating false positive alarms.

A number of third-party tools are available for intrusion detection. The vendors listed in Table 5.1 offer comprehensive IDS products and services.

3.3 Physical Security Measures

Physical security measures primarily control access to computers and networks and include devices for securing computers and peripherals from theft. As shown

Exhibit 5.5 *Common physical security measures*

Cable shielding

Lock for securing a computer

in Exhibit 5.5, common physical security measures can include the following:

- *Cable shielding*—Braided layers around the conductor cable protect it from electromagnetic interference (EMI), which could corrupt data or data transmissions.
- *Corner bolts*—An inexpensive way to secure a computer to a desktop or counter, they often have locks as additional protection against theft.
- *Electronic trackers*—These devices are secured to a computer at the power outlet. If the power cord is disconnected, a transmitter sends a message to an

alarm that sounds or to a camera that records what happens.

- *Identification (ID) badges*—These are checked against a list of authorized personnel, which must be updated regularly to reflect changes in personnel.
- *Proximity-release door openers*—An effective way to control access to the computer room. A small radio transmitter is placed in authorized employees' ID badges, and when they come within a predetermined distance of the computer room's door, a radio signal sends a key number to the receiver, which unlocks the door.
- *Room shielding*—A nonconductive material is sprayed in the computer room, which reduces the number of signals transmitted or confines the signals to the computer room.
- *Steel encasements*—These fit over the entire computer and can be locked.

With the increasing popularity of laptops, theft has become a major security risk. Laptops can store confidential data, so a variety of security measures should be used. For example, a cable lock on the laptop could be combined with a fingerprint scan to make sure only the laptop's owner can access files. The following information box discusses this security threat in more detail.

Table 5.1

IDS vendors

Vendor	URL
Enterasys Networks	www.enterasys.com
Cisco Systems	www.cisco.com
IBM Internet Security Systems	www.iss.net
Juniper Networks	www.juniper.net/us/en
Check Point Software Technologies	www.checkpoint.com

3.4 Access Controls

Access controls are designed to protect systems from unauthorized access in order to preserve data integrity. The following sections describe two widely used access controls: terminal resource security and passwords.

3.4.1 Terminal Resource Security

Terminal resource security is a software feature that erases the screen and signs the user off automatically after a specified length of inactivity. This method of access control prevents unauthorized users from using an unattended computer to access the network and data. Some programs also allow users to access data only during certain times, which reduces break-in attempts during off hours.

3.4.2 Passwords

A password is a combination of numbers, characters, and symbols that's entered to allow access to a system. A password's length and complexity determines its vulnerability to discovery by unauthorized users. For example, "p@s$w0rD" is much harder to guess than "password." The human element is one of the most notable weaknesses of password security, because users can forget passwords or give them to an unauthorized user (intentionally or unintentionally).

Access controls are designed to protect systems from unauthorized access in order to preserve data integrity.

To increase the effectiveness of passwords, follow these guidelines:

- Change passwords frequently.
- Passwords should be eight characters or longer.
- Passwords should be a combination of uppercase and lowercase letters, numbers, and special symbols, such as @ or $.
- Passwords should not be written down.
- Passwords shouldn't be common names, such as the user's first or last name, obvious dates (such as birthdays or anniversaries), or words that can be found in a dictionary.
- Passwords shouldn't be increased or decreased sequentially or follow a pattern (for example, 222ABC, 224ABC, 226ABC, etc.).
- Before employees are terminated, make sure their passwords have been deleted.

© Dmitriy Shironosov/Shutterstock.com

One of the oldest encryption algorithms, developed by Julius Caesar, is a simple substitution algorithm, in which each letter in the original message is replaced by the letter three positions farther in the alphabet.

3.5 Virtual Private Networks

A **virtual private network (VPN)** provides a secure "tunnel" through the Internet for transmitting messages and data via a private network (see Exhibit 5.6). It's often used so remote users have a secure connection to the organization's network. VPNs can also be used to provide security for extranets, which are networks set up between an organization and an external entity, such as a supplier (discussed in more detail in Chapter 7). Data is encrypted before it's sent through the tunnel with a protocol, such as Layer 2 Tunneling Protocol (L2TP) or Internet Protocol Security (IPSec). The cost of setting up a VPN is usually low, but transmission speeds can be slow, and lack of standardization can be a problem.

Typically, an organization leases the media used for a VPN on an as-needed basis, and network traffic can be sent over the combination of a public network (usually the Internet) and a private network. VPNs

A **virtual private network (VPN)** provides a secure "tunnel" through the Internet for transmitting messages and data via a private network.

Data encryption transforms data, called "plaintext" or "cleartext," into a scrambled form called "ciphertext" that can't be read by others.

Secure sockets layers (SSL) is a commonly used encryption protocol that manages transmission security on the Internet.

are an alternative to private leased lines or dedicated Integrated Services Digital Network (ISDN) lines and T1 lines.

3.6 Data Encryption

Data encryption transforms data, called "plaintext" or "cleartext," into a scrambled form called "ciphertext" that can't be read by others. The rules for encryption, known as the "encryption algorithm," determine how simple or complex the transformation process should be. The receiver then unscrambles the data by using a decryption key.

There are many different encryption algorithms used. One of the oldest encryption algorithms, developed by Julius Caesar, is a simple substitution algorithm, in which each letter in the original message is replaced by the letter three positions farther in the alphabet. For example, the word *top* is transmitted as *wrs*. Exhibit 5.7 shows a simple example of encryption with a substitution algorithm.

A commonly used encryption protocol is **secure sockets layer (SSL)**, which manages transmission security on the Internet. Next time you purchase an item online, notice that the *http* in the browser address bar changes to *https*. The *https* indicates a Secure HTTP connection

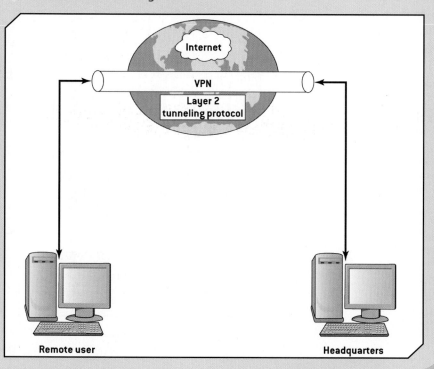

Exhibit 5.6 *A VPN configuration*

Internet

VPN

Layer 2 tunneling protocol

Remote user

Headquarters

over SSL. You might also see a padlock icon in the status bar at the bottom to indicate that your information has been encrypted and hackers can't intercept it. A more recent cryptographic protocol is **transport layer security (TLS)**, which ensures data security and integrity over public networks, such as the Internet. Similar to SSL, TLS encrypts the network segment used for performing transactions.

As mentioned, encryption algorithms use a key to encrypt and decrypt data. The key's size varies from 32 bits to 168 bits, and the longer the key, the harder it is to break. There are two main types of encryption: asymmetric (also called "public key encryption") and symmetric.

Asymmetric encryption uses two keys: a public key known to everyone and a private or secret key known only to the recipient. A message encrypted with a public key can be decrypted only with the same algorithm used by the public key and requires the recipient's private key, too. Anyone intercepting the message can't decrypt it, because he or she doesn't have the private key.

This encryption usually works better for public networks, such as the Internet. Each company conducting transactions or sending messages gets a private key and a public key; a company keeps its private key and publishes its public key for others to use. One of the first public key algorithms, RSA (named after its creators—Rivest, Shamir, and Adleman), is still widely used today. The main drawback of asymmetric encryption is that it's slower and requires a lot of processing power.

In **symmetric encryption** (also called "secret key encryption"), the same key is used to encrypt and decrypt the message. The sender and receiver must agree on the key and keep it secret. Advanced Encryption Standard (AES) is a symmetric encryption algorithm with a 56-bit key and is the encryption technique used by the U.S. government. The problem with symmetric encryption is that sharing the key over the Internet is difficult.

Encryption can also be used to create digital signatures that authenticate senders' identities and verify that the message or data hasn't been altered. Digital signatures are particularly important in online financial transactions. They also provide nonrepudiation, discussed in the next section. Here's how they work: You encrypt a message with your private key and use an algorithm that hashes the message and creates a message digest. The message digest can't be converted back to the original message, so anyone intercepting the message can't read it. Then you use your private key to encrypt the message digest, and this encrypted piece is called the "digital signature."

You then send the encrypted message and digital signature. The recipient has your public key and uses it

Exhibit 5.7 *Using encryption*

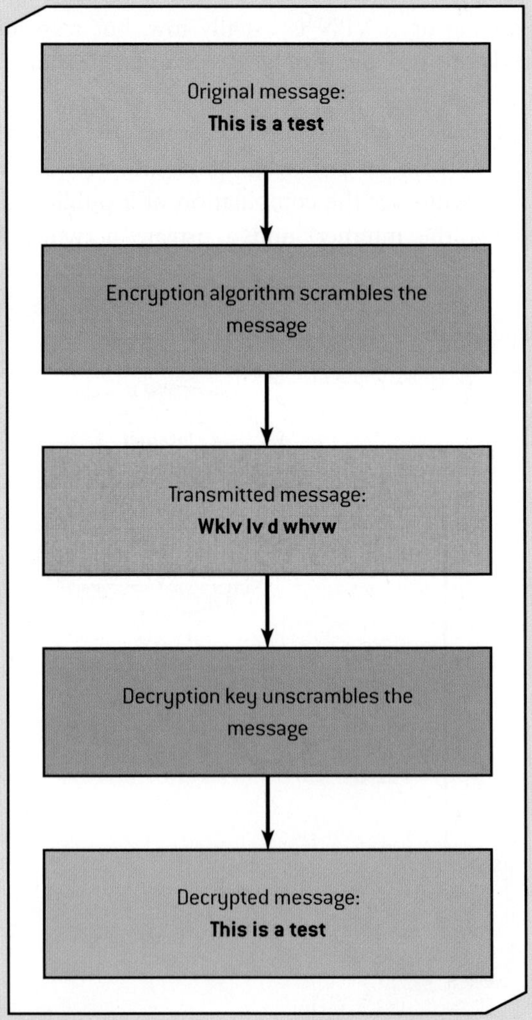

Original message:
This is a test

Encryption algorithm scrambles the message

Transmitted message:
Wklv lv d whvw

Decryption key unscrambles the message

Decrypted message:
This is a test

to decrypt the message, and then uses the same algorithm that you did to hash the message and create another version of the message digest. Next, the recipient uses your public key to decrypt your digital signature and get the message digest you sent. The recipient then compares the two message digests. If they match, the message wasn't tampered with and is the same as the one you sent.

3.7 E-Commerce Transaction Security Measures

In e-commerce transactions, three factors are critical for security: authentication, confirmation, and non-repudiation. Authentication is important because the person using a credit card number in an online transaction isn't necessarily the card's legitimate owner, for example. Two factors are important: what the receiver knows to be accurate and what the sender is providing. Passwords and personal information, such as your mother's maiden name, your Social Security number, or your date of birth, can be used for authentication. Physical proof, such as fingerprints or retinal scans, works even better.

Confirmation must also be incorporated into e-commerce transactions—to verify orders and receipt of shipments, for examples. When an electronic document, such as a payment, is sent from a customer to a vendor, a digitally signed confirmation with the vendor's private key is returned to verify that the transaction was carried out.

Nonrepudiation is needed in case a dispute over a transaction is raised. Digital signatures are used for this and serve to bind partners in a transaction. With this process, the sender receives proof of delivery, and the receiver is assured of the sender's identity. Neither party can deny sending or receiving the information.

E-commerce transaction security is concerned with the following issues:

- *Confidentiality*—How can you ensure that *only* the sender and intended recipient can read the message?
- *Authentication*—How can the recipient know that data is actually from the sender?
- *Integrity*—How can the recipient know that the data's contents haven't been changed during transmission?
- *Nonrepudiation of origin*—The sender can't deny having sent the data.
- *Nonrepudiation of receipt*—The recipient can't deny having received the data.

3.8 Computer Emergency Response Team

The Computer Emergency Response Team (CERT) was developed by the Defense Advanced Research Projects Agency (part of the Department of Defense) in response to the 1988 Morris worm attack, which disabled 10 percent of the computers connected to the Internet. Many organizations now follow the CERT model to form teams that can handle network intrusions and attacks quickly and effectively. Currently, CERT focuses on security breaches and DoS attacks and offers guidelines on handling and preventing these incidents. CERT also conducts a public awareness campaign and researches Internet security vulnerabilities and ways to improve security systems. Network administrators and e-commerce site managers should check the CERT Coordination Center for updates on protecting network and information resources. Exhibit 5.8 shows the CERT Coordination Center home page (*www.cert.org/certcc.html*).

In addition, the Office of Cyber Security at the Department of Energy offers a security service: Cyber Incident Response Capability (CIRC, *http://www.doecirc.energy.gov/aboutus.html*). CIRC's main function is to provide information on security incidents, including information systems' vulnerabilities, viruses, and malicious programs. CIRC also provides awareness training, analysis of threats and vulnerabilities, and other services.

4 Guidelines for a Comprehensive Security System

an organization's employees are an essential part of the success of any security system, so training employees about security awareness and security measures is important. Some organizations use a classroom setting for training, and others conduct it over the organization's intranet. Tests and certificates should be given to participants at the end of training sessions. In addition, making sure management supports security training is important to help promote security awareness throughout the organization.

Sarbanes-Oxley and Information Security

Section 404 of the Sarbanes-Oxley Act of 2002 requires IT professionals to document and test the effectiveness of security measures protecting information technology and systems, including general computer controls, application controls, and system software controls. IT professionals must be familiar with this act and incorporate it into their organizations' security policies. In addition, companies must set up a security system that protects vital records and data and prevents them from being destroyed, lost, corrupted, or altered by unauthorized users. The purpose of this act is to maintain data integrity and availability of business operations, which are particularly critical in financial organizations.[9]

Organizations should understand the principles of the Sarbanes-Oxley Act of 2002 (described in the information box) and conduct a basic risk analysis before establishing a security program. This analysis often makes use of financial and budgeting techniques, such as return on investment (ROI), to determine which resources are most important and should have the strongest protection. This information can also help organizations weigh the cost of a security system.

The following steps should be considered when developing a comprehensive security plan:[10]

1. Set up a security committee with representatives from all departments as well as upper management. The committee's responsibilities include the following:

   - Developing clear, detailed security policy and procedures

   - Providing security training and security awareness for key decision makers and computer users
   - Periodically assessing the security policy's effectiveness
   - Developing an audit procedure for logons and system use
   - Overseeing enforcement of the security policy
   - Designing an audit trail procedure for incoming and outgoing data

2. Post the security policy in a visible place, or post copies next to all workstations.

3. Raise employees' awareness of security problems.

4. Revoke terminated employees' passwords and ID badges immediately to prevent attempts at retaliation.

5. Keep sensitive data, software, and printouts locked up in secure locations.

6. Exit programs and systems promptly, and never leave logged-on workstations unattended.

7. Limit computer access to authorized personnel only.

8. Compare communication logs with communication billing periodically. The log should list all outgoing calls with users' names, call destinations, and time of calls. Investigate any billing discrepancies.

9. Install antivirus programs, and make sure they're updated automatically.

10. Install only licensed software purchased from reputable vendors.

11. Make sure fire protection systems and alarms are up to date, and test them regularly.

Exhibit 5.8 *CERT Coordination Center home page*

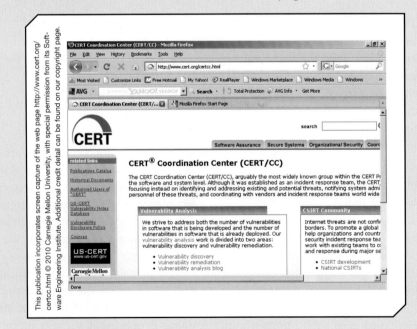

> An organization's employees are an essential part of the success of any security system, so training on security awareness and security measures is important.

12. Check environmental factors, such as temperature and humidity levels.
13. Use physical security measures, such as corner bolts on workstations, ID badges, and door locks.
14. Install firewalls and intrusion detection systems. If necessary, consider biometric security measures.

These steps should be used as a guideline. Not every organization needs to follow every step; however, some might need to include even more to fit their needs.

4.1 Business Continuity Planning

To lessen the effects of a natural disaster or a network attack or intrusion, planning the recovery is important. This should include **business continuity planning**, which outlines procedures for keeping an organization operational. A disaster recovery plan lists the tasks that must be performed to restore damaged data and equipment as well as steps to prepare for disaster, such as the following:

- Back up all files.
- Review security and fire standards for computer facilities periodically.
- Review information from CERT and other security agencies periodically.
- Make sure staff members have been trained and are aware of the consequences of possible disasters and steps to reduce the effects of disasters.
- Test the disaster recovery plan with trial data.
- Identify vendors of all software and hardware used in the organization, and make sure their mailing addresses, phone numbers, and Web site addresses are up to date.
- Document all changes made to hardware and software.
- Get a comprehensive insurance policy for computers and network facilities. Review the policy periodically to make sure coverage is adequate and up to date.
- Set up alternative sites to use in case of a disaster. Cold sites have the right environment for computer equipment (such as air conditioning and humidity controls), but no equipment is stored in them. Hot sites, on the other hand, have all the needed equipment and are ready to go.
- Investigate using a colocation facility, which is rented from a third party and usually contains telecommunication equipment.
- Check sprinkler systems, fire extinguishers, and halon gas systems.
- Keep backups in off-site storage, test data recovery procedures periodically, and keep a detailed record of machine-specific information, such as model and serial number. Backup facilities can be shared to reduce costs.

> **Business continuity planning** outlines procedures for keeping an organization operational in the event of a natural disaster or network attack.

- Keep a copy of the disaster recovery plan off site.
- Go through a mock disaster to assess response time and recovery procedures.

If disaster strikes, organizations should follow these steps to resume normal operations as soon as possible:

1. Put together a management crisis team to oversee the recovery plan.
2. Contact the insurance company.
3. Restore phone lines and other communication systems.
4. Notify all affected people, including customers, suppliers, and employees.
5. Set up a help desk to assist affected people.
6. Notify the affected people that recovery is underway.
7. Document all actions taken to regain normality so you know what worked and what didn't work; revise the disaster recovery plan, if needed.

The Industry Connection highlights McAfee Corporation, which offers several security products and services.

MCAFEE CORPORATION*

McAfee is a leading vendor of antivirus software and uses the Internet as a distribution medium for its products and services, although products can also be purchased through other outlets, such as retailers. In addition to antivirus software, McAfee offers network management software that includes virus scanning, firewalls, authentication, and encryption capabilities. McAfee also has an online bug-tracking system. The following list describes some popular McAfee products:

- *Internet Security*—Includes antivirus, antispyware, antispam, anti-phishing, identity protection, parental controls, data backup, and other features
- *VirusScan Plus*—Offers antivirus, antispyware, and firewall features, and Web site safety ratings
- *Total Protection*—Includes features for antivirus, antispyware, antispam, anti-phishing, two-way firewall, advanced Web site safety ratings, identity protection, parental controls, and data backup

McAfee also offers several free products and services, such as the following:

- *FreeScan*—Searches for the most recent viruses and displays a detailed list of any infected files
- *World Virus Map*—Shows where recent viruses are infecting computers worldwide
- *Virus Removal Tools*—Used to remove viruses and repair damage
- *Security Advice Center*—Offers tips and advice on keeping your computer and network safe and preventing attacker intrusions
- *Free PC and Internet Security Newsletter*—Includes virus alerts, special offers, and breaking news
- *Internet Connection Speedometer*—Tests your Internet connection to see how fast or slow it is

*This information was gathered from the company Web site (www.mcafee.com) and other promotional materials. For more information and updates, visit the Web site.

 5 Chapter Summary

his chapter discussed computer and network security risks and protective measures. Basic safeguards, including fault-tolerant systems, were covered, and an overview of intentional security threats was presented. We reviewed biometric, nonbiometric, and physical security measures as well as access controls, firewalls, and intrusion detection systems. We also discussed the importance of establishing a comprehensive security system and a business continuity plan and reviewed guidelines for both.

Key Terms

access controls (86)

asymmetric encryption (88)

availability (76)

backdoor (79)

biometric security measures (81)

blended threats (79)

business continuity planning (91)

callback modem (82)

confidentiality (76)

data encryption (87)

denial-of-service (DoS) attack (80)

fault-tolerant systems (77)

firewall (82)

integrity (76)

intrusion detection system (IDS) (84)

logic bomb (79)

physical security measures (84)

secure sockets layer (SSL) (87)

social engineering (80)

symmetric encryption (88)

transport layer security (TLS) (88)

Trojan program (79)

virtual private network (VPN) (87)

virus (78)

worm (79)

1. What are some examples of intentional threats?

2. How does a virus spread?

3. What effect does a DoS attack have?

4. List some major items that should be included in a disaster recovery plan.

5. Visit the following Web page, consult other sources, and write a one-page paper that summarizes the costs of cybercrime for businesses:

   http://www.infoworld.com/d/security-central/cybercrime-costs-businesses-each-38-million-year-732

6. Visit the following Web pages and write a one-page paper that summarizes iPad security issues and concerns:

   http://www.infoworld.com/d/security-central/ipad-buyers-targeted-windows-malware-948

   http://www.infoworld.com/t/data-security/hacker-group-apple-ipad-simply-not-safe-platform-154

7. Visit the following Web page, consult other sources, and write a one-page paper that summarizes insider security threats. How can organizations guard against this threat?

   http://www.infoworld.com/d/security-central/the-true-extent-insider-security-threats-281

8. The City of Oceanside, California, has implemented a biometric security that has significantly reduced the total cost of network operations and maintenance. Visit the following Web page and write a one-page paper that summarizes this savings. Can other cities and organizations learn from this experience?

   http://itmanagement.earthweb.com/secu/article. php/863861/Case-Study-Biometrics-Eases-Citys-Network-Access-Security-Woes.htm

9. Which of the following is considered a fault-tolerant system? (Choose all that apply.)

   a. Redundant array of independent disks (RAID)

   b. Disk operating system

   c. Uninterruptible power supply (UPS)

   d. Mirror disks

10. The first step an organization should take to resume normal operations after a disaster is contacting the insurance company. True or False?

casestudy

THE LOVE BUG VIRUS

The Love Bug virus that destroyed files and stole passwords first appeared in Hong Kong on May 11, 2000, and spread around the globe in two hours, three times faster than the Melissa virus. More than 45 million users in at least 20 countries were affected, as well as many organizations, including NASA and the CIA. The damages were estimated at $2 billion to $10 billion. Experts traced the origin of the virus to the Philippines, using information provided by an Internet service provider, the Philippines National Bureau of Investigation, and the FBI. The Philippines, however, had no cybercrime law, so creating and disseminating a virus wasn't considered a crime. Convincing the court to issue a warrant to search the suspect's apartment took several days, which gave the suspect plenty of time to destroy most of the evidence. Officials finally were able to search and seize evidence indicating that Onel de Guzman,

© Blazej Lyjak/Shutterstock.com

a former computer science student, was the person responsible for creating and spreading this virus.[11]

Answer the following questions:

1. How is the cost of the damage caused by a virus calculated? (*Hints:* Costs involve loss of sales, personnel time, replacing damaged files or programs, replacing equipment, and buying or upgrading equipment to protect against future attacks.)

2. Research U.S. laws for prosecuting hackers and describe two of them. For each law, include an example of a legal case involving that law.

3. How can organizations guard against viruses, such as the Love Bug?

DATA COMMUNICATION: DELIVERING INFORMATION ANYWHERE AND ANYTIME

learning outcomes

this chapter explains the role of data communication systems in delivering information for decision making, although you can see applications of data communication systems everywhere, from within your own home to within multinational corporations. We start with the basics of data communication systems, including components, processing configurations, and types of networks and topologies. We also cover important concepts in data communication, such as bandwidth, routing, routers, and the client/server model. Next, we give an overview of wireless and mobile networks and the technologies they use. Finally, we take a look at a growing phenomenon—the convergence of voice, video, and data—and its importance in the business world.

After studying this chapter, you should be able to:

LO1 Describe major applications of a data communication system.

LO2 Explain the major components of a data communication system.

LO3 Describe the major types of processing configurations.

LO4 Explain the three types of networks.

LO5 Describe the main network topologies.

LO6 Explain important networking concepts, such as bandwidth, routing, routers, and the client/server model.

LO7 Describe wireless and mobile technologies and networks.

LO8 Discuss the importance of wireless security and the techniques used.

LO9 Summarize the convergence phenomenon and its applications for business and personal use.

1 Defining Data Communication

data communication is the electronic transfer of data from one location to another. An information system's effectiveness is measured in part by how efficiently it delivers information, and a data communication system is what enables an information system to carry out this function. In addition, because most organizations collect and transfer data across large geographical distances, an efficient data communication system is critical. A data communication system can also improve the flexibility of data collection and transmission. For example, many workers use portable devices, such as laptops, personal digital assistants (PDAs), and other handheld devices, to communicate with the office at any time and from any location.

Data communication is also the basis of virtual organizations, discussed in Chapter 4. By using the capabilities of a data communication system, organizations aren't limited by physical boundaries. They can collaborate with other organizations, outsource certain functions to reduce costs, and provide customer services via data communication systems.

E-collaboration is another main application of data communication. Decision makers can be located throughout the world but can still collaborate with their colleagues no matter where they are.

1.1 Why Managers Need to Know About Data Communication

Data communication has become so woven into the fabric of corporate activity that separating an organization's core functions from the data communication systems that enable and support them is difficult. When a new product is introduced, for example,

> **Data communication** is the electronic transfer of data from one location to another.

MIS

Data communication systems…make virtual organizations possible, and these can cross geographic boundaries to develop products more quickly and effectively.

the executives who are the key decision makers might be scattered throughout the world in a multinational corporation. However, they can use data communication systems to collaborate and coordinate their efforts to introduce the new product in a timely manner.

Data communication applications can enhance decision makers' efficiency and effectiveness in many ways. For example, data communication applications support just-in-time delivery of goods, which reduces inventory costs and improves the competitive edge. As you've learned in previous chapters, many large corporations, such as Walmart, The Home Depot, and UPS, use data communication technologies to stay ahead of their competitors. As mentioned, data communication systems also make virtual organizations possible, and these can cross geographic boundaries to develop products more quickly and effectively.

Data communication systems also enable organizations to use e-mail and electronic file transfer to improve efficiency and productivity. A communication network, a crucial part of an organization's information system, shortens product and service development life cycles and delivers information to those who need it faster and more efficiently. Here are some of the ways data communication technologies affect the workplace:

- Online training for employees can be provided via virtual classrooms. In addition, employees get the latest technology and product information immediately.

- Internet searches for information on products, services, and innovation keep employees up to date.
- The Internet and data communication systems facilitate lifelong learning, which will be an asset for knowledge workers of the future.
- Boundaries between work and personal life are less clear-cut as data communication is more available in both homes and businesses. The increase in telecommuters is an example of this trend.
- Web and video conferencing are easier, which can reduce the costs of business travel.

Managers need a clear understanding of the following areas of data communication:

- The basics of data communication and networking
- The Internet, intranets, and extranets
- Wired and wireless networks
- Network security issues and measures
- Organizational and social effects of data communication
- Globalization issues
- Applications of data communication systems

E-collaborations and virtual meetings are other important applications of data communication systems for managers. These applications are cost effective and improve customer service. One example of an e-collaboration tool is WebEx, described in the following information box.

Example of an E-collaboration Tool

Cisco's WebEx (www.webex.com) is a Web-based service for collaboration, online meetings, Web conferencing, and video conferencing. One of its features enables users to share files on their computers with others while in a conference so that all participants can see the same thing. WebEx is particularly useful for companies with branch offices in many different locations, because it eliminates or reduces the need to travel for meetings. WebEx can help companies increase productivity and reduce costs. It can be used for many purposes, even sales presentations, product briefings, help desks, and online training.

An advantage of being Web based is that WebEx can be used on most current operating systems. In addition, WebEx sessions can be recorded and reused. Unlike other video conferencing services, WebEx sessions operate on a proprietary network, which increases security and reliability.

2 Basic Components of a Data Communication System

 typical data communication system includes the following components:

- Sender and receiver devices
- Modems or routers
- Communication medium (channel)

Before examining these components, you need to review some basic concepts in data communication. **Bandwidth** is the amount of data that can be transferred from one point to another in a certain time period, usually one second. It's often expressed as bits per second (bps). Other measurements include kilobits per second (Kbps), megabits per second (Mbps), and gigabits per second (Gbps). **Attenuation** is the loss of power in a signal as it travels from the sending device to the receiving device.

Data transmission channels are generally divided into two types: broadband and narrowband. In **broadband** data transmission, multiple pieces of data are sent simultaneously to increase the transmission rate.

Narrowband is a voice-grade transmission channel capable of transmitting a maximum of 56,000 bps, so only a limited amount of information can be transferred in a specific period of time.

Before a communication link can be established between two devices, they must be synchronized, meaning that both devices must start and stop communicating at the same point. Synchronization is handled with **protocols**, rules that govern data communication, including error detection, message length, and transmission speed. Protocols also help ensure compatibility between different manufacturers' devices.

2.1 Sender and Receiver Devices

A sender and receiver device can take various forms:

- An input/output device, or "thin client," is used only for sending or receiving information; it has no processing power.
- A smart terminal is an input/output device that can perform certain processing tasks, but it's not a full-featured computer. This type of device is often used on factory floors and assembly lines for collecting data and transmitting it to the main computer system.

- An intelligent terminal, a workstation, or a personal computer serves as an input/output device or as a stand-alone system. Using this type of device, a remote computer can perform certain processing tasks without the main computer's support. Generally, an intelligent terminal is considered a step up from a smart terminal.
- A netbook computer is a low-cost, diskless computer used to connect to the Internet or a LAN. It runs software off a server and saves data to a server. According to Forrester Research, however, the iPad and other tablet devices will significantly reduce the demand for netbooks.[1]
- Other types of computers, such as minicomputers, mainframes, and supercomputers.
- Many types of recent devices, such as smartphones, mobile phones, MP3 players, PDAs, and game consoles. Smartphones, briefly described in Chapter 1, are mobile phones with advanced capabilities, such as e-mail and Web-browsing, and most have a built-in keyboard or an external USB keyboard. A video game console is an electronic device for playing video games. It receives instructions from a game player and produces a video display signal on a monitor such as a television screen or a computer monitor.

2.2 Modems

A **modem** (short for "modulator-demodulator") is a device that connects a user to the Internet. Not all Internet connections require a modem; for example, wireless users connect via access points, and satellite users use a satellite dish. However, dial-up, a digital subscriber line (DSL), and cable access require modems to connect.

When phone lines are used for Internet connections, an analog modem is necessary to convert a computer's

---

Bandwidth is the amount of data that can be transferred from one point to another in a certain time period, usually one second.

Attenuation is the loss of power in a signal as it travels from the sending device to the receiving device.

In **broadband** data transmission, multiple pieces of data are sent simultaneously to increase the transmission rate.

Narrowband is a voice-grade transmission channel capable of transmitting a maximum of 56,000 bps, so only a limited amount of information can be transferred in a specific period of time.

Protocols are rules that govern data communication, including error detection, message length, and transmission speed.

A **modem** (short for "modulator-demodulator") is a device that connects a user to the Internet.

Digital subscriber line (DSL), a common carrier service, is a high-speed service that uses ordinary phone lines.

Communication media, or channels, connect sender and receiver devices. They can be conducted or radiated.

Conducted media provide a physical path along which signals are transmitted, including twisted pair cable, coaxial cable, and fiber optics.

digital signals to analog signals that can be transferred over analog phone lines. In today's broadband world, however, analog modems are rarely used. Instead, DSL or cable modems are common. **Digital subscriber line (DSL)**, a common carrier service, is a high-speed service that uses ordinary phone lines. With DSL connections, users can receive data at up to 7.1 Mbps and send data at around 1 Mbps, although the actual speed is determined by proximity to the provider's location. Also, different providers might offer different speeds. Cable modems, on the other hand, use the same cable connected to TVs for Internet connections and can usually reach transmission speeds of about 16 Mbps.

2.3 Communication Media

Communication media, or channels, connect sender and receiver devices. They can be conducted (wired or guided) or radiated (wireless), as shown in Exhibit 6.1.

Conducted media provide a physical path along which signals are transmitted, including twisted pair cable, coaxial cable, and fiber optics. Twisted pair cable consists of two copper lines twisted around each other and either shielded or unshielded; it's used in the telephone network and communication within buildings. Coaxial cables are thick cables that can be used for both data and voice transmissions. They are used mainly for long-distance telephone transmissions and local area networks. Fiber-optic cables are glass tubes (half the diameter of a human hair) surrounded by concentric layers of glass, called "cladding," to form a light path through wire cables. At the center is the core that is the central piece of glass that carries the light. Surrounding the core is the cladding, a second layer of glass, which keeps the light from escaping the core. And around both of these lies the buffer, an outer layer of plastic,

Exhibit 6.1 *Types of communication media*

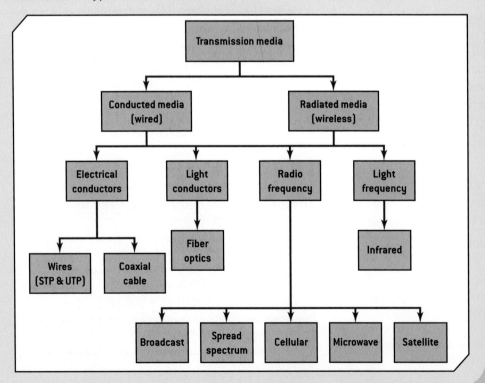

which provides protection and strength. Fiber-optic cables have a higher capacity, smaller size, lighter weight, lower attenuation, and higher security than other cable types and the highest bandwidth of any communication medium.

Radiated media use an antenna for transmitting data through air or water. Some of these media work based on "line of sight." They include broadcast radio, terrestrial microwave, and satellite. Satellites link ground-based microwave transmitters/receivers, known as Earth stations, and are commonly used in long-distance telephone transmissions and TV signals. Terrestrial microwave uses Earth-based transmitters and receivers and is often used for point-to-point links between buildings.

A communication medium can be a point-to-point or a multipoint system. In a point-to-point system, only one device at a time uses the medium. In a multipoint system, several devices share the same medium, and a transmission from one device can be sent to all other devices sharing the link.

3 Processing Configurations

ata communication systems can be used in several different configurations, depending on users' needs, types of applications, and responsiveness of the system. During the past 60 years, three types of processing configurations have emerged: centralized, decentralized, and distributed.

3.1 Centralized Processing

In a **centralized processing** system, all processing is done at one central computer. In the early days of computer technology, this type of processing was justified because data processing personnel were in short supply, hardware and software were expensive, and only large organizations could afford computers. The main advantage of this configuration is being able to exercise tight control on system operations and applications. The main disadvantage is lack of responsiveness to users' needs, because the system and its users could be located far apart from each other. This configuration isn't used much now.

3.2 Decentralized Processing

In **decentralized processing**, each user, department, or division has its own computer (sometimes called an "organizational unit") for performing processing tasks. A decentralized processing system is certainly more responsive to users than a centralized processing system. Nevertheless, decentralized systems have some drawbacks, including lack of coordination among organizational units, the high cost of having many systems, and duplication of efforts.

3.3 Distributed Processing

Distributed processing solves two main problems—the lack of responsiveness in centralized processing and the lack of coordination in decentralized processing—by maintaining centralized control and decentralizing operations. Processing power is distributed among several locations. For example, in the retail industry, each store's network does its own processing but is under the centralized control of the store's headquarters. Databases and input/output devices can also be distributed.

The advantages of distributed processing include the following:

- Accessing unused processing power is possible.
- Modular design means computer power can be added or removed, based on need.
- Distance and location aren't limiting.
- It's more compatible with organizational growth, because workstations can be added easily.
- Fault tolerance is improved because of the availability of redundant resources.
- Resources, such as high-quality laser printers, can be shared to reduce costs.
- Reliability is improved, because system failures can be limited to only one site.
- The system is more responsive to user needs.

> **Radiated media** use an antenna for transmitting data through air or water.
>
> In a **centralized processing** system, all processing is done at one central computer.
>
> In **decentralized processing**, each user, department, or division has its own computer (sometimes called an "organizational unit") for performing processing tasks.
>
> **Distributed processing** maintains centralized control and decentralizes operations. Processing power is distributed among several locations.

Disadvantages of distributed processing include: dependence on communication technology, incompatibility between equipment, and a more challenging network management.

3.4 Open Systems Interconnection Model

The **Open Systems Interconnection (OSI) model** is a seven-layer architecture for defining how data is transmitted from computer to computer in a network, from the physical connection to the network to the applications that users run. OSI also standardizes interactions between network computers exchanging information. Each layer in the architecture performs a specific task:

- *Application layer*—Serves as the window through which applications access network services. It performs different tasks, depending on the application, and provides services that support users' tasks, such as file transfers, database access, and e-mail.

- *Presentation layer*—Responsible for formatting message packets.

- *Session layer*—Establishes a communication session between computers.

- *Transport layer*—Generates the receiver's address and ensures the integrity of messages by making sure packets are delivered without error, in sequence, and with no loss or duplication. This layer provides methods for controlling data flow, ordering received data, and acknowledging received data.

- *Network layer*—Responsible for routing messages.

- *Data Link layer*—Oversees the establishment and control of the communication link.

- *Physical layer*—Specifies the electrical connections between computers and the transmission medium, and defines the physical medium used for communication. It's primarily concerned with transmitting binary data, or bits, over a communication network.

The **Open Systems Interconnection (OSI) model** is a seven-layer architecture for defining how data is transmitted from computer to computer in a network, from the physical connection to the network to the applications that users run. It also standardizes interactions between network computers exchanging information.

A **network interface card (NIC)** is a hardware component that enables computers to communicate over a network.

A **local area network (LAN)** connects workstations and peripheral devices that are in close proximity.

4 Types of Networks

here are three major types of networks: local area networks, wide area networks, and metropolitan area networks. In all these networks, computers are usually connected to the network via a **network interface card (NIC)**, a hardware component that enables computers to communicate over a network. A NIC, also called an "adapter card," is the physical link between a network and a workstation, so it operates at the OSI model's Physical and Data Link layers. NICs are available from many vendors, and the most common types of local area networks, Ethernet and token ring, can use NICs from almost any vendor. In addition, to operate a server in a network, a network operating system (NOS) must be installed, such as Windows Server 2008 or Novell Enterprise Server.

4.1 Local Area Networks

A **local area network (LAN)** connects workstations and peripheral devices that are in close proximity

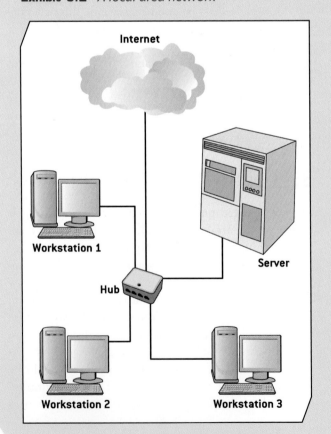

Exhibit 6.2 *A local area network*

Internet

Workstation 1

Server

Hub

Workstation 2

Workstation 3

© Francesco Bisignani/Shutterstock.com

ware and hardware devices used for building a LAN. An Ethernet cable is used to connect computers, hubs, switches, and routers to a network. Assessing information needs and careful planning are important in setting up a LAN.

4.2 Wide Area Networks

A **wide area network (WAN)** can span several cities, states, or even countries, and it's usually owned by several different parties (see Exhibit 6.3). The data transfer speed depends on the speed of its interconnections (called "links") and can vary from 28.8 Kbps to 155 Mbps. A WAN can be useful for a company headquartered in Washington, D.C., with 30 branch offices in 30 states, for example. The WAN makes it possible for all branch offices to communicate with headquarters and send and receive information.

A WAN can use many different communication media (coaxial cables, satellite, and fiber optics) and terminals of different sizes and sophistication (PCs, workstations, mainframes) and be connected to other networks.

(see Exhibit 6.2). Usually, a LAN covers a limited geographical area, such as a building or campus, and one company owns it. Its data transfer speed varies from 100 Mbps to 10 Gbps.

LANs are used most often to share resources, such as peripherals, files, and software. They're also used to integrate services, such as e-mail and file sharing. In a LAN environment, there are two key terms to remember: *Ethernet* and *Ethernet cable*. Ethernet is a standard communication protocol embedded in soft-

4.3 Metropolitan Area Networks

The Institute of Electrical and Electronics Engineers (IEEE) developed specifications for a public, independent, high-speed network that connects a variety of data communication systems, including LANs and WANs, in metropolitan areas. This network, called a **metropolitan area network (MAN)**, is designed to handle data communication for multiple organizations in a city and sometimes nearby cities, too (see Exhibit 6.4). The data transfer speed varies from 34 Mbps to 155 Mbps.

Table 6.1 compares these three types of networks.

Exhibit 6.3 *A wide area network*

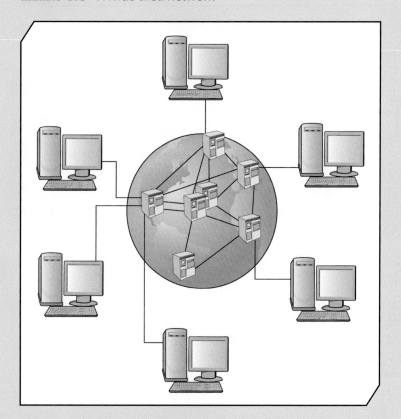

Exhibit 6.4 *Metropolitan area networks*

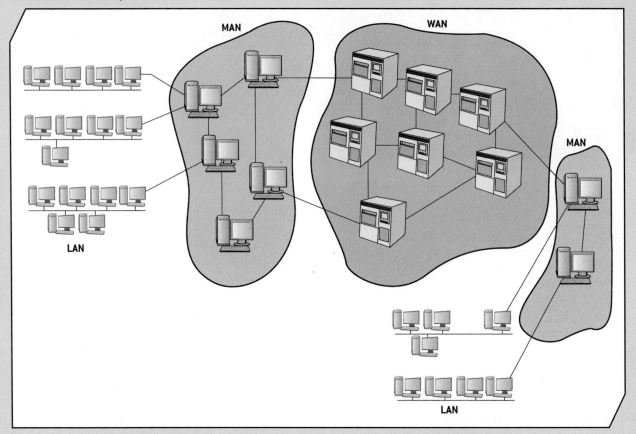

MAN

WAN

MAN

LAN

LAN

Table 6.1

Comparison of LAN, WAN, and MAN networks

Network type	Ownership	Data transfer speed	Scope
LAN	Usually one party	100 Mbps to 10 Gbps	A building or a campus
WAN	More than one party	28.8 Kbps to155 Mbps	Intercity to international
MAN	One to several parties	34 Mbps to155 Mbps	One city to several contiguous cities

 5 Network Topologies

 **network topology** represents a network's physical layout, including the arrangement of computers and cables. Five common topologies are discussed in the following sections: star, ring, bus, hierarchical, and mesh.

5.1 Star Topology

The **star topology** usually consists of a central computer (host computer, often a server) and a series of nodes (typically, workstations or peripheral devices). The host computer supplies the main processing power. A node failing doesn't affect the network's operation, but if the host computer fails, the entire network goes down.

In a ring topology, if any link between nodes is severed, the entire network is affected, and failure of a single node disrupts the entire network.

Advantages of this topology include the following:

- Cable layouts are easy to modify.
- Centralized control makes detecting problems easier.
- Nodes can be added to the network easily.
- It's more effective at handling heavy but short bursts of traffic.

Two main disadvantages are that having a central host means a single point of potential failure, and many cables are required, which increases cost.

5.2 Ring Topology

In a **ring topology**, no host computer is required, because each computer manages its own connectivity. Computers and devices are arranged in a circle so that each node is connected to two other nodes: its upstream neighbor and its downstream neighbor. Transmission is in one direction, and nodes repeat a signal before passing it to the downstream neighbor. If any link between nodes is severed, the entire network is affected, and failure of a single node disrupts the entire network. A token ring is a common implementation of the ring topology. It is a LAN protocol specified in the IEEE 802.5 where all stations are connected in a ring and each station can directly receive transmissions only from its immediate neighbor. Permission to transmit is granted by a message (token) that circulates around the ring. Modern ring topologies, such as Fibre Distributed Data Interface (FDDI), are capable of bidirectional transmission (clockwise and counterclockwise), which prevents the problems caused by a single node failure.

A ring topology needs less cable than a star topology, but it's similar to a star topology in that it's better for handling heavy but short bursts of traffic. Also, diagnosing problems and modifying the network are more difficult than with a star topology.

5.3 Bus Topology

The **bus topology** (also called "linear bus") connects nodes along a network segment, but the ends of the cable aren't connected, as they are in a ring topology. A hardware device called a terminator is used at each end of the cable to absorb the signal. Without a terminator, the signal would bounce back and forth along the length of the cable and prevent network communication.

Common speeds in a bus topology are 1, 2.5, 5, 10, 100 Mbps, 1 Gbps, and 10 Gbps (Gigabit Ethernet) with 10 Mbps, 100 Mbps, 1 Gbps, and 10 Gbps being the most widely used. A node failure has no effect on any other node. Advantages of this topology include the following:

- It's easy to extend.
- It's very reliable.
- The wiring layout is simple and uses the least amount of cable of any topology, which keeps costs down.
- It's ability to handle steady (even) traffic.

Two main disadvantages are that fault diagnosis is difficult and the bus cable can be a bottleneck when network traffic is heavy.

5.4 Hierarchical Topology

A **hierarchical topology** (also called a "tree topology") combines computers with different processing strengths in different organizational levels. For example, the bottom level might consist of workstations, with minicomputers in the middle and a server at the top. Companies that are geographically dispersed and organized hierarchically are good candidates for this type of network. Failure of nodes at the bottom might not have a big impact on network performance, but the middle

A **network topology** represents a network's physical layout, including the arrangement of computers and cables.

The **star topology** usually consists of a central computer (host computer, often a server) and a series of nodes (typically, workstations or peripheral devices).

In a **ring topology**, no host computer is required, because each computer manages its own connectivity.

The **bus topology** (also called "linear bus") connects nodes along a network segment, but the ends of the cable aren't connected, as they are in a ring topology.

A **hierarchical topology** (also called a "tree topology") combines computers with different processing strengths in different organizational levels.

nodes, and especially the top node (which controls the entire network) are crucial for network operation.

Traditional mainframe networks also use a hierarchical topology. The mainframe computer is at the top, front-end processors (FEPs) make up the next level, controllers and multiplexers are at the next level, and terminals and workstations make up the last level. A **controller** is a hardware and software device that controls data transfer from a computer to a peripheral device (examples are a monitor, a printer, or a keyboard) and vice versa. A **multiplexer** is a hardware device that allows several nodes to share one communication channel. The intermediate-level devices (FEPs and controllers) reduce the host's processing load by collecting data from terminals and workstations.

The hierarchical topology offers a great deal of network control and lower cost, compared with a star topology. Some disadvantages are that network expansion might pose a problem, and there could be traffic congestion at the root and higher-level nodes.

5.5 Mesh Topology

In a **mesh topology** (also called "plex" or "interconnected"), every node (which can differ in size and configuration from the others) is connected to every other node. This topology is highly reliable. Failure of one or a few nodes doesn't usually cause a major problem in network operation, because many other nodes are available. However, this topology is costly and difficult to maintain and expand.

6 Major Networking Concepts

t he following sections explain important networking concepts, including protocols, TCP/IP, routing, routers, and the client/server model.

6.1 Protocols

As mentioned, protocols are agreed-on methods and rules that electronic devices use to exchange information. People need a common language to communicate, and the same is true of computer and other electronic devices. Some protocols deal with hardware connections, and others control data transmission and file transfers. Protocols also specify the format of message packets sent between computers. In today's networks, multiple protocol support is becoming more important, as networks need to support protocols of computers running different operating systems, such as Mac OS, Linux/UNIX, and Windows. The following section describes the most widely used network protocol, TCP/IP.

6.2 Transmission Control Protocol/Internet Protocol

Transmission Control Protocol/Internet Protocol (TCP/IP) is an industry-standard suite of communication protocols. TCP/IP's main advantage is that it enables interoperability—in other words, it allows the linking of devices running on many different platforms. TCP/IP was originally intended for Internet communication, but because it addressed issues such as portability, it also became the standard protocol for UNIX network communication.

Two of the major protocols in the TCP/IP suite are Transmission Control Protocol (TCP), which operates at the OSI model's Transport layer, and Internet Protocol (IP), which operates at the OSI model's Network layer. TCP's primary functions are establishing a link between hosts, ensuring message integrity, sequencing and acknowledging packet delivery, and regulating data flow between source and destination nodes.

IP is responsible for packet forwarding. To perform this task, it must be aware of the available data link protocols and the optimum size of each packet. After it recognizes the size of each packet, it must be able to divide data into packets of the correct size. An IP address consists of 4 bytes in IPv4 or16 bytes in IPv6 (32 bits or 128 bits) and is divided into two parts: a network address and a node address. Computers on the same network must use the same network address, but each computer must have a unique node address. IP networks combine network and node addresses into one IP address; for example, 131.255.0.0 is a valid IP address.

A **controller** is a hardware and software device that controls data transfer from a computer to a peripheral device (examples are a monitor, a printer, or a keyboard) and vice versa.

A **multiplexer** is a hardware device that allows several nodes to share one communication channel.

In a **mesh topology** (also called "plex" or "interconnected"), every node (which can differ in size and configuration from the others) is connected to every other node.

Transmission Control Protocol/Internet Protocol (TCP/IP) is an industry-standard suite of communication protocols that enables interoperability.

Part 2: Data Communication, the Internet, E-Commerce and Global Information Systems

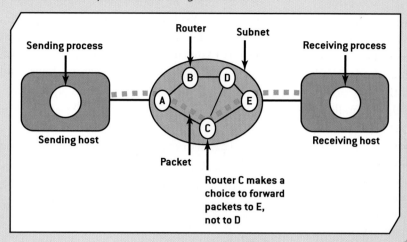

Exhibit 6.5 *A packet-switching network*

Sending process

Sending host

Router

Subnet

Receiving process

Receiving host

Packet

Router C makes a
choice to forward
packets to E,
not to D

6.3 Routing

To understand routing better, first examine packet switching, a network communication method that divides data into small packets and transmits them to an address, where they are reassembled. A **packet** is a collection of binary digits, including message data and control characters for formatting and transmitting, sent from computer to computer over a network. Packets are transmitted along the best route available between sender and receiver (see Exhibit 6.5). Any packet-switching network can handle multimedia data, such as text, graphics, audio, and video.

The path or route that data takes on a network is determined by the type of network and the software used to transmit data. The process of deciding which path that data takes is called **routing**. Routing is similar to the path you take from home to work. Although you probably take the same path most of the time, sometimes you have to change your path, depending on road and weather conditions, traffic, and time of day. Similarly, a packet's route can change each time a connection is made, based on the amount of traffic and the availability of the circuit. The decision about which route to follow is done in one of two ways: at a central location (centralized routing) or at each node along the route (distributed routing). In most cases, a **routing table**, generated automatically by software, is used to determine the best possible route for the packet. The routing table lists nodes on a network and the path to each node, along with alternate routes and the speed of existing routes.

In **centralized routing**, one node is in charge of selecting the path for all packets. This node, considered the network routing manager, stores the routing table,

and any changes to a route must be made at this node. All network nodes forward status information on the number of inbound, outbound, and processed messages to the network routing manager periodically. The network routing manager, therefore, has an overview of the network and can determine whether any part of it is underused or overused. As with all centralized configurations, there are disadvantages to having control at one node. For example, if the network routing manager is at a point far from the network's center, many links and paths that make up the network are far from the central node. Status information sent by other nodes to initiate changes to the routing table have to travel a long distance to the central node, causing a delay in routing some data and reducing network performance. In addition, if the controlling node fails, no routing information is available.

Distributed routing relies on each node to calculate the best possible route. Each node contains its own routing table with current information on the status of adjacent nodes so that the best possible route can be followed. Each node also sends status messages periodically so that adjacent nodes can update their tables. Distributed routing eliminates the problems caused by having the routing table at a centralized site. If one node isn't operational, routing tables at other nodes are updated, and the packet is sent along a different path.

A **packet** is a collection of binary digits, including message data and control characters for formatting and transmitting, sent from computer to computer over a network.

Routing is the process of deciding which path to take on a network. This is determined by the type of network and the software used to transmit data.

A **routing table**, generated automatically by software, is used to determine the best possible route for the packet.

In **centralized routing**, one node is in charge of selecting the path for all packets. This node, considered the network routing manager, stores the routing table, and any changes to a route must be made at this node.

Distributed routing relies on each node to calculate its own best possible route. Each node contains its own routing table with current information on the status of adjacent nodes so that the best possible route can be followed.

6.4 Routers

A **router** is a network connection device containing software that connects network systems and controls traffic flow between them. The networks being connected can be operating on different protocols, but they must use a common routing protocol. Routers operate at the Network layer of the OSI model and handle routing packets on a network. Cisco Systems and Juniper Networks are two major router vendors.

A router performs the same functions as a bridge but is a more sophisticated device. A bridge connects two LANs using the same protocol, and the communication medium doesn't have to be the same on both LANs.

Routers can also choose the best possible path for packets based on distance or cost; routers can prevent network jams that delay packet delivery and handle packets of different sizes. A router can also be used to isolate a portion of the LAN from the rest of the network; this process is called "segmenting." For example, you might want to keep information about new product development or payroll information isolated from the rest of the network, for confidentiality reasons.

There are two types of routers: static and dynamic. A **static router** requires the network routing manager to give it information about which addresses are on which network. A **dynamic router** can build tables that identify addresses on each network. Dynamic routers are used more often now, particularly on the Internet.

6.5 Client/Server Model

In the **client/server model**, software runs on the local computer (the client) and communicates with the remote server to request information or services. A server is a remote computer on the network that provides

information or services in response to client requests. For example, on your client computer, you make this request: "Display the names of all marketing majors with a GPA greater than 3.8." The database server receives your request, processes it, and returns the following names: Alan Bidgoli, Moury Jones, and Jasmine Thomas.

In the most basic client/server configuration, the following events usually take place:

1. The user runs client software to create a query.

2. The client accepts the request and formats it so that the server can understand it.

3. The client sends the request to the server over the network.

4. The server receives and processes the query.

5. The results are sent to the client.

6. The results are formatted and displayed to the user in an understandable format.

The main advantage of the client/server architecture is its scalability, meaning its capability to grow. Client/server architectures can be scaled horizontally or vertically. Horizontal scaling means adding more workstations (clients), and vertical scaling means migrating the network to larger, faster servers.

To understand client/server architecture better, you can think of it in terms of these three levels of logic:

- Presentation logic
- Application logic
- Data management logic

Presentation logic, the topmost level, is concerned with how data is returned to the client. The Windows graphical user interface (GUI) is an example of presentation software. An interface's main function is to translate tasks and convert them to something users can understand. Application logic is the software processing requests for users. Data management logic is responsible for handling data management and storage operations. The real challenge in a client/server architecture is how to divide these three logics between the client and server. The following sections describe some typical architectures used for this purpose.

6.5.1 Two-Tier Architecture

In the **two-tier architecture**, known as the traditional client/server model, a client (tier one) communicates directly with the server (tier two), as shown in Exhibit 6.6. The presentation logic is always on the client, and the data

---

A **router** is a network connection device containing software that connects network systems and controls traffic flow between them.

A **static router** requires the network routing manager to give it information about which addresses are on which network.

A **dynamic router** can build tables that identify addresses on each network.

In the **client/server model**, software runs on the local computer (the client) and communicates with the remote server to request information or services. A server is a remote computer on the network that provides information or services in response to client requests.

In the **two-tier architecture**, known as the traditional client/server model, a client (tier one) communicates directly with the server (tier two).

Exhibit 6.6 *A two-tier client/server architecture*

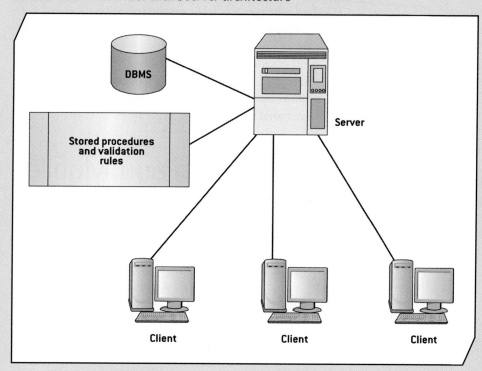

management logic is on the server. The application logic can be on the client, on the server, or split between them, although it's usually on the client side.

This architecture is effective in small workgroups (that is, groups of 50 clients or less). Because application logic is usually on the client side, a two-tier architecture has the advantages of application development speed, simplicity, and power. On the downside, any changes in application logic, such as stored procedures and validation rules for databases, require major modifications on clients, resulting in upgrade and modification costs. However, this depends on the application.

Exhibit 6.7 *An n-tier architecture*

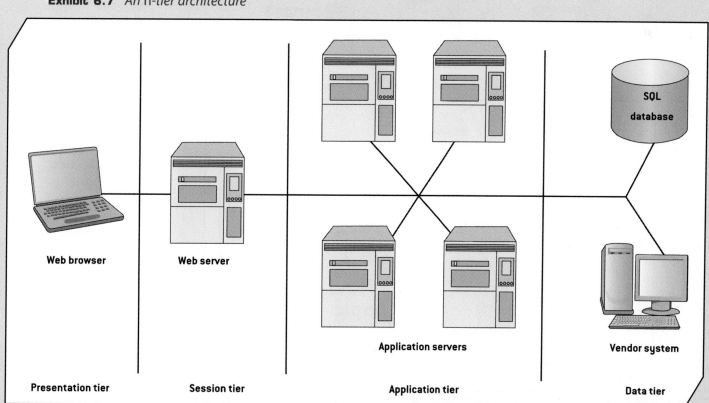

6.5.2 N-Tier Architectures

In a two-tier architecture, if the application logic is modified, it can affect the processing workload. For example, if application software is placed on the client, changing the data management software requires modifying the software on all clients. An *n*-tier architecture attempts to balance the workload between client and server by removing application processing from both the client and server and placing it on a middle-tier server, as shown in Exhibit 6.7. The most common *n*-tier architecture is the three-tier architecture. This arrangement leaves the presentation logic on the client and the data management logic on the server (see Exhibit 6.8).

Improving network performance is a major advantage of this architecture. However, network management is more challenging because there's more network traffic, and testing software is more difficult in an n-tier architecture because more devices must communicate to respond to a user request.

> An *n*-tier architecture attempts to balance the workload between client and server by removing application processing from both the client and server and placing it on a middle-tier server.
>
> A **wireless network** is a network that uses wireless instead of wired technology.
>
> A **mobile network** (also called a cellular network) is a network operating on a radio frequency (RF), consisting of radio cells, each served by a fixed transmitter, known as a cell site or base station.

7 Wireless and Mobile Networks

a wireless network is a network that uses wireless instead of wired technology. A **mobile network** (also called a cellular network) is a network operating on a radio frequency (RF), consisting of radio cells, each served by a fixed transmitter, known as a cell site or base station (discussed later, in "Mobile Networks"). These cells are used to provide radio coverage over a wider area.

Wireless and mobile networks have the advantages of mobility, flexibility, ease of installation, and low cost. These systems are particularly effective when no infrastructure (such as communication lines or established wired networks) is in place, which is common in many developing nations and in old buildings that don't have the necessary wiring for a network. Drawbacks of mobile and wireless networks include the following:

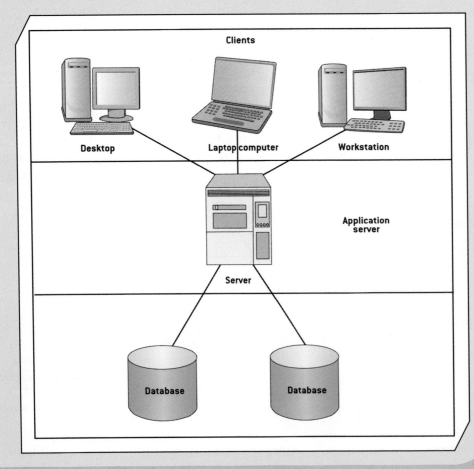

Exhibit 6.8 *A three-tier architecture*

Clients

Desktop | Laptop computer | Workstation

Application server

Server

Database | Database

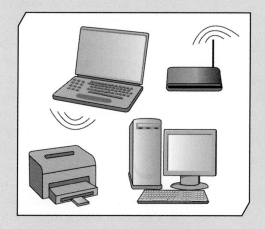

Exhibit 6.9 *A wireless notebook connecting to a wired LAN*

- *Limited throughput*—**Throughput** is similar to bandwidth. It's the amount of data transferred or processed in a specified time, usually one second. As other disadvantages in this list are solved, throughput is expected to increase in the future.

- *Limited range*—The distance a signal can travel without losing strength is more limited in mobile and wireless networks. For example, a Wi-Fi (Wireless Fidelity) network can have a range of 120 feet indoors and 300 feet outdoors. Chapter 14 covers range specifications for mobile and wireless networks in more detail.

- *In-building penetration problems*—Wireless signals might not be able to pass through certain building materials or might have difficulty passing through walls.

- *Vulnerability to frequency noise*—Interference from other signals, usually called "noise," can cause transmission problems. Common sources of noise include thunderstorms and lightning, which creates radio waves (the same waves used by wireless networks); transformers; high-voltage cables; and fluorescent lights.

- *Security*—Wireless network traffic can be captured with sniffers. (Security is discussed in more detail later in this chapter.)

There are different definitions of mobile and wireless computing. Mobile computing might simply mean using a laptop away from the office or using a modem to access the corporate network from a client's office. Neither activity requires wireless technology. Wireless LANs usually refer to proprietary LANs, meaning they use a certain vendor's specifications, such as Apple Computer's LAN protocol for linking Macintosh devices. The term is also used to describe any wireless network. Wireless LANs have the same features and characteristics as wired LANs, except they use wireless media, such as infrared (IR) light and RF.

Exhibit 6.9 shows a wireless notebook connecting to a wired LAN. The transceiver on the laptop establishes radio contact with the wired LAN (although the figure doesn't show the entire wired network). The transceiver/receiver can be built in, attached to the notebook, or mounted on a wall or placed on a desk next to the notebook.

There are many applications of wireless networks. For example, health care workers are more productive because they can use handheld or notebook computers with wireless capabilities to get patient information quickly. Instead of writing notes on the patient's condition on paper and transcribing them into an electronic form later, health care workers can enter information directly into handheld wireless devices. Because the information can be sent to and saved on a centralized database, it's available to other workers instantly. In addition, entering notes directly prevents errors that are common during the transcription process, which improves the quality of information.

7.1 Wireless Technologies

In a wireless environment, portable computers use small antennas to communicate with radio towers in the surrounding area. Satellites in near-Earth orbit pick up low-powered signals from mobile and portable network devices. The wireless communication industry has many vendors and rapid changes, but wireless technologies generally fall into two groups:

- *Wireless LANs (WLANs)*—These networks are becoming an important alternative to wired LANs in many companies. Like their wired counterparts, WLANs are characterized by having one owner and covering a limited area.

- *Wireless WANs (WWANs)*—These networks cover a broader area than WLANs and include the following devices: cellular networks, cellular digital packet data (CDPD), paging networks, personal communication systems (PCS), packet radio networks, broadband personal communications systems (BPCS), microwave networks, and satellite networks.

Throughput is similar to bandwidth. It's the amount of data transferred or processed in a specified time, usually one second.

Table 6.2

WLANs versus WWANs

	WLANs	WWANs
Coverage	About 100 m	Much wider area; capable of a regional, nationwide, or international range
Speed	With the 802.11b wireless standard, data transfer rate up to 11 Mbps, 802.11a up to 54 Mbps, and 802.11n up to 100 Mbps	Depending on the technology, could vary from 115 Kbps to 14 Mbps
Data security	Usually lower than WWANs	Usually higher than WLANs

To improve the efficiency and quality of digital communications, **Time Division Multiple Access (TDMA)** divides each channel into six time slots. Each user is allocated two slots: one for transmission and one for reception. This method increases efficiency by 300%, as it allows carrying three calls on one channel.

These technologies enable computing devices to communicate with other devices or networks at any time and from any location. They use a protocol for communication between a cellular network's transmitters and receivers and users' devices. Both WLANs and WWANs rely on the RF spectrum as the communication medium, but they differ in the areas outlined in Table 6.2.

Note that 802.11a and 802.11b have been largely replaced by the current wireless standards: 802.11g and 802.11n. 802.11g uses the 2.4 GHz frequency and has a data transfer rate of 54 Mbps; 802.11n uses the same frequency but increases the data transfer rate to 100 Mbps.

7.2 Mobile Networks

Mobile networks have a three-part architecture, shown in Exhibit 6.10:

- Base stations send and receive transmissions to and from subscribers.
- Mobile telephone switching offices (MTSOs) transfer calls between national or global phone networks and base stations.
- Subscribers (users) connect to base stations by using mobile communication devices.

Mobile devices register by subscribing to a carrier service (provider) licensed for certain geographic areas. When a mobile unit is outside its provider's coverage area, roaming occurs. Roaming is using a cellular phone outside of a carrier's limited service area. By doing this, users are extending the connectivity service in a location that is different from the home location where the service was first registered.

To improve the efficiency and quality of digital communications, two technologies have been developed: Time Division Multiple Access and Code Division Multiple Access. **Time Division Multiple Access (TDMA)**

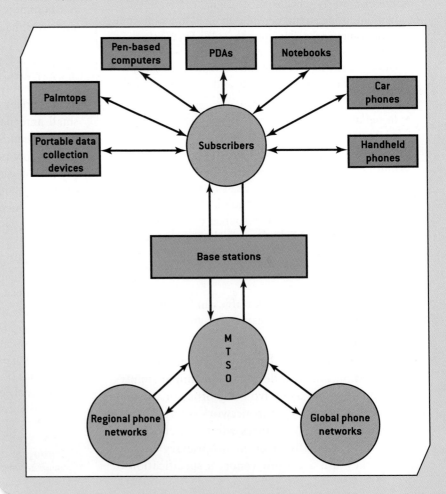

Exhibit 6.10 *Mobile network architecture*

In the past, separate networks were used to transmit data, voice, and video, but as the demand for integrated services increased, technology has developed to meet this demand.

Table 6.3

Generations of cellular networks

Generation	Description
1G	Analog transmission with limited bandwidth
2G	Support for voice, data, paging, and fax services added
2.5G	Added packet-switching technology, which transmits data packets over radio signals; different from the phone system's circuit-switching technology, which transmits data as a continuous stream of bits
3G	Supports transmission of high-quality multimedia data, including data, voice, and video
4G	An advanced version of 3G that provides broadband, large-capacity, high-speed data transmission, and high-quality interactive multimedia services

divides each channel into six time slots. Each user is allocated two slots: one for transmission and one for reception. This method increases efficiency by 300%, as it allows carrying three calls on one channel. **Code Division Multiple Access (CDMA)** transmits multiple encoded messages over a wide frequency and then decodes them at the receiving end.

Advanced Mobile Phone System (AMPS) is the analog mobile phone standard developed by Bell Labs and introduced in 1983. Digital technologies, however, are more widely used because of higher data capacities, improved voice quality, encryption capabilities, and integration with other digital networks. As of February 18, 2008, U.S. carriers were no longer required to support AMPS. Companies such as AT&T and Verizon have discontinued this service permanently. Table 6.3 describes generations of cellular networks. (Bluetooth, a wireless networking technology, is discussed in Chapter 14.)

Many businesses use wireless and mobile networks to improve customer service and reduce operational costs. The following information box gives an overview of the Apple iPhone and its business applications.

To improve the efficiency and quality of digital communications, **Code Division Multiple Access (CDMA)** transmits multiple encoded messages over a wide frequency and then decodes them at the receiving end.

Mobile Computing in Action: The Apple iPhone

Apple released its first version of the iPhone in 2007. In 2008, the 3G version, which allowed fast 3G networking, was introduced. In 2009, the iPhone 3GS, which was much faster than the earlier versions and also allowed video recording, was introduced. In 2010, the iPhone 4 was released; it enabled video conferencing, enhanced e-mail, multitasking among all apps, and much more. There are numerous business applications for the iPhone. Here are some of the business benefits of using the iPhone[2,3]:

- *It integrates with Microsoft Exchange so that you can check your e-mails, your contacts, and your calendar, staying up to date no matter where you are.*
- *The Safari browser gives you access to your company's resources anytime and anywhere.*
- *It includes innovative apps like Maps, Voice Memos, and Voice Control.*
- *You can send SMS messages to multiple recipients.*
- *You can check stocks from anywhere or get a quick weather report before heading off on your next business trip.*

8 Wireless Security

Security is important in any type of network, but it is especially important in a wireless network, since anyone walking or driving within the range of an access point (AP) (even outside a home or office) can use the network. An AP is the part of a wireless LAN (WLAN) that connects it to other networks. Finding WLANs is an easy task. A user can simply walk or drive around office buildings or homes with a WLAN-equipped computer and try to pick up a signal. Wireless signals can also be intercepted, and they are susceptible to the same DoS attacks that wired networks are susceptible to, as discussed in Chapter 5.

There are several techniques for improving the security of a wireless network:

1. SSID (Service Set Identifier)—All client computers that try to access the AP are required to include a SSID in all of their packets. A packet without a SSID is not processed by the AP. The major weakness of using SSID is that it can be picked up by other devices within the range, given the right software.

2. WEP (Wired Equivalent Privacy)—A key must be manually entered into the AP and the client computer. The key encrypts the message before transmission. Because this manual process is complex and time consuming, the WEP technique is not suitable for large networks.

3. EAP (Extensible Authentication Protocol)—EAP keys are dynamically generated based on the user's ID and password. When the user logs out of the system, the key is discarded. A new key is generated when the user logs back into the network.

4. WPA (Wi-Fi Protected Access)—This technique combines the strongest features of WEP and EAP: Keys are fixed, as in WEP, or dynamically changed, as in EAP. However, the WPA key is longer than the WEP key; therefore, it is more difficult to break. Also, the key is changed for each frame (a distinct and identifiable data set) before transmission.

5. WPA2 or 802.11i—This technique uses EAP to obtain a master key. With this master key, a user's computer and the AP negotiate for a key that will be used for a session. After the session is terminated, the key is discarded. This technique uses Advanced Encryption Standard, which is more complex than WPA and much harder to break.

9 Convergence of Voice, Video, and Data

In data communication, **convergence** refers to integrating voice, video, and data so that multimedia information can be used for decision making. In the past, separate networks were used to transmit voice, video, and data, but as the demand for integrated services has increased, technology has developed to meet this demand.

Convergence required major network upgrades, because video requires much more bandwidth. This has changed, however, with the availability of high-speed technologies, such as Asynchronous Transfer Mode (ATM), Gigabit Ethernet, 3G and 4G networks, and more demand for applications using these technologies. Gigabit Ethernet is a LAN transmission standard capable of 1 Gbps and 10 Gbps data transfer speeds. ATM is a packet-switching service that operates at 25 Mbps, 622 Mbps, with maximum speed of up to 10 Gbps. As mentioned earlier, the 3G network is the third generation of mobile networking and telecommunications.

> In data communication, **convergence** refers to integrating voice, video, and data so that multimedia information can be used for decision making.

Telepresence: A New Use of Data Communication and Convergence

Telepresence has attracted a lot of attention recently, particularly because of the economic downturn; companies are finding that they can reduce business travel and business meeting expenses. This technology integrates audio and video conferencing into a single platform, and recent improvements have resulted in higher quality, greater ease of use, and better reliability.

Telepresence systems can be used to record meetings for later use or to incorporate multimedia technologies into presentations. They can also offer plug-and-play collaboration applications. Some products offer a telepresence room, with customized lighting and acoustics as well as large high-density (HD) screens that can be configured for up to 20 users. Others are on a smaller scale, with a single HD screen. Major vendors of telepresence products include Cisco, Polycom, Tandberg, Teliris, and HP.[4]

© B Busco/Getty Images

It features a wider range of services and advanced network capacity than the 2G network. The 3G network has increased the rate of information transfer, its quality, video and broadband wireless data transfers, and the quality of Internet Telephony or Voice over Internet Protocol (VoIP). It has also made possible streaming video and much faster uploads and downloads. The 4G network will further enhance all these features.

More content providers, network operators, telecommunication companies, and broadcasting networks, among others, have moved toward convergence. Even smaller companies are now taking advantage of this fast-growing technology by offering multimedia product demonstrations and using the Internet for multimedia presentations and collaboration. Convergence is possible now because of a combination of technological innovation, changes in market structure, and regulatory reform. Common applications of convergence include the following:

- E-commerce
- Entertainment (the number of available TV channels will increase substantially, and movies and videos on demand will become available)
- Increased availability and affordability of video and computer conferencing
- Consumer products and services, such as virtual classrooms, telecommuting, and virtual reality

The Internet, as a tool for delivering services, is an important contributor to the convergence phenomenon. Advances in digital technologies are helping to move convergence technologies forward, and when standards

Industry Connection

CISCO SYSTEMS, INC.*

The main goal of Cisco Systems, Inc., the largest vendor of networking equipment, is to make it easier to connect different computers. Cisco offers a wide variety of products, including routers, switches, network management tools, optical networking, security software, VPNs, firewalls, and collaboration and telepresence products. The variety of products makes it possible for organizations to get everything they need for networking solutions from one vendor. Cisco's products and services include the following:

- **PIX Firewall series**—Allows corporations to protect their internal networks from outside intruders.
- **Network Management Tools**—Allow network managers to automate, simplify, and integrate their networks to reduce operational costs and improve productivity.
- **Identity Management Tools**—Protect information resources through identity policies, access control, and compliance features.
- **TelePresence Network Management**—Integrates audio, high-definition video, and interactive features to deliver face-to-face collaboration capabilities.
- **Cisco 800 Series routers**—Offer built-in security, including content filtering, WAN connection with multiple access options, four 10/100 Mbps Fast Ethernet managed switch ports, and more. The routing system released in 2010, CRS-3 (Cisco Carrier Routing System-3), delivers Internet speeds of up to 322 terabits per second, which can offer video and other content 12 times faster than its rivals' systems.
- **LinkSys**—This router is popular for home and small office use; it connects devices to each other and to the Internet.

*This information has been gathered from the company Web site (www.cisco.com) and from other promotional materials. For more information and updates, visit the Web site.

in data collection, processing, and transmission become more available and acceptable, their use should increase even further.

One notably growing application of data communication and convergence is telepresence. Discussed in more detail in Chapter 14, telepresence refers to using related technologies, such as digital networks and virtual reality, to create a real-life communication experience in an audiovisual environment. In other words, it seems as though participants in a meeting are sitting across the table from one another instead of appearing on a screen. Telepresence is being used for many purposes, such as teleconferencing, education and training, advertising, and entertainment. See the information box on Telepresence for an explanation of how this technology is being used by organizations to reduce costs, generate revenue, and improve customer service.

The Industry Connection box highlights Cisco Systems, Inc., which offers many products and services used in data communication systems.

10 Chapter Summary

this chapter has provided an overview of data communication systems and networking concepts. It has discussed the basic components of a data communication system, processing configurations, types of networks, and network topologies as well as important networking concepts, such as routing and the client/server model. It also covered wireless and mobile networks, as well as future trends in data communication: convergence and telepresence technologies.

Key Terms

attenuation (97)

bandwidth (97)

broadband (97)

bus topology (103)

centralized processing (99)

centralized routing (105)

client/server model (106)

Code Division Multiple Access (CDMA) (111)

communication media (98)

conducted media (98)

controller (104)

convergence (112)

data communication (95)

decentralized processing (99)

digital subscriber line (DSL) (98)

distributed processing (99)

distributed routing (105)

dynamic router (106)

hierarchical topology (103)

local area network (LAN) (100)

mesh topology (104)

metropolitan area network (MAN) (101)

mobile network (108)

modem (97)

multiplexer (104)

narrowband (97)

network interface card (NIC) (100)

network topology (103)

n-tier architecture (108)

Open Systems Interconnection (OSI) model (100)

packet (105)

protocols (97)

radiated media (99)

ring topology (103)

router (106)

routing (105)

routing table (105)

star topology (103)

static router (106)

throughput (109)

Time Division Multiple Access (TDMA) (110)

two-tier architecture (106)

Transmission Control Protocol/Internet Protocol (TCP/IP) (104)

wide area network (WAN) (101)

wireless network (108)

Problems, Activities, and Discussions

1. How is bandwidth measured?

2. What are some examples of network topologies?

3. Describe the client/server model.

4. Visit the following Web pages, consult other sources, and write a one-page paper on wireless security. How is it different from wired security? Why is it more challenging to secure a wireless network than a wired one?

   http://www.practicallynetworked.com/support/wireless_secure.htm

   http://www.ehow.com/about_5244116_definition-wireless-security.html

5. Dimdim (*http://dimdim.com/*) is an example of an e-collaboration tool. Using a Web browser, it allows use rs to share documents, Web pages, whiteboards, audio, video, as well as record a meeting. Compare and contrast this tool with WebEx (introduced in the chapter). What are some of the advantages of each?

6. Visit the following Web pages, consult other sources, and write a one-page paper that highlights mobile phone security. How could the security of a mobile device be improved?

   http://www.infoworld.com/t/hacking/defcon-hackers-target-cell-phone-security-742?source=IFWNLE_nlt_blogs_2010-07-27

   http://www.infoworld.com/d/mobilize/mobile-phone-security-dos-and-donts-516?source=IFWNLE_nlt_sec_2010-06-08

7. Visit the following Web page, consult other sources, and write a one-page paper that outlines the applications of mobile computing in the healthcare industry.

   http://webcache.googleusercontent.com/search?q=cache:cof3t0a95ikJ:subs.emis.de/LNI/Proceedings/Proceedings15/GI-Proceedings.15-22.pdf+what+companies+are+using+Business+applications+of+Mobile+computing%3F&cd=35&hl=en&ct=clnk&gl=us

8. Visit the following Web page and summarize the applications of data communication technologies in Ford's MyTouch system.

   http://www.crm-daily.com/news/On-the-Go-CRM—New-Developments/story.xhtml?story_id=023001ISUN55

9. Which of the following isn't a communication medium?

   a. Network computer

   b. Twisted pair cable

   c. Fiber optics

   d. Satellite

10. Mesh is the most reliable network topology. True or False?

casestudy

DATA COMMUNICATION AT WALMART

Walmart has made several changes in its data communication systems to improve its suppliers' access to sales and inventory data. For example, the company added a customized Web site for its suppliers, such as Mattel, Procter & Gamble, and Warner-Lambert. Walmart's goal is to improve efficiency in order to keep prices low and maintain a high level of customer service. With Walmart's network, suppliers can access sales, inventory, and forecasting data over extremely fast connections. To ensure confidentiality of data, a sophisticated security system has been implemented to prevent suppliers from accessing data about one another's products.

Walmart has also added Web-based access to its RetailLink system so that suppliers can use information in the database. Other data communication applications at Walmart include automated distribution, computerized routing, and electronic data interchange (EDI).[5]

Walmart uses the latest in wireless technology in its operations for warehouse management systems (WMS) to track and manage the flow of goods through its distribution centers. Another application of wireless technology is for controlling and monitoring forklifts and industrial vehicles

that move merchandise inside its distribution centers. The Vehicle Management System (VMS) is the latest application of data communications in Walmart. Among other features, VMS includes a two-way text messaging system that enables management to effectively divert material-handling resources to where they are needed the most. (VMS works effectively with RFID systems.) According to Walmart, VMS has improved safety and has also significantly improved the productivity of its operations.[6]

Answer the following questions:

1. How has Walmart improved its data communication systems for suppliers?

2. What are some typical data communication applications in Walmart?

3. What are some of the applications of wireless technology in Walmart?

4. What are some of the features and capabilities of VMS?

© Anton Gvozdikov/Shutterstock.com

SPEAK UP!

MIS2 was built on a simple principle: to create a new teaching and learning solution that reflects the way today's faculty and students teach and learn. Through conversations, focus groups, surveys, and interviews, we collected data that drove the creation of the version of **MIS2** that you are using today.

But it doesn't stop there – in order to make **MIS2** an even better learning experience, we'd like you to SPEAK UP and tell us how **MIS2** works for you.

What do you like about it? What would you change? Do you have additional ideas that would help us build a better product for next year's Management Information Systems students?

Speak Up! **Go to CourseMate for MIS2. Access at login.cengagebrain.com.**

THE INTERNET, INTRANETS, AND EXTRANETS

t his chapter gives you an introduction to the Internet and Web technologies as well as an overview of the Domain Name System and types of Internet connections. You also learn about how navigational tools, search engines, and directories are used on the Internet, and you get a brief survey of common Internet services and Web applications. This chapter also explains intranets and extranets and how they're used. Finally, you learn about trends of Web 2.0 and 3.0 eras and Internet2.

The Internet started in 1969 as a U.S. Department of Defense project.

learning outcomes

After studying this chapter you should be able to:

LO1 Describe the makeup of the Internet and the World Wide Web.

LO2 Discuss navigational tools, search engines, and directories.

LO3 Describe common Internet services.

LO4 Summarize widely used Web applications.

LO5 Explain the purpose of intranets.

LO6 Explain the purpose of extranets.

LO7 Summarize new trends in the Web 2.0 and 3.0 eras.

1 The Internet and the World Wide Web

the **Internet** is a worldwide collection of millions of computers and networks of all sizes. The term *Internet* was derived from the term *internetworking*, which means connecting networks. Simply put, the Internet is a network of networks. No one actually owns or runs the Internet, and each network is administered and funded locally.

The Internet started in 1969 as a U.S. Department of Defense project called **Advanced Research Projects Agency Network (ARPANET)** that connected four nodes: University of California at Los Angeles, University of California at Santa Barbara, Stanford Research Institute at Stanford, and University of Utah at Salt Lake City. Other nodes composed of computer networks from universities and government laboratories were added to the network later. These connections were linked in a three-level hierarchical structure: backbones, regional networks, and local area networks.

> The **Internet** is a worldwide collection of millions of computers and networks of all sizes. It is a network of networks.
>
> The **Advanced Research Projects Agency Network (ARPANET)**, a project started in 1969 by the U.S. Department of Defense, was the beginning of the Internet.

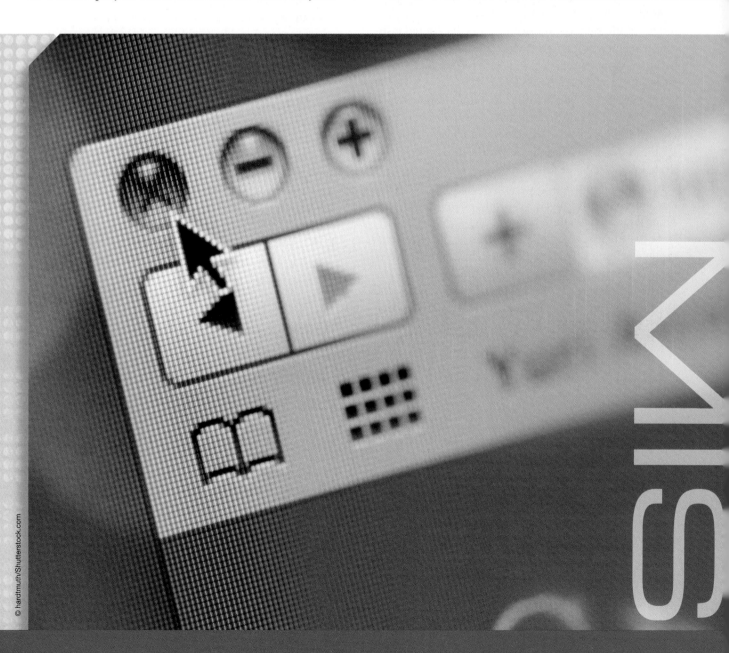

© hardtmuth/Shutterstock.com

Exhibit 7.1 *IBM's backbone*

IBM Global Network
Advantis
45 Mbps DS-3
Backbone

Courtesy of International Business Machines Corporation, © IBM.

ARPANET evolved into the National Science Foundation Network (NSFNET) in 1987, which is considered the first Internet backbone. The NSF initially restricted Internet use to research and educational institutions; commercial use wasn't allowed. Eventually, because of increased demand, other backbones were allowed to connect to NSFNET.

The **Internet backbone** is a foundation network linked with fiber-optic cables that can support very high bandwidth. It's called a "backbone" because it supports all the other networks that form the Internet, just as the human backbone is the foundation of the nervous system. The Internet backbone is made up of many interconnected government, academic, commercial, and other high-capacity data routers.

Several private companies operate their own Internet backbones that interconnect at network access points (NAPs). You can find a list of Internet backbones in the United States at *www.nthelp.com/maps.htm*. Exhibit 7.1 shows IBM's backbone (*www.nthelp.com/images/ibm.jpg*). NAPs determine how traffic is routed over the Internet. As you learned in Chapter 6, local

area networks (LANs) serve as localized Internet connections, and they use NAPs to connect to the Internet backbone.

The World Wide Web (WWW, or "the Web") changed the Internet in 1989 by introducing a graphical interface to the largely text-based Internet. The Web was proposed by Tim Berners-Lee at CERN. CERN is the European Organization for Nuclear Research, the world's largest particle physics center. It stands for *Conseil Europeen pour la Recherche Nucleaire*.

The Web organizes information by using **hypermedia**, meaning documents that include embedded references to audio, text, images, video, and other documents. Composed of billions of hypermedia documents, the Web constitutes a large portion of the Internet. The embedded references in hypermedia documents are called **hypertext**; they consist of links users can click to follow a particular thread (topic). By using hypertext links, users can access files, applications, and other computers in any order they like (unlike in paper documents) and retrieve information with the click of a button. In essence, hypertext is an approach to data management, in which data is stored in a network of nodes connected by links. Data in these nodes is accessed with an interactive browsing system, meaning the user determines the order in which information is accessed.

Any computer that stores hypermedia documents and makes them available to other computers on the Internet is called a server or Web server, and computers requesting these documents are called clients. A client can be a home computer or a node in an organization's LAN. The most exciting feature of the Web is that hypermedia documents can be stored anywhere in the world, so users can jump from a site in the United States to a site in Paris, France, all in a few milliseconds. The following information box summarizes major events in the development of the Internet.

1.1 Domain Name System

Domain names, such as IBM.com or whitehouse.gov, are unique identifiers of computer or network addresses on the Internet. Each computer or network also has an Internet Protocol (IP) address, such as 208.77.188.166, that's assigned by the Internet Corporation for Assigned Names and Numbers (ICANN). These numbers are difficult to remember, however, so English-like domain names are used more often to access Web sites. When information is transferred from one network to another, domain names are converted to IP addresses by the protocol **Domain Name System (DNS)**. Servers using this

The **Internet backbone** is a foundation network linked with fiber-optic cables that can support very high bandwidth. It is made up of many interconnected government, academic, commercial, and other high-capacity data routers.

With **hypermedia**, documents include embedded references to audio, text, images, video, and other documents.

The embedded references in hypermedia documents are called **hypertext**; they consist of links users can click to follow a particular thread (topic).

When information is transferred from one network to another, domain names are converted to IP addresses by the protocol **Domain Name System (DNS)**. Servers using this protocol (called DNS servers) maintain lists of computers' and Web sites' addresses and their associated IP addresses.

protocol (called DNS servers) maintain lists of computers' and Web sites' addresses and their associated IP addresses. DNS servers translate all domain names into IP addresses.

You see domain names used in **uniform resource locators (URLs)**, also called "universal resource locators," to identify a Web page. A URL is the address of a document or site on the Internet. For example, in the URL *http://www.csub.edu*, the domain name is *csub.edu*. Every domain name has a suffix indicating the top-level domain (TLD) it belongs to. In this example, the suffix is .edu, which stands for educational institutions. Combinations of letters and numerals 0 through 9 as well as hyphens can be used in domain names, too. Spaces are not allowed.

The TLD denotes the type of organization or country the address specifies. TLDs are divided into organizational domains (generic top-level domains, gTLDs) and geographic domains (country code top-level domains, ccTLDs). Table 7.1 lists common gTLDs.

Many new gTLDs have been proposed, including .aero (aviation industry), .museum, .law, and .store. Some are already in use, such as .info for organizations providing information services, .biz for businesses, and .news for news-related sites.

In addition, most countries have geographic domains. These ccTLDs include .au for Australia, .ca for Canada, .fr for France, .jp for Japan, and .uk for the United Kingdom. You can find a complete list of ccTLDs at *http://www.thrall.org/domains.htm*.

> **Uniform resource locators (URLs)**, also called "universal resource locators," identify a Web page. A URL is the address of a document or site on the Internet.

Major Events in the Development of the Internet

September 1969: *ARPANET is born.*

1971: *Ray Tomlinson of BBN invents an e-mail program to send messages across a network.*

January 1983: *Transition from Network Control Protocol (NCP) to Transmission Control Protocol/Internet Protocol (TCP/IP), the protocol for sending and receiving packets.*

1987: *The National Science Foundation creates a backbone to the National Research and Education Network called NSFNET; it signifies the birth of the Internet.*

November 1988: *A worm attacks more than 6,000 computers, including those at the Department of Defense.*

1989: *The World Wide Web is developed at CERN.*

February 1991: *The Bush administration approves Senator Al Gore's idea to develop a high-speed national network, and the term "information superhighway" is coined.*

November 1993: *Pacific Bell announces a plan to spend $16 billion on the information superhighway.*

January 1994: *MCI announces a six-year plan to spend $20 billion on an international communication network.*

October 1994: *The first beta version of Netscape Navigator is available.*

April 1995: *Netscape becomes the most popular graphical navigator for surfing the Web.*

August 1995: *Microsoft releases the first version of Internet Explorer.*

April 1996: *Yahoo! goes public.*

June 1998: *A U.S. appellate court rules that Microsoft can integrate its browser with its operating system, allowing the company to integrate almost any application into the Microsoft OS.*

February 2000: *A denial-of-service attack (DoS) shuts down several Web sites, including Yahoo!, Ameritrade, and Amazon.com, for several hours.*

2004: *A worm called "MyDoom" or "Novarg" spreads through Web servers. About 1 in 12 e-mail messages are infected.*

2005: *YouTube.com is launched.*

2008: *Microsoft offers to buy Yahoo! for $44.6 billion, and San Francisco federal judge Jeffrey S. White orders the disabling of Wikileaks.org, a Web site that discloses confidential information.*

2009: *The number of worldwide Internet users surpasses one billion per month.*[1]

2010: *The number of worldwide Facebook users tops 500 million.*

To understand the parts of a URL more clearly, examine the URL

http://www.csub.edu/~hbidgoli/books.html.
A brief explanation from left to right follows:

- *http*—Stands for Hypertext Transfer Protocol, the protocol used for accessing most Web sites.
- *www.csub.edu*—WWW, World Wide Web, or the Web, the subdomain, is the network consisting of millions of Internet sites that offer text, graphics, sounds, animations, and other multimedia resources. It organizes information by using hypermedia. The *csub* stands for California State University at Bakersfield, and the *.edu* is the suffix for educational institutions. Together, *csub.edu* identifies this Web site uniquely.
- */~hbidgoli*—This part is the name of the directory in which files pertaining to the books are stored. A server can be divided into directories for better organization.
- *books.html*—This part is the document itself. The .html extension means it's a Hypertext Markup Language (HTML) document. (See the information box for more on HTML.) Servers that don't support long extensions display *.htm*, and other servers display *.html*.

What Is HTML?

Hypertext Markup Language (HTML) *is the language used to create Web pages. It defines a page's layout and appearance by using tags and attributes. A tag delineates a section of the page, such as the header or body; an attribute specifies a value for a page component, such as a font color. (Note that HTML codes are not case sensitive.) The most recent version of HTML is HTML5, which competes with Adobe Flash. Flash is a multimedia platform used to add animation, video, and interactivity to Web pages. A typical structure for an HTML document is as follows:*

<HTML>

<HEAD>

(Enter the page's description.)

</HEAD>

<BODY>

(Enter the page's content.)

</BODY>

</HTML>

Table 7.1

Generic top-level domains

gTLD	Purpose
.com	Commercial organizations (such as Microsoft)
.edu	Educational institutions (such as California State University)
.int	International organizations (such as the United Nations)
.mil	U.S. military organizations (such as the U.S. Army)
.gov	U.S. government organizations (such as the Internal Revenue Service)
.net	Backbone, regional, and commercial networks (for example, the National Science Foundation's Internet Network Information Center)
.org	Other organizations, such as research and nonprofit organizations (for example, the Internet Town Hall)

1.2 Types of Internet Connections

As you learned in Chapter 6, there are several methods for connecting to a network, including the Internet. These methods include dial-up and cable modems as well as Digital Subscriber Line (DSL). Several types of DSL services are available, including the following:

- *Symmetric DSL (SDSL)*—Has the same data transmission rate to and from the phone network (called "upstream" and "downstream"), usually up to 1.5 Mbps (millions bits per second) in both directions.

- *Asymmetric DSL (ADSL)*—Has a lower transmission rate upstream (3.5 Mbps); downstream rates are typically 24 Mbps (e.g., the ITU G.992.5 Annex M standard).

- *Very High-Speed DSL (VDSL)*—Has downstream/upstream transmission rate up to 100 Mbps over short distances (e.g., ITU G.993.2 standard).

Organizations often use T1 or T3 lines. These are provided by the telephone company and are capable of transporting the equivalent of 24 conventional telephone lines, using only two pairs of copper wires. T1 uses two pairs of copper wires to carry up to 24 simultaneous conversations ("channels") and has a transmission rate of 1.544 Mbps; it's more widely used than T3. (In other countries, T1 is called E1 and has a transmission rate of 2.048 Mbps.) A T3 line is a digital communication link that supports transmission rates of about 43 to 45 Mbps. A T3 line actually consists of 672 channels, each one supporting rates of 64 Kbps.

2 Navigational Tools, Search Engines, and Directories

ow that you know what the Internet is and how to connect to it, you'll need tools to get around it and find what you're looking for. These tools can be divided into three general categories:

1. **Navigational tools** are used to travel from site to site ("surf") the Internet.

2. Search engines, which give you an easy way to look up information and resources on the Internet by entering keywords related to your topic of interest.

3. Directories, which are indexes of information based on keywords in documents, make it possible for search engines to find what you're looking for. Some Web sites, such as Yahoo!, also use directories to organize content into categories.

Originally, Internet users used text-based commands for simple tasks such as downloading files or sending e-mails. It's tedious to type commands at the command line. The user would also need certain programming knowledge to be able to use these systems. The graphical browsers changed all of this by providing menus and graphical tools that allow point-and-click techniques. These systems make the user-system interface easier and more convenient. These graphical browsers also support multimedia information, such as images and sound.

The following sections describe the three main tools available for getting around the Internet and finding information.

2.1 Navigational Tools

Many graphical Web browsers are available, such as Microsoft Internet Explorer (IE), Mozilla Firefox, Google Chrome, Apple Safari, and Opera. Typically, these browsers have menu options you've seen in word-processing programs, such as File, Edit, and Help. They also include options for viewing your browsing history, bookmarking favorite Web sites, and setting viewing preferences as well as navigation buttons to move backward and forward in Web pages you've visited. With some browsers, you can also set up specialized toolbars for accessing frequently visited sites or conducting searches.

Hypertext Markup Language (HTML) is the language used to create Web pages. It defines a page's layout and appearance by using tags and attributes. A tag delineates a section of the page, such as the header or body; an attribute specifies a value for a page component, such as a font color.

Navigational tools are used to travel from site to site ("surf") the Internet.

Directories are indexes of information based on keywords in documents, making it possible for search engines to find what you're looking for.

2.2 Search Engines and Directories

A **search engine**, such as Google.com, Bing.com or Ask.com, is an information system that enables users to retrieve data from the Web by searching for information using search terms. All search engines follow this three-step process:

1. *Crawling the Web.* Search engines use software called "crawlers," "spiders," "bots," and other similar names. These automated modules search the Web continuously for new data. When you post a new Web page, crawlers find it (if it's public), and when you update it, crawlers find the new data. Crawlers also check to see what links are on your page and make sure they work; if a link is broken, crawlers identify it and include this information as part of the data about that page. In addition, crawlers can also go through other pages that are part of your Web site, as long as there are links to these pages. All the gathered data is sent back to the search engine's data center so that the search engine always has the most current information on the Web.

2. *Indexing.* Search engines use server farms to index data coming in from crawlers by using keywords. Every keyword has an index entry that's linked to all Web pages containing the keyword. For example, a company selling picture frames has the term *picture frame* on its Web site several times. The indexing process recognizes the frequency of use and creates an index entry for the term *picture frame*. This index entry is linked to the company's Web site, along with all other sites containing the term *picture frame*. Indexing makes it possible for search engines to retrieve all related Web pages when you enter a search term.

3. *The search process.* When you enter a search term, the search engine uses the index created in Step 2 to look up the term. If the term exists in the index, the search engine identifies all Web pages linked to the term. However, it needs some way of prioritizing Web pages based on how close each one is to the search term. For example, say your Aunt Emma makes picture-frame cookies and has a Web site for selling

them. Someone searching on the term *picture frames* might see Aunt Emma's site listed, too. Because search engines are programmed to try to differentiate different types of search requests, however, they can use other terms, such as *posters*, *photos*, and *images*, to give a higher priority to Web pages containing these additional terms, along with the search term *picture frames*, and a lower priority to Web pages containing terms such as *cookies* or *baked goods* along with *picture frames*. Search engines vary in intelligence, which is why you can use the same search term and get different results with two different search engines.

Directories organize information into categories. There are two kinds of directories on the Web. The first is the automated, or crawler-based, directory that search engines use; it creates indexes of search terms and collects these terms automatically by using crawlers. Google, Yahoo!, Ask.com, and others fall into this category. When your Web page changes, for example, these directories update their indexes and databases automatically to reflect the change. The second type of directory is the human-powered directory. If you want your Web page to be listed in a search engine's results, you have to manually submit keywords to a human-powered directory. It doesn't use crawlers to collect data; instead, it relies on users to supply the data. After keywords are submitted, they're indexed with search engines and can then be listed in search results. The main difference is that if your Web page changes, the directory doesn't have the updated content until you submit changes to the directory. Open Directory is an example of a human-powered directory. However, Google has made many directories obsolete, and directories in general are not as relevant as they used to be.

Crawler-based directories are based on index terms, just as the phone book's white pages are based on the last names and first names of people. Some search engines, in addition to their index-term-based directory, offer directories based on popular categories such as business, sports, entertainment, travel, and dining. Each category can have subcategories; for example, an Entertainment category might contain Movies, Music, and Theater subcategories. Yahoo! Travel, Yahoo! Business, and Yahoo! RealEstate are some categories in the Yahoo! directory, which is considered top-of-the-line by both users and experts.

A **search engine**, such as Google.com or Ask.com, is an information system that enables users to retrieve data from the Web by searching for information using search terms.

3 Internet Services

many services are available via the Internet, and most are made possible by the TCP suite of protocols in the Application layer (introduced in Chapter 6). For instance, TCP/IP provides several useful e-mail protocols, such as Simple Message Transfer Protocol (SMTP) for sending e-mails and Post Office Protocol (POP) for retrieving messages. Popular services include e-mail, newsgroups, and discussion groups; Internet Relay Chat (IRC) and instant messaging; and Internet telephony, discussed in the following sections.

3.1 E-mail

E-mail is one of the most widely used services on the Internet. In addition to personal use, many companies use e-mail for product announcements, payment confirmations, and newsletters. There are two main types of e-mail. Web-based e-mail enables you to access your e-mail account from any computer and, in some cases, store your e-mails on a Web server. MSN Hotmail and Google Gmail are two examples of free Web-based e-mail services. The other type is client-based e-mail, which consists of an e-mail program you install on your computer; e-mail is downloaded to and stored locally on your computer. Examples of client-based e-mail programs include Microsoft Outlook, Mozilla Thunderbird, and Apple Mail.

Most e-mail programs include a system for organizing your e-mails in folders as well as address books to store e-mail addresses. Many address books include an auto-completion feature, too, so that you can just type a recipient's name and the e-mail address is filled in automatically. You can also set up distribution groups for sending an e-mail to several people at the same time. Other commonly available features are spell checkers and delivery notifications. You can also attach documents and multimedia files to e-mails.

3.2 Newsgroups and Discussion Groups

The Internet serves millions of people with diverse backgrounds and interests. Discussion groups and newsgroups are a great way for people with similar interests to find each other. Although newsgroups and discussion groups are alike in many ways, **discussion groups** are usually for exchanging opinions and ideas on a specific topic, usually of a technical or scholarly nature. Group members post messages or articles that others in the group can read. **Newsgroups** are typically more general in nature and can cover any topic; they allow people to get together for fun or for business purposes. For example, you could join a newsgroup for people interested in ancient civilizations or one that's used to help people write and debug computer programs. Newsgroups can also serve as an effective advertising medium in e-commerce.

3.3 Instant Messaging

Internet Relay Chat (IRC) enables users in chat rooms to exchange text messages with people in other locations in real time. Think of it as a coffee shop where people sit around a table and chat, except each person could be in a different country. In addition, the entire conversation is "recorded," so you can scroll back to see something you missed, for example. You can find chat rooms on a variety of topics, such as gardening, video games, and relationships.

Discussion groups are usually for exchanging opinions and ideas on a specific topic, usually of a technical or scholarly nature. Group members post messages or articles that others in the group can read.

Newsgroups are typically more general in nature and can cover any topic; they allow people to get together for fun or for business purposes.

Internet Relay Chat (IRC) enables users in chat rooms to exchange text messages with people in other locations in real time.

© Dusan Jankovic/Shutterstock.com

Instant messaging (IM) is a service for communicating with others via a private "chat room" on the Internet. Many IM applications are available, such as Windows Messenger, Yahoo! Messenger, and Google Chat, and the capabilities and features vary depending on the application. For example, some IM applications notify you when someone on your chat list comes online; others have features for audio or video conversations.

3.4 Internet Telephony

Internet telephony is using the Internet rather than the telephone network to exchange spoken conversations. The protocol used for this capability is **Voice over Internet Protocol (VoIP)**. To use VoIP, you need a high-speed Internet connection and usually a microphone or headset. Some companies have special adapters that connect to your high-speed modem and allow you to use your regular phone. Because access to the Internet is available at local phone connection rates, international and other long-distance calls are much less expensive. Many businesses use VoIP to offer hotlines, help desks, and other services at far lower cost than with telephone networks.

VoIP is also used to route traffic starting and ending at conventional public switched telephone network (PSTN) phones. The only drawback is the call quality, which isn't as good as with regular phone lines. However, the quality has been improving steadily. In addition to cost savings, VoIP offers the following advantages:

- Users don't experience busy lines.
- Voicemails can be received on the computer.
- Users can screen callers, even if the caller has caller ID blocked.
- Users can have calls forwarded from anywhere in the world.
- Users can direct calls to the correct departments and take automated orders.

4 Web Applications

Several service industries use the Internet and its supporting technologies to offer services and products to a wide range of customers at more competitive prices and with increased convenience. In the current economic downturn, the Internet is playing an important role in helping organizations reduce expenses, because Web applications can be used with minimum costs. The following sections describe how a variety of service industries use Web applications.

4.1 Tourism and Travel

The tourism and travel industry has benefited from e-commerce Web applications. For example, the Tropical Island Vacation (*www.tropicalislandvacation.com*) home page directs prospective vacationers to online brochures after vacationers respond to questions about the type of vacation they want to take. They can click appealing photographs or phrases to explore further. Many travel Web sites allow customers to book tickets for plane trips and cruises, and make reservations for hotels and rental cars. On some sites, such as InfoHub.com (*www.infohub.com/*), specialty

Instant messaging (IM) is a service for communicating with others via a private "chat room" on the Internet.

Internet telephony is using the Internet rather than the telephone network to exchange spoken conversations.

Voice over Internet Protocol (VoIP) is the protocol used for Internet telephony.

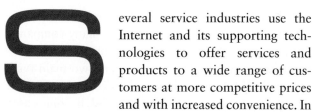

Many businesses use VoIP to offer hotlines, help desks, and other services at far lower cost than with telephone networks.

© Guy Shapira/Shutterstock.com

travel adventures are offered covering activities such as artist workshops, yoga retreats, hunting, scuba diving, and more. Expedia.com, Travel.com, Travelocity.com, Priceline.com, Hotels.com, and Yahoo! Travel are other examples of sites that offer all types of tourism and travel services.

4.2 Publishing

Many major publishers in the United States and Europe have Web sites that offer descriptions of forthcoming books, post sample chapters, offer online ordering, and include search features for looking up books on certain topics or by specific authors. Some publishers even offer books that can be read online free for 90 days or allow you to buy e-book versions or even selected chapters. Exhibit 7.2 shows the home page of this book's publisher, Cengage Learning (*www.cengage.com*).

Exhibit 7.2 *The Cengage Learning home page*

© Cengage Learning

4.3 Higher Education

Most universities have Web sites with information about departments, programs, faculty, and academic resources. Some even offer virtual tours of the campus for prospective students, and more universities are creating virtual divisions that offer entire degree programs via the Internet. Online degree programs help colleges and universities facing an enrollment decline, because they make it possible for students who couldn't attend school otherwise to enroll in classes. With online classes, universities can also have renowned experts give lectures or seminars, usually at a reduced cost, because travel expenses aren't a factor. In addition, many professional certification programs are offered through the Internet, which is convenient for people who live in remote areas or can't attend regular classes.

4.4 Real Estate

Real estate Web sites provide millions of up-to-date listings of homes for sale. Buyers can review neighborhoods, schools, and local real estate prices, and customers can use these sites to find realtors and brokerage firms and learn home-buying tips. Some sites have virtual tours of houses for sale, which is convenient for buyers moving to another state, for example. (Chapter 14 covers virtual reality technologies.) Other services include appraisals, neighborhood and school profiles, financing options, home improvement advice, and more. Major real estate Web sites include Remax (*www.remax.com*), Century 21 (*www.century21.com*), Prudential (*www.prudential.com*), and ERA (*www.era.com*).

4.5 Employment

Employment services are widely available on the Internet. You might be familiar with Monster.com, for example. These sites offer comprehensive services for job seekers, such as the following:

- Expert advice and tools for managing your career
- Resume assistance, including tools for creating professional-looking resumes
- Job search tutorials
- Posting and distributing resumes
- Job alerts
- Searches by company, industry, region, or category
- Announcements of job fairs

- Career tests to see what career is right for you
- Salary calculators

4.6 Financial Institutions

Almost all U.S. and Canadian banks and credit unions, and many others worldwide, offer online banking services and use e-mail to communicate with customers and send account statements and financial reports. E-mail helps banks reduce the time and costs of communicating via phone (particularly long-distance calls) and postal mail. Customers can get more up-to-date account information and check balances at any time of the day or night. Despite all these advantages, consumer acceptance has been slow. Measures are being taken, however, to ensure that a secure nationwide electronic banking system is in place, which should help ease consumers' concerns. For example, digital signatures (discussed in Chapter 5) are a key technology, because they provide an electronic means of guaranteeing the authenticity of involved parties and verifying that encrypted documents haven't been changed during transmission.

The following list describes some banking services available via the Internet:

- Access to customer service by e-mail around the clock
- Viewing current and old transactions
- Online mortgage applications
- Interactive tools for designing a savings plan, choosing a mortgage, or getting insurance quotes online
- Finding loan status and credit card account information online
- Paying bills and credit card accounts
- Transferring funds
- Viewing digital copies of checks

4.7 Software Distribution

Many vendors distribute software on the Internet as well as drivers and patches. For example, most antivirus vendors make updates available for download to keep up with new viruses and worms. Typically, patches, updates, and small programs such as new browser versions are fast and easy to download. Trying to download large programs, such as Microsoft Office Suite, takes too long, so these types of programs aren't usually distributed via the Internet.

Developing online copyright-protection schemes continues to be a challenge. If users need an encryption code to "unlock" software they've downloaded, making backups might not be possible. Despite these challenges, online software distribution provides an inexpensive, convenient, and fast way to sell software.[2, 3]

4.8 Health Care

With patient records stored on the Internet, health care workers can order lab tests and prescriptions, admit patients to hospitals, and refer patients to other physicians more easily; also, test and consultation results can be directed to the right patient records automatically. All patient information can be accessible from one central location, so that finding critical health information is faster and more efficient, especially if a patient falls ill while away from home. However, these systems have potential problems involving information privacy, accuracy, and currency.

There are other uses for health care Web sites. Telemedicine (*http://telemedtoday.com*), for example, enables medical professionals to conduct remote consultation, diagnosis, and conferencing, which can save on office overhead and travel costs. In addition,

© Todd Pearson/Getty Images

personal health information systems (PHISs) can make interactive medical tools available to the public. These systems use public kiosks (often in shopping malls) equipped with Internet-connected computers and a diagnosis procedure that prompts patients with a series of questions. These systems can be useful in detecting early onset of diseases.[4]

In addition, virtual medicine on the Internet enables specialists at major hospitals to operate on patients remotely. Telepresence surgery, as it's called, allows surgeons to operate all over the world without physically traveling anywhere. A robot performs the surgery based on the digitized information sent by the surgeon via the Internet. These robots have stereoscopic cameras to create three-dimensional images for the surgeon's virtual reality goggles and tactical sensors that provide position information to the surgeon.

Already, prescription drugs are sold online, and several Web sites offer medical services.[5] For example, WebMD (*www.webmd.com*) offers a variety of health information, such as prevention tips, proper nutrition, and common symptoms of diseases.

4.9 Politics

Most political candidates now make use of Web sites in campaigns. The sites are a helpful tool for announcing candidates' platforms, publicizing their voting records, posting notices of upcoming appearances and debates, and even raising campaign funds. President Barack Obama's fund-raising efforts in his first presidential campaign are a good example of how successful Web sites can be in this area.

Some claim the Internet has helped empower voters and revitalize the democratic process. Being well informed about candidates' stances on political issues is much easier with Web sites, for example, and online voting may make voting easier for people who in the past couldn't make it to polling sites. In addition, there's the possibility of legislators being able to remain in their home states, close to their constituents, and voting on bills via an online system. However, a stringent ID system would have to be in place, one that most likely

would use biometric security measures. Currently, the U.S. House of Representatives is attempting to put all pending legislation online, and presidential documents, executive orders, and other materials are available on the White House's Web site. You can also find full-text versions of speeches, proclamations, press briefings, daily schedules, the proposed federal budget, health care reform documents, and the Economic Report of the President. The information box below highlights some of the possible features of the Internet in 2020.

The Internet in 2020

There are some radical predictions for the future of the Internet having to do with architectural changes, IP addresses, DNS, and routing tables. However, most experts agree that the Internet in 2020 will have the following characteristics[6]:

1. *More people will use it—as many as 5 billion, as compared to 1.7 billion today.*
2. *It will be more geographically dispersed, stretching to more remote corners of the globe. (It will also support more languages.)*
3. *It will be more a network of devices—for example, sensors on buildings to monitor security—than a network of computers.*
4. *It will carry much more content—a veritable "exaflood," which is the term coined by researchers for the onrushing flow of exabytes.*
5. *It will be wireless. By 2014, as many as 2.5 billion people worldwide will subscribe to mobile broadband, according to one estimation.*
6. *More Internet services will be offered in cloud as cloud-computing will continue to grow, generating $45.5 billion in annual revenue as early as 2015.*
7. *It will be greener, thanks to more energy-efficient technologies.*
8. *Network management will be more automated.*
9. *Connectivity will become less important, thanks to techniques that will tolerate delays or forward communications from one user to another.*
10. *There will be more hackers.*

5 Intranets

many of the applications and services made possible with the Internet can be offered to an organization's users by establishing an intranet. An **intranet** is a network within an organization that uses Internet protocols and technologies (for example, TCP/IP, which includes File Transfer Protocol [FTP], SMTP, and others) for collecting, storing, and disseminating useful information that supports business activities, such as sales, customer service, human resources, and marketing. Intranets are also called "corporate portals." You might wonder what the difference is between a company's Web site and its intranet. The main difference is that the company Web site is usually public; an intranet is for internal use by employees. However, many companies also allow trusted business partners to access their intranets, usually with a password or another authentication method to protect confidential information.

An intranet uses Internet technologies to solve organizational problems that have been solved in the past by proprietary databases, groupware, scheduling, and workflow applications. An intranet is different from a LAN, although it uses the same physical connections. An intranet is an application or service that uses an organization's computer network. Although intranets are physically located in an organization, they can span the globe, allowing remote users to access the intranet's information. However, defining and limiting access carefully is important for security reasons, so intranets are typically set up behind a firewall.

In a typical intranet configuration (Exhibit 7.3), users in the organization can access all Web servers, but the system administrator must define each user's level of access. Employees can communicate with one another and post information on their departmental Web servers.

> An **intranet** is a network within an organization that uses Internet protocols and technologies (for example, TCP/IP, which includes File Transfer Protocol [FTP], SMTP, and others) for collecting, storing, and disseminating useful information that supports business activities, such as sales, customer service, human resources, and marketing.

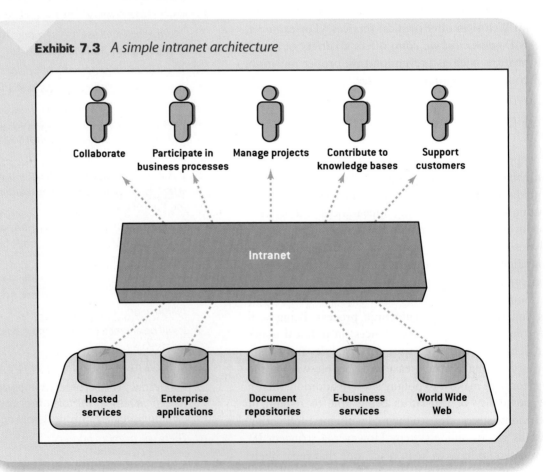

Exhibit 7.3 *A simple intranet architecture*

Collaborate · Participate in business processes · Manage projects · Contribute to knowledge bases · Support customers

Intranet

Hosted services · Enterprise applications · Document repositories · E-business services · World Wide Web

Table 7.2 *The Internet versus intranets*

Key feature	Internet	Intranet
User	Anybody	Approved users only
Geographical scope	Unlimited	Limited to unlimited
Speed	Slower than an intranet	Faster than the Internet
Security	Less than an intranet's	More than the Internet's; user access is restricted more

Departmental Web servers can be used to host Web sites. For example, the human resources department might have a separate Web site containing information that employees need to access frequently, such as benefits information or 401K records. Similarly, the marketing department could have a Web site with the latest product information. Employees can also bookmark important sites in the intranet.

5.1 The Internet versus Intranets

The Internet is a public network; an intranet is a private network. Any user can access the Internet, but access to an intranet is only for certain users and must be approved. Table 7.2 summarizes the major differences between the Internet and intranets.

Despite these differences, both use the same protocol, TCP/IP, and both use browsers for accessing information. Typically, they use similar languages for developing applications, such as Java, and offer files in similar formats.

Another advantage of an intranet is that because the organization can control which browser is used, it can specify a browser that supports the technologies the organization uses, such as Internet telephony or video conferencing. In addition, the organization knows documents will be displayed the same way in all users' browsers; on the Internet, there's no assurance that a Web page will be displayed the same way for every user who views it. Intranets also enable organizations to share software, such as an office suite or a DBMS.

5.2 Applications of an Intranet

A well-designed intranet can make the following types of information, among others, available to the entire organization in a timely manner to improve an organization's efficiency and effectiveness:[7]

- *Human resources management*—401K plans, upcoming events, company's mission statement and policies, job postings, medical benefits, orientation materials, online training sessions and materials, meeting minutes, vacation time
- *Sales and marketing*—Call tracking, information on competitors, customer information, order tracking and placement, product information
- *Production and operations*—Equipment inventory, facilities management, industry news, product catalog, project information
- *Accounting and finance*—Budget planning, expense reports

Intranets can also help organizations move from a calendar- or schedule-based document-publishing strategy to one that's based on events or needs. In the past, for example, a company usually published an employee handbook once a year, and it wasn't updated until the following year, even if major changes happened that required updating, such as organizational restructuring. The company might have sent single pages occasionally as updates, and employees had to insert these pages in a binder. Needless to say, these binders were often out of date and difficult to use. With an intranet, however, a company can make updates as soon as they're needed, in response to company events rather than a set schedule.

Intranets reduce the costs and time of document production, too. In the past, document production went through several steps, such as creating content, producing and revising drafts, migrating content to desktop publishing, duplicating, and distributing. Intranets eliminate the duplication and distribution steps, and often the step of migrating to a publishing application can be streamlined or eliminated.

6 Extranets

an **extranet** is a secure network that uses the Internet and Web technologies to connect intranets of business partners so that communication between organizations or between consumers is possible. Extranets are considered a type of interorganizational system (IOS). These systems facilitate information exchange among business partners. Some of these systems, such as electronic funds transfer (EFT) and e-mail, have been used in traditional businesses as well as in e-commerce. Electronic data interchange (EDI) is another common IOS.

> An **extranet** is a secure network that uses the Internet and Web technologies to connect intranets of business partners so that communication between organizations or between consumers is possible.

As mentioned, some organizations allow customers and business partners to access their intranets for specific purposes. For example, a supplier might want to check the inventory status, or a customer might want to check an account balance. Often an organization makes only a portion of its intranet accessible to external parties as its extranet. Comprehensive security measures must ensure that access is granted only to authorized users and trusted business partners. Exhibit 7.4 shows a simple extranet. In this exhibit, DMZ refers to the demilitarized zone, an area of the network that's separate from the organization's LAN, where the extranet server is placed. Table 7.3 compares the Internet, intranets, and extranets.[8]

There are numerous applications of extranets. For example, Toshiba America, Inc., has designed an extranet for timely order-entry processing. Using this extranet, more than 300 dealers can place orders for

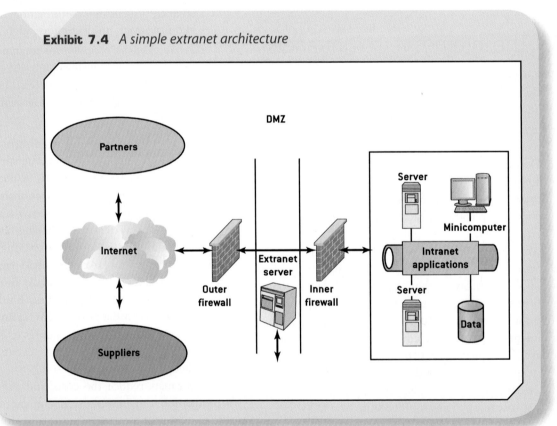

Exhibit 7.4 *A simple extranet architecture*

Extranets not only allow companies to reduce internetworking costs, they give companies a competitive advantage, which can lead to increased profits.

Table 7.3 *Comparison of the Internet, intranets, and extranets*

	Internet	Intranet	Extranet
Access	Public	Private	Private
Information	General	Typically confidential	Typically confidential
Users	Everybody	Members of an organization	Groups of closely related companies, users, or organizations

parts until 5 p.m. for next-day delivery. Dealers can also check accounts receivable balances and pricing arrangements, read press releases, and more. This secure system has decreased costs and improved customer service.[9]

Another example of an extranet is the Federal Express Tracking System (*www.fedex.com*). Federal Express uses its extranet to collect information and make it available to customers over the Internet. Customers can enter a package's tracking number and locate any package still in the system as well as prepare and print shipping forms, get tracking numbers, and schedule pickups.

Extranets not only allow companies to reduce internetworking costs, they give companies a competitive advantage, which can lead to increased profits. A successful extranet requires a comprehensive security system and management control, however. The security system should include access control, user-based authentication, encryption, and auditing and reporting capabilities.

An extranet offers an organization the same benefits as an intranet but provides other advantages, such as the following:[10]

- *Coordination*—Improves coordination between business partners, such as suppliers, distributors, and customers. Critical information can be made available quickly so that decisions can be made without delays. For example, a manufacturer can coordinate production by checking the inventory status.

- *Feedback*—Provides instant feedback from customers and other business partners to an organization and gives consumers an opportunity to express opinions on new products or services before they're introduced to the market.

- *Customer satisfaction*—Links customers to an organization so they can get more information on products and services. Customers can also order products online, expediting B2B (business-to-business). E-commerce is a major benefit of an extranet.

- *Cost reduction*—Reduces inventory costs by providing information to participants in a supply network program. For example, Mobil Corporation designed an extranet that allows its distributors to submit purchase orders, which has increased efficiency and expedited delivery of goods and services.[11]

- *Expedited communication*—Improves communication by linking intranets for access to critical information. A traveling salesperson can get the latest product information remotely before going to a sales meeting, for example.

7 New Trends: The Web 2.0 and 3.0 Eras

Web 2.0 refers to the trend toward Web applications that are more interactive than traditional Web applications. Collaboration or e-collaboration is one of the key components of Web 2.0. Table 7.4 compares various Web 1.0 and Web 2.0 applications,[11] and the following sections describe some Web 2.0 applications in more detail.

Most experts agree that Web 3.0, or the Semantic Web, provides personalization that allows users to access the Web more intelligently. Computers, not their users, will perform the tedious work involved in finding, combining, and acting upon information on the Web. For example if a user searches for the word *spring* in a sentence, the search engine figures out if the user is searching for the season, a coil, or *jump*, and then it provides the most relevant search results.

Web 2.0 refers to the trend toward Web applications that are more interactive than traditional Web applications. Collaboration or e-collaboration is one of its key components.

Table 7.4 *Web 1.0 versus Web 2.0*

Web 1.0	Web 2.0
DoubleClick (used for online marketing)	Google AdSense
Ofoto (sharing digital photos)	Flickr
Akamai (streaming media services)	BitTorrent
mp3.com	iTunes
Britannica Online	Wikipedia
Personal Web sites	Blogging
eVite (type of wiki for event planning)	Upcoming.org and Events and Venues Database (EVBD)
Domain name speculation	Search engine optimization
Page views	Cost per click
Content management systems	Wikis
ERoom and Groove (collaboration software)	Collaboration portals, such as IBM Quickr and Microsoft Sharepoint
Posting a movie file on a personal Web page	YouTube

The main focus of Web 2.0 has been on social networking and collaboration. Web 3.0, on the other hand, focuses on "intelligent" Web applications using various artificial intelligent technologies (discussed in Chapter 13). These include natural language processing, artificial neural networks, and intelligent agents. The goal is to tailor online searching and requests to users' specific search patterns, preferences, and needs.

One part of Web 3.0 could be the semantic Web proposed by World Wide Web inventor Tim Berners-Lee. According to Berners-Lee, the Web can be made more useful by using methods (such as content tags) that will enable computers to understand what they are displaying and communicate more effectively with each other. Using the semantic Web, computers will be able to read Web sites as easily as humans read them. Nova Spivack's Twine (*http://www.slideshare.net/novaspivack/web-evolution-nova-spivack-twine*) is one of the first online services to use Web 3.0 technologies. Twine automatically organizes information, learns about users' specific interests and search patterns, and makes recommendations based on this information.[12]

A **blog** (short for "Weblog") is a journal or newsletter that's updated frequently and intended for the general public. Blogs reflect their authors' personalities and often include philosophical reflections and opinions on social or political issues.

A **wiki** is a type of Web site that allows users to add, delete, and sometimes modify content.

7.1 Blogs

A **blog** (short for "Weblog") is a journal or newsletter that's updated frequently and intended for the general public. Blogs reflect their authors' personalities and often include philosophical reflections and opinions on social or political issues. Sometimes, they're simply used as a way for families or groups of friends to keep in touch with one another. Automated tools have made creating and maintaining blogs easy, so even people with very little technical background can have blogs. Many sites, such as Blogger.com, offer free space for blogs and even include posting photos.

You can also find blogs on Web sites dedicated to a particular topic or organization; these are periodically updated with the latest news and views. For example, on the CNN Web site, you can find blogs written by Anderson Cooper. Blogs are becoming a popular source of online publication, too, especially for political information, opinions, and alternative news coverage; some examples can be found at *www.HuffingtonPost.com* and *www.Slate.com*.

7.2 Wikis

A **wiki** is a type of Web site that allows users to add, delete, and sometimes modify content. One of the best-known examples is the online encyclopedia Wikipedia. What's unique about wikis is that an information user can also be an information provider. The most serious problem with wikis is the quality of information, because allowing anyone to modify content affects its accuracy. Wikipedia is currently working on methods to verify the credentials of users contributing to the site because of past problems with contributors falsifying credentials.

Wikis have caught on at many companies, too. For example, an Intel employee developed Intelpedia as a way for employees around the world to share information on company history, project progress, and more. However, some employees don't like their content being edited by others. For this reason, "corporate wikis" were developed; these include tighter security and access controls. Corporate wikis are used for a variety of purposes, such as posting news about product development. Many open-source software packages for creating wikis are available, such as MediaWiki and TWiki. Companies are also creating wikis to give customers information. For example, Motorola and T-Mobile have set up wikis about their products that function as continually updated user guides, and eBay has formed eBay Wiki, where buyers and sellers can share information on a wide range of topics.[13]

7.3 Social Networking Sites

Social networking refers to a broad class of Web sites and services that allow users to connect with friends, family, and colleagues online as well as meet people with similar interests or hobbies. More than 100 social networks are available on the Internet. Two of the most popular are Facebook and MySpace. In addition, LinkedIn is a professional networking site, where you can connect with professional contacts and exchange ideas and job opportunities with a large network of professionals. Many people now use both LinkedIn and Facebook to keep their professional and social contacts separate.

Social networking sites are also popular for business use. For example, many companies use Twitter to keep track of customer opinions on their products. Comcast, Dell, General Motors, H&R Block, and Kodak are a few examples.[14] Companies also use social networking sites for advertising; they might include links to their company Web sites or use pay-per-click

(PPC) features. PPC is an Internet advertising method used on Web sites, in which advertisers pay their host only when their ad is clicked.

As mentioned in Chapter 1, Twitter is extremely popular, and the term *tweet* is often used for a response or comment no longer than 140 characters, the maximum allowed length of a Twitter post. Even this book has a Twitter account! Follow us at *twitter.com/4LTRPress_MIS*.

7.4 RSS Feeds

RSS feeds are a fast, easy way to distribute Web content in Extensible Markup Language (XML) format. RSS, which stands for Really Simple Syndication, is a subscription service you sign up for, and new content from Web sites you've selected is delivered via a feed reader. The content all goes to one convenient spot where you can read "headlines." With this service, you don't have to keep checking a site for updates.

XML, a subset of the Standard Generalized Markup Language (SGML), is a flexible method for creating common formats for information. Unlike HTML tags that specify layout and appearance, XML tags represent the kind of content being posted and transmitted. Although HTML contains some layout and appearance features, these "presentational attributes" are deprecated by the W3C, which suggests that HTML only be used for creating structured documents through markup. Layout and appearance should be handled by CSS (Cascading Style Sheets). Data can be meaningless without a context for understanding it.

For example, consider the following: "Information Systems, Smith, John, 357, 2009, Cengage, 45.00,

Social networking refers to a broad class of Web sites and services that allow users to connect with friends, family, and colleagues online as well as meet people with similar interests or hobbies.

Really Simple Syndication (RSS) feeds are a fast, easy way to distribute Web content in Extensible Markup Language (XML) format. It is a subscription service you sign up for, and new content from Web sites you've selected is delivered via a feed reader to one convenient spot.

As you can see, each piece of data is defined with its context by using tags, which makes the data much easier to interpret. Although both HTML and XML are tag-based languages, they have different purposes. XML was designed to improve interoperability and data sharing between different systems, which is why RSS feeds are in XML. Any system can interpret the data in an RSS feed the correct way because it's based on the data's meaning, not its format and layout.

7.5 Podcasting

A **podcast** is an electronic audio file, such as an MP3 file, that's posted on the Web for users to download to their mobile devices—iPhones, iPods, and Zune, for example—or even their computers. Users can also listen to it over the Web. A podcast has a specific URL and is defined with an XML item tag.

Podcasts are usually collected by an "aggregator," such as iTunes or iPodder. You can also subscribe to various podcasts; NPR, *The Economist*, and ESPN all offer podcast subscriptions. What differentiates a podcast from a regular audio file is that users can subscribe to it. Each time a new podcast is available, the aggregator collects it automatically, using the URL, and makes it available for subscribers. Subscribers can then "sync" the podcast with their mobile devices and listen to it whenever they want.

This subscription model makes podcasts more useful and popular and increases their accessibility. Syndication feeds are one way of announcing a podcast's availability. Organizations use podcasts to update people on their products and services, new trends, changes in organizational structure, and merger/acquisition news. Financial institutions, for example, offer podcasts to inform customers about investment strategies, market performance, and trading. When multimedia information is involved, the terms *video podcast*, *vodcast*, or *vidcast* are sometimes used.

02-139-4467-X." From this string of data, you might make an educated guess that this refers to a book, including the title, author name, number of pages, year of publication, publisher, price, and ISBN. A computer, however, might interpret this same data as indicating that Smith John spoke about information systems for 357 minutes at a conference organized by Cengage in 2009 and received 45 Euros as compensation, with a transaction ID of 02-139-4467-X for the payment.

XML prevents this kind of confusion by defining data with a context, so that you would format the preceding data string as follows:

```
<book>
<title>Information Systems</title>
<authorlastname>Smith</authorlastname>
<authorfirstname>John</authorfirstname>
<pages>357</pages>
<yearofpub>2009</yearofpub>
<publisher>Cengage</publisher>
<priceinS>45.00</priceinS>
<isbn>02-139-4467-X </isbn>
</book>
```

A **podcast** is an electronic audio file, such as an MP3 file, that's posted on the Web for users to download to their mobile devices—iPhones, iPods, and Zune, for example—or even their computers.

7.6 The Internet2

Another recent development is **Internet2 (I2)**, a collaborative effort involving more than 200 U.S. universities and corporations (including AT&T, IBM, Microsoft, and Cisco Systems) to develop advanced Internet technologies and applications for higher education and academic research.

The I2 project started in 1987 and has been planned as a decentralized network, where universities in the same geographic region can form an alliance to create a local connection point-of-presence called a **gigapop**. Gigapops connect a variety of high-performance networks, and a gigapop's main function is the exchange of I2 traffic with a specified bandwidth. One major objective of the I2 project is to develop new applications that can enhance researchers' ability to collaborate and conduct scientific experiments.

Internet2 relies on the NSFNET and MCI's very high-speed backbone network service (vBNS), which was designed in 1995 as a high-bandwidth network for research applications. This nationwide network operates at 622 Mbps, using MCI's advanced switching and fiber-optic transmission technologies. Some applications of Internet2 include the following:

- *Learningware*—This suite of applications is intended to make education more accessible, targeting distance learning and self-education. The proposed Instructional Management System (IMS) provides an environment that enables students to learn in an "anytime, anywhere" fashion. This technology also gives instructors access to a broad range of teaching materials for online classes. Some software, such as WebEx and Elluminate Live, is already available.

- *Digital Library*—This initiative, started in the 1990s, aimed to create an electronic repository of educational resources, such as textbooks and journals. The goal was to include rare books and documents, such as the Dead Sea Scrolls and the Magna Carta, although this goal hasn't been achieved yet. The bottom line is that researchers can have access to everything they need without leaving their offices, including access to experts who can guide them in their work. Imagine learning the theory of relativity directly from Albert Einstein, for example.

- *Teleimmersion*—A teleimmersion system allows people in different locations to share a virtual environment created on the Web. Virtual reality has important applications in education, science, manufacturing, and collaborative decision making.

> **Internet2 (I2)** is a collaborative effort involving more than 200 U.S. universities and corporations to develop advanced Internet technologies and applications for higher education and academic research.
>
> A **gigapop** is a local connection point-of-presence that connects a variety of high-performance networks, and its main function is the exchange of I2 traffic with a specified bandwidth.

For instance, a pediatric cardiologist in Los Angeles can work with a surgeon in India to explain how to perform an operation transposing the pulmonary artery and vein in an infant born with an abnormality. The cardiologist can train the surgeon in a virtual reality environment with an imaginary infant. This technology can be used in many other training settings, such as putting out oil-rig fires, repairing complex machinery, and even co-teaching a virtual class. Oil companies have used this technology to conduct an exploratory dig and evaluate situations such as high-pressure natural gas in the vicinity or too much water in the dig. In this case, a company builds its own teleimmersion center, and all participants are in the same location.

- *Virtual laboratories*—These are environments designed specifically for scientific and engineering applications, and they allow a group of researchers connected to Internet2 to work on joint projects, such as large-scale simulations and global databases.

The Industry Connection for this chapter highlights the role of Google, Inc., as a leader in search technology and Web advertising platforms.

8 Chapter Summary

this chapter has given you an overview of the Internet and the World Wide Web, starting with a brief overview of the Internet's development. You also learned how navigational tools, search engines, and directories are used, you reviewed widely used Internet services, such as e-mail, newsgroups, and instant messaging, and you saw how Web applications are used in parts of the service industry. This chapter also explained intranets and extranets and how they're used in organizations. Finally, you got an overview of Web 2.0 and 3.0 innovations and Internet2 applications.

Industry Connection

GOOGLE, INC. *

Google, founded in 1998, is one of the most widely used search engines in the world. It also offers products and services in the following categories:

- *Search*—Google offers more than 20 search categories, including Web Search, Blog Search (by blog name, posts on a certain topic, a specific date range, and more), Catalogs (for mail-order catalogs, many previously unavailable online), Images, and Maps. There is also Google Earth, an exciting feature that combines searching with the viewing of satellite images, terrain maps, and 3-D models.

- *Ads*—AdSense, AdWords, and Analytics are used for displaying ads on your Web site or with Google search results. For example, with AdWords, you create an ad and choose keywords related to your business. When people search Google with one of your keywords, your ad is displayed next to the search results, so you're reaching an audience that's already interested in your product or service.

- *Applications*—Google has many applications for account holders, such as Gmail, Google Talk (instant messaging and voice calls), Google Groups, YouTube, Blogger, Google Checkout (for shopping online with just a single

Google sign-in), and Google Docs, a free, Web-based word-processing and spreadsheet program.

- *Enterprise*—Google has applications for organizations, such as Google Maps for Enterprise, which is particularly useful for planning operations and logistics, and SketchUp Pro, a 3-D modeling tool for architects, city planners, game developers, and others that can be used in presentations and documents.

- *Mobile*—Many of Google's services and applications, such as Blogger, YouTube, and Gmail, are available on mobile devices. You can also use text messaging to get real-time information from Google on a variety of topics, including weather reports, sports scores, and flight updates. Google also offers Android OS for smartphone and mobile devices.

- *Google Wave*— Is a collaboration and communication tool designed to consolidate features from e-mail, instant messaging, blogging, wikis, and document sharing while offering a variety of social networking features.

* This information has been gathered from the company Web site (*www.google.com*) and other promotional materials. For more information and updates, visit the Web site.

Key Terms

Advanced Research Projects Agency Network (ARPANET) (119)

blog (134)

directories (123)

discussion groups (125)

Domain Name System (DNS) (120)

extranet (132)

gigapop (137)

Hypertext Markup Language (HTML) (123)

hypermedia (120)

hypertext (120)

instant messaging (IM) (126)

Internet (119)

Internet backbone (120)

Internet Relay Chat (IRC) (125)

Internet telephony (126)

Internet2 (I2) (137)

intranet (130)

navigational tools (123)

newsgroups (125)

podcast (136)

RSS feeds (135)

search engine (124)

social networking (135)

uniform resource locators (URLs) (121)

Voice over Internet Protocol (VoIP) (126)

Web 2.0 (133)

wiki (134)

Problems, Activities, and Discussions

1. What is the Domain Name System?

2. What is Voice over Internet Protocol (VoIP)? What are some of the advantages and disadvantages of using the Internet for phone conversation?

3. What are some of the popular applications of intranets?

4. Log on to the Web site for Kiva.org (*http://www.kiva.org*). What is this site's mission? How could this site help alleviate poverty?

5. Google is planning to build an experimental, ultra-high-speed broadband network called Google Fiber for Communities. What is the purpose of this initiative? How many people will be served by this network?

6. Visit the following Web pages, consult other sources, and write a one-page paper that summarizes the applications of social networking sites for selling products and services:

   http://online.wsj.com/article/SB10001424052748704 59650457527285046301 9656.html

   http://www.newsfactor.com/story.xhtml?story_ id=74343

7. Visit the following Web page, consult other sources, and write a one-page paper that describes issues and concerns related to net neutrality:

   http://www.infoworld.com/d/adventures-in-it/net-neutrality-numbers-dont-add-804

8. Chrome, Firefox, Internet Explorer, Opera, and Safari are the top Web browsers on the market. Visit the following Web page, consult other sources, and write a one-page paper that compares and contrasts these Web browsers. List an advantage and a disadvantage of each.

   http://www.infoworld.com/d/applications/the-best-web-browser-chrome-firefox-internet-explorer-opera-or-safari-516

9. Which of the following is an example of a navigational tool? (Choose all that apply.)
   a. Chrome (Google)
   b. Mozilla (Firefox)
   c. Safari (Apple)
   d. Yahoo!

10. Podcasting is not one of the features of Web 2.0. True or False?

case study

IBM'S INTRANET

In 2006, IBM's intranet was selected as one of the 10 best intranets by the Nielsen Norman Group, a consulting firm specializing in user-focused design for products and services. The company's intranet was cited for its consistent design and for personalizing the dissemination of information to its employees in a timely fashion. It's also used as a productivity and collaboration tool by IBM employees. Its features include the following:[15]

- Personalized news, customized to employees' jobs and interests and based on profiles they create

- Job-specific portlets for different departments, such as sales or finance

- IBM Blue Pages, an employee directory that helps with collaboration

- BlogCentral, which consists of blogs through which

© Yuri Arcurs/Shutterstock.com

employees share ideas as well as RSS feeds that allow employees to subscribe to other employees' blogs

- An accessible design for people with disabilities, including motor disabilities, memory or literacy issues, and poor vision

Answer the following questions:

1. To design an intranet such as IBM's, where should an organization start? Who should participate in the design process?

2. Which features of this intranet do you think would be most useful? Why?

3. What could other organizations learn from IBM's intranet?

4. What are some challenges in designing an intranet such as IBM's?

E-COMMERCE

this chapter provides an overview of e-commerce and value chain analysis, then compares e-commerce with traditional commerce. It explains e-commerce business models and the major categories of e-commerce. Along the way, it shows what the major activities are in the business-to-consumer e-commerce cycle and the major models of business-to-business e-commerce as well as mobile- and voice-based e-commerce. Finally, it provides an overview of electronic payment systems and Web marketing, two supporting technologies for e-commerce operations.

learning outcomes

After studying this chapter, you should be able to:

LO1 Define e-commerce and describe its advantages, disadvantages, and business models.

LO2 Explain the major categories of e-commerce.

LO3 Describe the business-to-consumer e-commerce cycle.

LO4 Summarize the major models of business-to-business e-commerce.

LO5 Describe mobile- and voice-based e-commerce.

LO6 Explain two supporting technologies for e-commerce.

1 Defining E-Commerce

E-commerce and e-business differ slightly. **E-business** encompasses all the activities a company performs in selling and buying products and services using computers and communication technologies. In broad terms, e-business includes several related activities, such as online shopping, sales force automation, supply chain management, electronic procurement (e-procurement), electronic payment systems, Web advertising, and order management. **E-commerce** is buying and selling goods and services over the Internet. In other words, e-commerce is part of e-business. However, the two terms are often used interchangeably.

E-business includes not only transactions that center on buying and selling goods and services to generate revenue, but also transactions that support revenue generation by generating demand for goods and services, offering sales support and customer service, and facilitating communication between business partners.

E-commerce builds on traditional commerce by adding the flexibility that networks offer and the availability of the Internet. The following are common business applications that use the Internet:

- Buying and selling products and services
- Collaborating with other companies

> **E-business** encompasses all the activities a company performs in selling and buying products and services using computers and communication technologies.
>
> **E-commerce** is buying and selling goods and services over the Internet.

- Communicating with business partners
- Gathering business intelligence on customers and competitors
- Providing customer service
- Making software updates and patches available
- Offering vendor support
- Publishing and disseminating information

1.1 The Value Chain and E-Commerce

One way to examine e-commerce and its role in the business world is through value chain analysis. Michael Porter introduced the **value chain** concept in 1985.[1] It consists of a series of activities designed to meet business needs by adding value (or cost) in each phase of the process. A typical business or division within a business designs, produces, markets, delivers, and supports its products or services. Each activity adds cost and value to the product or service delivered to the customer (see Exhibit 8.1).

In Exhibit 8.1, the top four components—organizational infrastructure, human resource management, technological development, and procurement (gathering input)—are supporting activities. The "margin" represents the value added by supporting primary activities (the components at the

> A **value chain** is a series of activities designed to meet business needs by adding value (or cost) in each phase of the e-commerce process.

bottom). The following list describes primary activities:

- *Inbound logistics*—Movement of materials and parts from suppliers and vendors to production or storage facilities; includes tasks associated with receiving, storing, and converting raw materials to finished goods.
- *Operations*—Processing raw materials into finished goods and services.
- *Outbound logistics*—Moving and storing products, from the end of the production line to end users or distribution centers.
- *Marketing and sales*—Activities for identifying customer needs and generating sales.
- *Service*—Activities to support customers after the sale of products and services.

For instance, by having superior relationships with suppliers (through prompt payments, electronic ordering, loyalty, and so forth), the company can ensure timely delivery and high quality of raw materials. These, in turn, add value for customers by providing a high-quality product at a lower cost. If good quality and lower costs are top priorities for customers, the company knows which parts of the value chain to focus on (e.g., better suppliers to ensure quality and reduce costs, superior operations to ensure quality, better distribution to reduce costs, better after-sales service to ensure quality with warranties). So the value chain is really about understanding what aspects of an organization's business add value for customers and then maximizing those aspects.

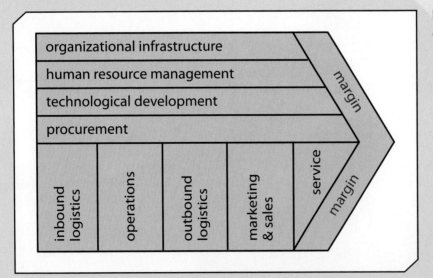

Exhibit 8.1
Michael Porter's value chain

> The value chain is really about understanding what aspects of an organization's business add value for customers and then maximizing those aspects.

A furniture manufacturing company, for example, buys raw materials (wood) from a logging company, converts the raw materials into finished products (chairs), ships the finished products to retailers, distributors, or customers, and markets these products. The value chain can continue after delivering chairs to furniture stores by offering other products and services. Value chain analysis can also help the furniture company spot manufacturing opportunities. For example, it could cut costs of raw materials if it owned or formed a partnership with a lumber company. In any industry, a company is part of a value chain when it buys goods or services from suppliers, adds features to increase value, and sells the goods or services to customers. As another example, a computer vendor can buy components from different vendors and then assemble them into complete PCs. The vendor has added value by assembling the components and, therefore, can charge a higher price than the total combined cost of the components.

E-commerce, its applications, and its supporting technologies, as discussed in this chapter, are an example of using Porter's value chain concept. The Internet can increase the speed and accuracy of communication between suppliers, distributors, and customers. Moreover, the Internet's low cost means companies of any size can take advantage of value-chain integration, which is the process of multiple companies in a shared market working together to plan and manage the flow of goods, services, and information from manufacturers to consumers. This process optimizes the value chain's efficiency, thus creating a competitive advantage for all companies involved.

E-commerce can enhance a value chain by offering new ways to reduce costs or improve operations. The following are some examples:

- Using e-mail rather than regular mail to notify customers of upcoming sales can reduce costs.
- Selling to customers via the company Web site can generate new sources of revenue, particularly from customers who live far away from the company's headquarters or physical store.
- Offering online customer service can make products or services more appealing to customers.

As you learn throughout this book, many companies have taken advantage of the Web and e-commerce to reduce costs, increase revenue, and improve customer service. For example, Dell Computer generates a large portion of its revenue through the Web and eliminates the middleman in the process. Similarly, Cisco Systems sells networking hardware and software over the Web, and customers can track packages on the Internet with United Parcel Service (UPS) and FedEx. The information box below describers how Twitter, a popular social networking site, can help businesses.

1.2 E-Commerce vs. Traditional Commerce

Although the goal of e-commerce and traditional commerce is the same—selling products and services to generate profit—they do it quite differently. In

Twitter Helps Businesses to Find Customers

In addition to its personal and social applications, Twitter can be used by businesses as a promotional tool and as a way to find sales leads. John Pepper, CEO and cofounder of Boloco, a burrito restaurant chain, posted a photo of a coupon on Twitter and invited customers to bring in any image of the coupon—a photocopy, a printout, or even an image on a mobile phone—to get the discount. The promotion was a big success, given that 900 customers redeemed the coupon (including by bringing their mobile phones) as opposed to the usual 350.

Similarly, Dell Computer announced that it had generated $3 million in sales on Twitter during a two-year period by posting the coupon numbers for discounts on Dell Outlet items. Before long, the company's Twitter account, @DellOutlet, had attracted over 700,000 followers. Twitter also helps businesses to find customers who are looking for their products or services, and it allows businesses to deal with unhappy customers quickly.[2]

Table 8.1

E-commerce vs. Traditional commerce

Activity	Traditional commerce	E-commerce
Product information	Magazines, flyers	Web sites, online catalogs
Business communication	Regular mail, phone calls	E-mail
Check product availability	Phone calls, faxes, letters	E-mail, Web sites, and extranets
Order generation	Printed forms	E-mail, Web sites
Product acknowledgements	Phone calls, faxes	E-mail, Web sites, and EDI
Invoice generation	Printed forms	Web sites

e-commerce, the Internet and telecommunication technologies play a major role. Often, there's no physical store, and the buyer and seller don't see each other. Many companies now operate as a mix of traditional commerce and e-commerce, however, and have some kind of e-commerce presence. These companies, referred to as **click-and-brick e-commerce**, capitalize on the advantages of online interaction with their customers yet retain the benefits of having a physical store location. For example, customers can buy items from the company's Web site but take them to the physical store if they need to return items. Table 8.1 compares e-commerce and traditional commerce.

1.3 Advantages and Disadvantages of E-Commerce

Businesses of all sizes use the Internet and e-commerce applications to gain a competitive edge. For example, IBM does business with more than 12,000 suppliers over the Web and uses the Web for sending purchase orders, receiving invoices, and paying suppliers. In addition, IBM uses the Internet and Web technologies for its transaction-processing network.

Similar to traditional businesses, e-commerce has many advantages and disadvantages. If e-commerce is based on a sound business model (discussed in the next section), its advantages outweigh its disadvantages. Advantages of e-commerce include the following:

Click-and-brick e-commerce mixes traditional commerce and e-commerce. It capitalizes on the advantages of online interaction with customers yet retains the benefits of having a physical store location.

- Creating better relationships with suppliers, customers, and business partners

- Creating "price transparency," meaning all market participants can trade at the same price
- Being able to operate around the clock and around the globe
- Gathering more information about potential customers
- Increasing customer involvement (for example, offering a feedback section on the company Web site)
- Improving customer service
- Increasing flexibility and ease of shopping
- Increasing the number of customers
- Increasing opportunities for collaboration with business partners
- Increasing return on investment because inventory needs are reduced
- Offering personalized services and product customization
- Reducing administrative and transaction costs

E-commerce also has the following disadvantages, although many of these should be eliminated or reduced in the near future:

- Bandwidth capacity problems (in certain parts of the world)
- Security issues
- Accessibility (not everybody is connected to the Web yet)
- Acceptance (not everybody accepts this technology)

1.4 E-Commerce Business Models

The Internet has improved productivity in many organizations, but this improvement must be converted to profitability. As the fall of many dot.com companies in

> The Internet has improved productivity in many organizations, but this improvement must be converted to profitability.

© Worldpics/Shutterstock.com

2000 and 2001 shows, just improving productivity isn't enough. The companies that survive have a sound business model governing how they plan to make a profit and sustain a business for future growth.

To achieve profitability, e-commerce companies focus their operations in different parts of the value chain, discussed earlier. To generate revenue, for example, an e-commerce company might decide to sell only products or services or cut out the middleman in the link between suppliers and consumers. Many business-to-consumer (B2C) business models do the latter by using the Web to deliver products and services to customers, which helps reduce prices and improve customer service. As you learned in the discussion of Michael Porter's differentiation strategies in Chapter 1, by differentiating themselves from their competitors in this fashion, these companies can increase their market shares as well as their customer loyalty.

The products that e-commerce companies sell could be traditional products, such as books and apparel, or digital products, such as songs, software, and e-books. Similarly, e-commerce models can be traditional or "digital." Traditional e-commerce models are usually an extension or revision of traditional business models, such as advertising or merchant, or a new type suitable for Web implementation, such as informediary (described in the following list). The following are the most widely used business models in e-commerce:[3,1]

- *Merchant*—The **merchant model** transfers the old retail model to the e-commerce world by using the medium of the Internet. In the most common type of merchant model, an e-commerce company uses Internet technologies and Web services to sell goods

and services over the Web. Companies following this model offer good customer service and lower prices to establish a presence on the Web. Amazon.com uses this model, but traditional businesses, such as Dell, Cisco, and Hewlett-Packard, have adopted this model to eliminate the middleman and reach new customers.

- *Brokerage*—Using the **brokerage model** brings sellers and buyers together on the Web and collects commissions on transactions between these parties. The best example of this model is online auction sites, such as eBay (*www.ebay.com*) and Swoopo. com (*www.swoopo.com*). Auction sites can generate additional revenue by selling banner advertisements. Other examples of the brokerage model are online stockbrokers, such as TDAmeritrade.com and Schwab.com, which generate revenue by collecting commissions from buyers and sellers of securities.

- *Advertising*—The **advertising model** is an extension of traditional advertising media, such as radio and television. Directories such as Yahoo! provide content (similar to radio and TV) to users for free. By creating more traffic with this free content, they can charge companies for placing banner ads or leasing spots on their sites. Google, for example, generates revenue from AdWords, which offers pay-per-click (PPC) advertising and site-targeted advertising for both text and banner ads.

- *Mixed*—The **mixed model** refers to generating revenue from more than one source. For example, ISPs such as AOL generate revenue from advertising and from subscription fees for Internet access. An auction site can also generate revenue from commissions collected from buyers and sellers and from advertising.

The **merchant model** transfers the old retail model to the e-commerce world by using the medium of the Internet.

Using the **brokerage model** brings sellers and buyers together on the Web and collects commissions on transactions between these parties.

The **advertising model** is an extension of traditional advertising media, such as radio and television. Directories such as Yahoo! provide content (similar to radio and TV) to users for free. By creating more traffic with this free content, they can charge companies for placing banner ads or leasing spots on their sites.

The **mixed model** refers to generating revenue from more than one source.

- *Informediary*—E-commerce sites that use the **informediary model** collect information on consumers and businesses and then sell this information to other companies for marketing purposes. For example, Bizrate.com collects information about the performance of other sites and sells this information to advertisers.

- *Subscription*—Using the **subscription model**, e-commerce sites sell digital products or services to customers. For example, the *Wall Street Journal* and *Consumer Reports* offer online subscriptions, and antivirus vendors use this model to distribute their software and updates.

© boumen&japet/Shutterstock.com

 2 Major Categories of E-Commerce

Several categories of e-commerce, based on the nature of their transactions, are in use: business-to-consumer (B2C), business-to-business (B2B), consumer-to-consumer (C2C), consumer-to-business (C2B), and government. Table 8.2 summarizes these categories, and the following sections describe them in more detail.

Table 8.2 *Major E-Commerce Categories*

	Consumer	Business	Government
Consumer	C2C	C2B	C2G
Business	B2C	B2B	B2G
Government	G2C	G2B	G2G

2.1 Business-to-Consumer E-Commerce

Business-to-consumer (B2C) companies, such as Amazon.com, Barnesandnoble.com, and Onsale.com, sell directly to consumers. As discussed later in the Industry Connection, Amazon.com and its business partners sell a wide array of products and services, including books, DVDs, prescription drugs, clothing, and household products. Amazon.com is an example of a pure-play company, meaning it has no physical store. Companies that do have a physical store—called brick-and-mortar companies—have also entered the virtual marketplace by establishing comprehensive Web sites and virtual storefronts. Walmart, the Gap, and Staples are some examples of companies active in B2C e-commerce. In these cases, e-commerce supplements traditional commerce. Some experts believe that these companies could be more successful than pure-play companies because of the advantages a physical space can offer, such as customers being able to visit a store for returns.

2.2 Business-to-Business E-Commerce

Business-to-business (B2B) e-commerce involves electronic transactions between businesses. These transactions have been around for many years in the form of electronic data interchange (EDI) and electronic funds transfer (EFT). In recent years, the Internet has increased the number of B2B transactions and made B2B the fastest growing segment of e-commerce. As discussed in Chapter 7, extranets have been used effectively for B2B operations, as companies rely on other companies for supplies, utilities, and services. Companies using B2B applications for purchase orders, invoices, inventory status, shipping logistics, business contracts, and other operations report millions of dollars in savings by increasing transaction speed, reducing errors, and eliminating manual tasks. Walmart is a major player in B2B e-commerce, and suppliers such as Proctor and Gamble, Johnson & Johnson, and others sell products to Walmart electronically, meaning all transactions are handled electronically. These suppliers can check the inventory status in each store and replenish products in a timely manner.

2.3 Consumer-to-Consumer E-Commerce

Consumer-to-consumer (C2C) e-commerce involves business transactions between users, such as consumers selling to other consumers via the Internet. For example, people can use online classified ads (Craigslist. org, for example) or online auction sites, such as eBay. com. People can also advertise products and services on organizations' intranets (discussed in Chapter 7) and sell them to other employees.

2.4 Consumer-to-Business E-Commerce

Consumer-to-business (C2B) e-commerce involves people selling products or services to businesses, such as a service for creating online surveys for a company to use. In other cases, people might search for sellers of a product and service, such as using Priceline.com to make travel arrangements.

Business-to-business (B2B) e-commerce involves electronic transactions between businesses.

Consumer-to-consumer (C2C) e-commerce involves business transactions between users, such as consumers selling to other consumers via the Internet.

Consumer-to-business (C2B) e-commerce involves people selling products or services to businesses, such as a service for creating online surveys for a company to use.

E-government applications can include government-to-citizen, government-to-business, government-to-government, and government-to-employee transactions. Services include tax filing, online voter registration, disaster assistance, and e-training for government employees.

2.5 Government and Nonbusiness E-Commerce

Many government and other nonbusiness organizations are using e-commerce applications now, including the Department of Defense, the Internal Revenue Service, and the Department of Treasury. These applications are broadly called **e-government** (or just "e-gov") applications and are divided into these categories:

- *Government-to-citizen (G2C)*—Tax filing and payments; completing, submitting, and downloading forms; requests for records; online voter registration
- *Government-to-business (G2B)*—Sales of federal assets, license applications and renewals
- *Government-to-government (G2G)*—Disaster assistance and crisis response
- *Government-to-employee (G2E)*—E-training

Exhibit 8.2 shows the home page of USA.gov, a Web site for delivering government-related information to users.

Exhibit 8.2 *The USA.gov home page*

Organizational or **intra-business e-commerce** involves e-commerce activities that take place inside an organization, typically via the organization's intranet. These activities can include exchange of goods, services, or information among employees.

Universities are an example of nonbusiness organizations that use e-commerce applications; for example, many universities use Web technologies for online classes, registration, and grade reporting. In addition, nonprofit, political, and social organizations use e-commerce applications for activities such as fund-raising, political forums, and purchasing.

2.6 Organizational or Intrabusiness E-Commerce

Organizational or **intrabusiness e-commerce** involves e-commerce activities that take place inside an organization, typically via the organization's intranet (discussed in Chapter 7). These activities can include exchange of goods, services, or information among employees (such as the C2C e-commerce discussed previously). Other examples are conducting training programs and offering human resource services. Some of these activities, although not specifically selling and buying, are considered supporting activities in Porter's value chain. For example, a human resources department supports personnel involved in producing and distributing the company's products.

 3 A B2C E-Commerce Cycle

s Exhibit 8.3 shows, five major activities are involved in conducting B2C e-commerce:

- *Information sharing*—A B2C e-commerce company can use a variety of methods to share information with its customers, such as company Web sites, online catalogs, e-mail, online advertisements, video conferencing, message boards, and newsgroups.
- *Ordering*—Customers can use electronic forms or e-mail to order products from a B2C site.
- *Payment*—Customers have a variety of payment options, such as credit cards, e-checks, and e-wallets.

Electronic payment systems are discussed later in "E-commerce Supporting Technologies."

- *Fulfillment*—Delivering products or services to customers is multifaceted, depending on whether physical products (books, videos, CDs) or digital products (software, music, electronic documents) are being delivered. For example, delivery of physical products can take place via air, sea, or ground at varying costs and with different options. Delivering digital products is more straightforward, usually involving just downloading, although products are typically verified with digital certificates. (Refer to Chapter 5.) Fulfillment also varies depending on whether the company handles its own fulfillment operations or outsources them to third parties. Fulfillment often includes delivery address verification and digital warehousing, which maintains digital products on storage media until they're delivered. Several third-party companies are available to handle fulfillment functions for e-commerce sites.
- *Service and support*—Service and support are even more important in e-commerce than in traditional commerce, given that e-commerce companies don't

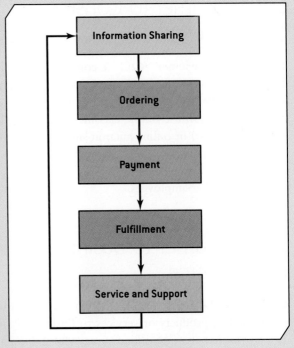

Exhibit 8.3 *Major activities in B2C e-commerce*

- Information Sharing
- Ordering
- Payment
- Fulfillment
- Service and Support

The main objective of e-procurement is to avoid buying from suppliers not on the approved list of sellers and to eliminate the processing costs of purchases.

have a physical location to help maintain current customers. Because maintaining current customers is less expensive than attracting new customers, e-commerce companies should make an effort to improve customer service and support by using some of these methods: e-mail confirmations and product updates, online surveys, help desks, and guaranteed secure transactions.

The activities listed in Exhibit 8.3 are the same in traditional commerce and probably occur in the same sequence, too. However, each stage has been transformed by Web technologies and the Internet.

4 B2B E-Commerce: A Second Look

he B2B e-commerce model uses a cycle similar to the one in Exhibit 8.3; however, B2Bs use these additional technologies extensively: intranets, extranets, virtual private networks, electronic data interchange (EDI, discussed in Chapter 11), and electronic funds transfer (EFT). B2B e-commerce reduces delivery time, inventory, and prices, and it helps business partners share relevant, accurate, and timely information. The end result is improved supply chain management among business partners.

B2B e-commerce lowers production costs and improves accuracy by eliminating many labor-intensive tasks, such as creating invoices and tracking payments manually. In addition, the information flow with business partners is improved by creating a direct online connection in the supply chain network, which also reduces delivery time. In other words, raw materials are received faster, and information related to customers' demands is transferred faster. Improved electronic communication between business partners improves overall communication, which results in better inventory management and control.

4.1 Major Models of B2B E-Commerce

There are three major models of B2B e-commerce, based on who controls the marketplace: seller, buyer, or intermediary (third party). As a result, the following marketplace models have been created: seller-side marketplace, buyer-side marketplace, and third-party exchange marketplace. A fourth model, called trading partner agreements, which facilitates contracts and negotiations among business partners, is also gaining popularity. The following sections explain these models.

4.1.1 Seller-Side Marketplace

The **seller-side marketplace** model is the most popular B2B model. In this model, sellers who cater to specialized markets, such as chemicals, electronics, and auto components, come together to create a common marketplace for buyers—sort of a one-stop shopping model. Sellers can pool their market power, and buyers' search for alternative sources is simplified.

A popular application of this model is **e-procurement**, which enables employees in an organization to order and receive supplies and services directly from suppliers. Typically, a company negotiates reduced prices with suppliers ahead of time. E-procurement streamlines the traditional procurement process by using Web technologies, which reduce costs, improve the timeliness of procurement processes, and improve relationships between suppliers and participating organizations.

E-procurement applications often have purchase-approval procedures that allow users to connect only to company-approved e-catalogs that give employees

The **seller-side marketplace** model is the most popular B2B model. In this model, sellers who cater to specialized markets, such as chemicals, electronics, and auto components, come together to create a common marketplace for buyers—sort of a one-stop shopping model.

E-procurement enables employees in an organization to order and receive supplies and services directly from suppliers.

© Alexey Pustoshilov/Shutterstock.com

prenegotiated prices. The main objectives of e-procurement is to prevent purchases from suppliers that are not on the approved list of sellers and to eliminate the processing costs of purchases. Not following this process can be costly for the receiving partner, because it can result in paying higher prices for supplies.[4] E-procurement can also qualify customers for volume discounts or special offers.

E-procurement applications can also automate some buying and selling activities, which reduces costs

and improves processing speeds. Companies using these applications expect to control inventory more effectively, reduce purchasing overhead, and improve the manufacturing production cycle. E-procurement will likely be integrated into standard business systems with the trend toward computerized supply-chain management (discussed in Chapter 11).

Major vendors of e-commerce and B2B solutions include I2 Technologies, IBM, Oracle, and SAP. The information box highlights e-procurement applications at Schlumberger.

4.1.2 Buyer-Side Marketplace

Large corporations (such as General Electric or Boeing) as well as consortiums of large companies use the **buyer-side marketplace** model. Here's how it works: A buyer, or a group of buyers, opens an electronic marketplace and invites sellers to bid on announced products or requests for quotation (RFQs). Using this model, buyers can manage the procurement process more efficiently, lower administrative costs, and implement uniform pricing.[5] Companies are investing in buyer-side marketplaces with the goal of establishing new sales channels that increase their market presence and lower the cost of each sale. By participating in buyer-side marketplaces, sellers can do the following:

- Conduct sales transactions
- Automate the order management process
- Conduct postsales analysis
- Automate the fulfillment function
- Improve understanding of buying behaviors
- Provide an alternative sales channel
- Reduce order placement and delivery time

In a **buyer-side marketplace** model, a buyer, or a group of buyers, opens an electronic marketplace and invites sellers to bid on announced products or requests for quotation (RFQs). Using this model, buyers can manage the procurement process more efficiently, lower administrative costs, and implement uniform pricing.

The **third-party exchange marketplace** model isn't controlled by sellers or buyers. Instead, it's controlled by a third party, and the marketplace generates revenue from the fees charged for matching buyers and sellers. These marketplaces are usually active in vertical or horizontal markets.

A **vertical market** concentrates on a specific industry or market. Utility companies, the beef and dairy industries, and medical products are a few examples of vertical markets.

E-Procurement at Schlumberger

Schlumberger, an oil field services provider, developed an e-procurement system for order processing that has reduced the cost per order. The new system accomplished reduced cost by streamlining the paperwork that was required both to route purchase orders for approval and for other administrative tasks. The old centralized electronic data interchange (EDI) procurement system was replaced with a Web-based system that enables employees to contact any approved supplier directly from their workstations. The system has an easy-to-use, flexible interface that has allowed Schlumberger to conduct business with a more diverse group of suppliers. The Internet connection for the system is inexpensive and fast, and the system's open platform has been an advantage.[4]

4.1.3 Third-Party Exchange Marketplace

The **third-party exchange marketplace** model isn't controlled by sellers or buyers. Instead, it's controlled by a third party, and the marketplace generates revenue from the fees charged for matching buyers and sellers. These marketplaces are usually active in vertical or horizontal markets. A **vertical market** concentrates on a specific industry or market. Utility companies, the beef and dairy industries, and medical products are a few examples of vertical markets. A **horizontal market** concentrates on a specific function or business process and automates this function or process for different industries. Employee-benefits administration and media buying are examples of horizontal markets.

This model offers suppliers a direct channel of communication to buyers through online storefronts. The interactive procedures in the marketplace have features such as product catalogs, requests for information (RFI), rebates and promotions, broker contacts, and product sample requests. The Worldbid.com information box describes an example of a third-party exchange marketplace.

4.1.4 Trading Partner Agreements

The main objectives of **trading partner agreements** are to automate negotiating processes and enforce contracts between participating businesses. Using this model, business partners can send and receive bids, contracts, and other information needed when offering and purchasing products and services. In addition, this model will become more common with the development of electronic business Extensible Markup Language (ebXML), a worldwide project for standardizing the exchange of e-commerce data via XML, including electronic contracts and trading partner agreements.

Using this model enables customers to submit, via the Internet, electronic documents that previously required hard copies with signatures. The Digital Signature Act of 1999 gives digital signatures the same legal validity as handwritten signatures. Accepting an electronic trading agreement binds the parties to all its terms and conditions.

With XML, contracts can be transmitted electronically, and many processes between trading partners can be performed electronically, including inventory status, shipping logistics, purchase orders, reservation systems,

and electronic payments. The main advantage of XML over HTML is that you can assign data type definitions to information in a page, so Web browsers can select only the data requested in a search. This feature makes data transfer easier because not all data is transferred, just the data needed in a particular situation. It is particularly useful in m-commerce (mobile commerce), because loading only the necessary data in a browser makes searches more efficient. This process reduces traffic on the Internet and helps prevent delays during peak usage hours.

5 Mobile and Voice-Based E-Commerce

obile commerce (m-commerce), based on the Wireless Application Protocol (WAP), has been around for several years, particularly in European countries. M-commerce is using handheld devices, such as smartphones or PDAs, to conduct business transactions, such as making stock trades

© Colin Anderson/Getty Images

with an online brokerage firm. Supporting technologies for m-commerce applications include wireless wide area networks and 3G and 4G networks (discussed in Chapter 6) as well as short-range wireless communication technologies, such as Wi-Fi, WiMAX, Bluetooth, and RFID (discussed in Chapter 14).

Many telecommunication companies offer Web-ready cell phones. A wide variety of m-commerce applications are available; among the most popular are iPhone apps, which include games, entertainment, news, and travel information. Some are free, and others must be purchased via iTunes.

In addition, Microsoft has a wireless version of Internet Explorer called Internet Explorer Mobile, and many e-commerce companies are developing the simple, text-based interface required by current cell phones. For example, Google offers features for searching, news, map, and Gmail to mobile Internet users (*www.google.com/mobile/*). MSN Mobile (*http://mobile.msn.com*) provides a special browser for accessing e-mail, news, sports, entertainment, maps, and Windows Live services, such as Hotmail and Windows Live Messenger. Other applications of m-commerce include banking, traffic updates, tourism services, shopping, and video conferencing. Mobile user-to-user applications, such as sharing games and pictures, are also popular.

You can already use a mobile phone to access a Web site and order a product. The next step is **voice-based e-commerce**, which will rely on voice recognition and text-to-speech technologies that have improved dramatically in the past decade.[6,7] For example, you'll be able to simply speak the name of the Web site or service you want to access and use voice commands to search a database by product name and find the merchant with the most competitive prices. Voice-based e-commerce would be suitable for applications, such as making stock trades, looking up sports scores, reserving movie tickets, or getting directions to a restaurant. Also, the iPhone 3GS and 4G supports voice-based Google searches.

One method of conducting voice-based e-commerce is using e-wallets, covered in the next section, "E-Commerce Supporting Technologies." In addition to storing financial information, e-wallets could store information such as the customer's address and driver's license number. Security features for voice-based e-commerce are expected to include the following:

- Call recognition, so that calls have to be placed from specific mobile devices
- Voice recognition, so that authorizations have to match a specific voice pattern
- Shipping to a set address that can't be changed by voice commands

Several voice portals are already available, including Nuance.com, InternetSpeech.com, and Tellme.com. You can visit these sites to discover the latest in voice-based e-commerce.

6 E-Commerce Supporting Technologies

 number of technologies and applications support e-commerce activities. The following sections explain these widely used supporting technologies: electronic payment systems, Web marketing, and search engine optimization.

6.1 Electronic Payment Systems

Electronic payment refers to money or scrip that is exchanged only electronically. It usually involves use of the Internet, other computer networks, and digitally stored value systems. Payment cards are the most popular instrument for electronic payment transactions and include credit cards, debit cards, charge cards, and smart cards. **Smart cards** have been used in Europe, Asia, and Australia for many years and are slowly gaining acceptance in the United States because of their multipurpose functions. A smart card is about the size of a credit card and contains an embedded microprocessor chip for storing important financial and personal information. The chip can be loaded with information and updated periodically.

Voice-based e-commerce relies on voice recognition and text-to-speech technologies.

Electronic payment refers to money or scrip that is exchanged electronically. It usually involves use of the Internet, other computer networks, and digitally stored value systems. It includes credit cards, debit cards, charge cards, and smart cards.

A **smart card** is about the size of a credit card and contains an embedded microprocessor chip for storing important financial and personal information. The chip can be loaded with information and updated periodically.

> Payment cards are the most popular instrument for electronic payment transactions and include credit cards, debit cards, charge cards, and smart cards.

E-cash, a secure and convenient alternative to bills and coins, complements credit, debit, and charge cards and adds convenience and control to everyday cash transactions. E-cash usually works with a smart card, and the amount of cash stored on the chip can be "recharged" electronically.

An **e-check**, the electronic version of a paper check, offers high security, speed, and convenience for online transactions. Many utility companies offer e-checks for customers' payments, and most banks have e-checks for online bill paying. E-checks are a good solution when other electronic payment systems are too risky or not appropriate.

E-wallets are available for most handheld devices and offer a secure, convenient, and portable tool for online shopping. As described previously, they store personal and financial information, such as credit card numbers, passwords, and PINs. E-wallets can be used for micropayments (discussed later in this section) and are useful for frequent online shoppers because personal and financial information doesn't have to be reentered each time you place an order.

You're probably familiar with **PayPal**, a popular online payment system used in many online auction sites. Users with a valid e-mail address can set up a PayPal account and use it for secure payments of online transactions, using their credit cards or bank accounts.

Micropayments are used for very small payments on the Web. They began as a method for advertisers to pay for cost per view or cost per click, which is typically one-tenth of a cent. These fractional amounts are difficult to handle with traditional currency methods, and electronic micropayments reduce the cost of handling them for financial institutions. Payment amounts are accumulated for customers until they're large enough to offset the transaction fee, and then the account deduction or charge is submitted to the bank. Of course, micropayment systems charge a fee for tracking and processing the transactions. However, the World Wide Web Consortium (w3C; *www.w3c.org*), which defines standards for Web-related technologies, has canceled support for micropayments and is no longer working on standards for them. Google is developing an online subscription and micropayment system that will assist online content providers to more efficiently charge for their content.[8]

6.2 Web Marketing

Web marketing uses the Web and its supporting technologies to promote goods and services. Although traditional media, such as radio and TV, are still used for marketing, the Web offers many unique capabilities, such as message boards for customers to post questions and newsgroups for sending information to customers. To better understand Web marketing, review the following list of terms:

- *Ad impression*—One user viewing one ad.
- *Banner ads*—Usually placed on frequently visited Web sites, these ads are around 468 x 60 pixels and have simple animation. Clicking a banner ad displays a short marketing message or transfers the user to another Web site.
- *Click*—The opportunity for a user to click a URL or a banner ad and be transferred to another Web

E-cash, a secure and convenient alternative to bills and coins, complements credit, debit, and charge cards and adds convenience and control to everyday cash transactions.

An **e-check,** the electronic version of a paper check, offers high security, speed, and convenience for online transactions.

E-wallets are available for most handheld devices and offer a secure, convenient, and portable tool for online shopping. They store personal and financial information, such as credit card numbers, passwords, and PINs.

PayPal is a popular online payment system used in many online auction sites. Users with a valid e-mail address can set up an account and use it for secure payments of online transactions, using their credit cards or bank accounts.

Micropayments are used for very small payments on the Web. They began as a method for advertisers to pay for cost per view or cost per click.

Web marketing uses the Web and its supporting technologies to promote goods and services.

site or see a marketing message, as recorded by the Web server. For a user, defining a "click" is simple. However, because of monetary and advertising factors associated with a click, the definition varies depending on who's defining it. Each time a keyword used for searching takes a user to a particular Web page, for example, the site owner (advertiser) pays the search engine a cost per click (discussed later in this list). A consortium of Yahoo!, Microsoft, Google, and the Interactive Advertising Bureau has formed the Click Measurement Working Group, which is trying to define what a "legal and valid" click ought to be.

- *Cost per thousand (CPM)*—Most Web and e-mail advertising is priced based on the cost per thousand (*M* stands for *mille*, which means thousand) ad impressions. For example, a $125 CPM means it costs $125 for 1000 ad impressions.
- *Cost per click (CPC)*—The cost of every click on an ad. For example, $1.25 CPC means that for every click an advertiser gets, the advertiser pays $1.25 to the sponsoring Web site. MIVA (*www.miva.com/*) is an example of a cost-per-click network. It serves over 150 million clicks per month and displays text ads over a wide variety of platforms.
- *Click-through rate (CTR)*—This rate is computed by dividing the number of clicks an ad gets by the total impressions bought. For example, if an advertiser buys 100,000 impressions and gets 20,000 clicks, the CTR is 20% (20,000/100,000 = 20%).
- *Cookie*—Information a Web site stores on the user's hard drive so that it can be remembered for a later visit, such as greeting a visitor by name. This information is also used to record user preferences and browsing habits.
- *Hit*—Every element of a Web page (including text, graphics, and interactive items) is counted as a hit to a server. Hits aren't the preferred unit of measurement for site traffic, because the number of hits per page can vary widely, depending on the number of graphics, type of browser uses, and page size.
- *Meta tag*—This HTML tag doesn't affect how a Web page is displayed; it simply provides information about a Web page, such as keywords that represent the page content, the Web designer, and frequency of page updates. Search engines use this information, particularly keywords, to create indexes.

- *Page view (PV)*—One user viewing one Web page.
- *Pop-up ad*—Opens a new window to display ads.
- *Pop-under ad*—Opens a new window hidden under the active window. Pop-unders don't interrupt users as much as pop-ups.
- *Splash screen*—A Web page displayed when the user first visits the site; it's designed to capture the user's attention and motivate the user to browse through the site. The splash screen can display the company logo, for example, as well as a message about any requirements for viewing the site, such as installing plug-ins.
- *Spot leasing*—Search engines offer space that companies can purchase for advertising purposes. Spots have an advantage over banner ads because their placement is permanent; banner ad placement can change from visit to visit. However, spots can be more expensive than banner ads, especially on high-traffic sites, such as Google.

Intelligent agents (discussed in Chapter 13) and push technology (discussed in Chapter 14) are also used as Web marketing tools. Briefly, intelligent agents are an application of artificial intelligence that can be used for Web marketing. For example, product-brokering agents could alert customers about a new product. Push technology is the opposite of pull technology, in which users search the Web to find (pull) information. With push technology, information is sent to users based on their previous inquiries, interests, or specifications. This technology can be used to send and update marketing information, product and price lists, and product updates.

6.3 Search Engine Optimization

Search engine optimization (SEO) is a method for improving the volume or quality of traffic to a Web site. A higher ranking in search results should generate more revenue for a Web site. For example, if you conduct a search on the keywords *digital camera* for information on buying a camera, the search engine might list hundreds or thousands of Web sites, but you probably visit only the top 5 or 10 and ignore the rest.

A comprehensive Web marketing campaign should use a variety of methods, and SEO is another method that can help improve business. Some companies offer SEO services. Unlike Web marketing methods that involve paying for listings on search engines, SEO aims at increasing a Web site's performance on search engines in a natural (and free) fashion. As you learned in Chapter 7, a typical search engine, such as Google or Bing, uses a crawler or spider to find a Web site and

Search engine optimization (SEO) is a method for improving the volume or quality of traffic to a Web site. A higher ranking in search results should generate more revenue for a Web site.

then, based on the site's contents and relevance, indexes it and gives it a ranking. Optimizing a Web site involves editing a site's contents and HTML code to increase its relevance to specific keywords. SEO includes techniques that make it easier for search engines to find and index a site for certain keywords. The following are some common recommendations for optimizing a Web site:

- *Keywords*—Decide on a few keywords that best describe the Web site and use them consistently throughout the site's contents.
- *Page title*—Make sure the page title reflects the site and its contents accurately.
- *Inbound links*—Get people to comment on your Web site, using one of your top keywords.

This chapter's Industry Connection focuses on Amazon.com as one of the leaders in e-commerce.

 7 Chapter Summary

n this chapter, you've learned about the role of e-commerce in Michael Porter's value chain, you've compared e-commerce with traditional commerce, and you've seen the advantages and disadvantages of e-commerce. You've also learned about the major e-commerce business models as well as the main categories of e-commerce. In addition, you've reviewed important activities in the B2C e-commerce cycle, various B2B e-commerce business models, and applications of m-commerce and voice-based e-commerce. Finally, you've learned about electronic payment systems and Web marketing as supporting technologies for e-commerce operations.

Industry Connection

AMAZON.COM*

Amazon.com, a leader in B2C e-commerce, offers a variety of products and services, including books, CDs, videos, games, free e-cards, online auctions, and other shopping services and partnership opportunities. By using customer accounts, shopping carts, and its 1-Click feature, Amazon.com makes shopping fast and convenient, and it uses e-mail for order confirmation and customer notifications of new products tailored to customers' shopping habits. In addition, Amazon.com has created an open forum with customers by posting customers' book and product reviews and allowing customers to rate products on a scale of one to five stars. Most recently, Amazon.com has become one of the major players in the cloud computing environment. Here are some of the activities customers can do on Amazon.com:

- Search for books, music, and many other products and services.
- Browse in hundreds of product categories, from audio books, jazz, and video documentaries to coins and stamps available for auction.
- Get personalized recommendations based on previous purchases.

- Sign up for an e-mail subscription service to get the latest reviews of new titles in categories that interest the customer.
- Create wish lists that can be saved for later viewing.
- Search book content on specific keywords and view selected pages in some books.

Amazon.com is known for its personalization system, which is used to recommend goods, and its collaborative filtering, which is used to improve customer service. (Both features are discussed in Chapter 11). Amazon.com also collaborates with business partners via Amazon zShops, where other merchants can sell their products through Amazon.com.

* This information has been gathered from the company Web site (*www.amazon.com*) and other promotional materials. For more information and updates, visit the Web site.

Key Terms

advertising model (145)

brokerage model (145)

business-to-business (B2B) (147)

business-to-consumer (B2C) (146)

buyer-side marketplace (150)

click-and-brick e-commerce (144)

consumer-to-business (C2B) (147)

consumer-to-consumer (C2C) (147)

e-business (141)

e-cash (153)

e-check (153)

e-commerce (141)

e-government (147)

electronic payment (152)

e-procurement (149)

e-wallets (153)

horizontal market (151)

informediary model (146)

merchant model (145)

micropayments (153)

mixed model (145)

mobile commerce (m-commerce) (151)

organizational or intrabusiness e-commerce (148)

PayPal (153)

search engine optimization (SEO) (154)

seller-side marketplace (149)

smart card (152)

subscription model (146)

third-party exchange marketplace (150)

trading partner agreements model (151)

value chain (142)

vertical market (150)

voice-based e-commerce (152)

Web marketing (153)

Problems, Activities, and Discussions

1. Why did so many dot.com companies go out of business in 2000 and 2001?

2. What are some advantages of e-procurement?

3. What are some popular applications of m-commerce?

4. Alibaba.com (*www.alibaba.com*) is considered a global trade platform for importers and exporters. Log on to this site and find out its offerings. In which country does Alibaba.com do most of its business?

5. Visit the following Web page, consult other sources, and write a one-page paper that explains how reverse auctions can reduce cost.
   www.cio.com/article/29717/

6. Visit the following Web page, consult other sources, and write a one-page paper that explains how Facebook has helped 1-800-Flowers.com raise revenue.
   www.internetnews.com/ec-news/article.php/3832436

7. Visit the following Web page, consult other sources, and write a one-page paper that explains how language translation could be incorporated into search optimization strategy.
   www.imediaconnection.com/content/21902.asp

8. Visit the following Web page, consult other sources, and write a one-page paper that explains how Web sites can help streamline your shipping.
   www.inc.com/magazine/20091101/websites-to-help-streamline-your-shipping.html

9. Which of the following is a disadvantage of e-commerce? (Choose all that apply.)
   a. Bandwidth capacity problems
   b. Fragmented markets
   c. Security issues
   d. Accessibility

10. Craigslist is an example of C2B e-commerce. True or False?

casestudy

E-COMMERCE APPLICATIONS IN ONLINE TRAVEL

People use online travel agents (OTAs) to book hotels, airline tickets, cruises, and more. Currently, Expedia, Travelocity, and Orbitz are the top three OTAs, and they have transferred the traditional travel agency model to the Internet, using Web features and technologies to sell travel services. OTAs are considered the first generation of travel services. Most OTAs charge a fee for performing these services, but customers get confirmed travel arrangements as well as guarantees in case flights are canceled, among other things. Travel suppliers, such as airlines, hotels, and car rental companies, prefer that customers book through the suppliers' own Web sites, because they often have to pay high fees to be included on an OTA. JetBlue and InterContinental Hotels, for example, don't list their services on OTAs.

© Supri Suharjoto/Shutterstock.com

The second generation of travel services is represented

by travel search engines, which use Web technologies to help customers make travel arrangements quickly and efficiently; major ones include SideStep, Kayak, and Mobissimo. Customers can use travel search engines to search for the best deal and then click to book flights, hotels, and cars directly with the travel supplier. Using these services is less expensive than using OTAs, but it's also less convenient.[9]

Answer the following questions:

1. What are the differences between online travel agents, such as Orbitz, and travel search engines, such as Mobissimo?

2. Find examples of other travel suppliers that don't list their services on OTAs.

3. From the perspective of travel suppliers, what are the advantages of travel search engines?

GLOBAL INFORMATION SYSTEMS

n this chapter, you review the reasons organizations should go global and adopt global information systems, including the rise in e-business and the growth of the Internet. Global information systems are a growing application of telecommunications and networking, and you learn the requirements and components of these systems as well as the types of organizational structures used with global information systems. Offshore outsourcing, as one beneficiary of global information systems, is also discussed. Finally, you explore some obstacles to using global information systems.

learning outcomes

After studying this chapter, you should be able to:

LO1 Discuss the reasons for globalization and for using global information systems, including e-business and Internet growth.

LO2 Describe global information systems and their requirements and components.

LO3 Explain the types of organizational structures used with global information systems.

LO4 Discuss obstacles to using global information systems.

1 Why Go Global?

the global economy is creating customers who demand integrated worldwide services, and the expansion of global markets is a major factor in developing global information systems to handle these integrated services. To understand the need for integrated worldwide services, consider the example of a U.S.-based shoe company that procures leather and has the upper parts of its shoes produced in Italy because of the high quality of leather and the expertise in shoe stitching available there. The uppers are then shipped to China, where they're attached to soles, thereby taking advantage of the inexpensive manufacturing labor available in that country. The shoes are then shipped to Ireland for testing because of Ireland's high concentration of high-tech facilities. Finally, the shoes are shipped to a variety of retail outlets in the United States, where they're sold. The entire supply-chain logistics—from Italy to China to Ireland—must be managed and coordinated from the U.S. headquarters. This example shows why companies choose other countries for different manufacturing processes and how important integration is in making sure all these processes are coordinated.

Many companies have become international. In 2008, for example, the Coca-Cola Company generated more than 80 percent of its revenue from outside the United States. Major corporations, such as Procter & Gamble, IBM, McDonald's, Unilever, Nestle, and Motorola, have been prime users of global information

> In 2008, the Coca-Cola Company generated more than 80 percent of its revenue from outside the United States.

systems. Because today's multinational corporations operate in a variety of markets and cultures, a clear understanding of factors such as customs, laws, technological issues, and local business needs and practices is a prerequisite to the success of a global information system.

Airline reservation systems are considered the first large-scale interactive global system; hotels, rental car companies, and credit card services also now require worldwide databases to serve their customers more efficiently and effectively.[1] Global products, which are products or services that have been standardized for all markets, are becoming increasingly important in international marketing efforts. In addition, a manufacturer might "regionalize" operations—that is, move them to another country—because of advantages available in certain regions. For example, raw materials might be less expensive in Indonesia than in Singapore, and specialized skills needed for production might be available in India but not in Brazil.

The growing trend toward global customers and products means globalization has also become an important factor in purchasing and the supply chain. Worldwide purchasing gives suppliers the incentive to consider foreign competition as well as domestic competition. Furthermore, large global organizations can reduce costs in purchasing, manufacturing, and distribution because they have access to cheaper labor and can sell products and services locally as well as internationally.[2,3] The following information box highlights the use of global information systems at Rohm & Haas.

1.1 E-Business: A Driving Force

E-business is a major factor in the widespread use of global information systems. As discussed in Chapter 8, e-business includes transactions that support revenue generation as well as those that focus on buying and selling goods and services. These revenue-generating transactions include generating demand for goods and services, offering sales support and customer service, and facilitating communication between business partners. An effective global information system can support all these activities.

E-business builds on the advantages and structures of traditional business by adding the flexibility that networks offer. By generating and delivering timely and relevant information supported by networks, e-business creates new opportunities for conducting commercial activities. For example, by using online information for commercial activities, e-business makes it easier for different groups to cooperate. Branches of a multinational company can share information to plan a new marketing campaign, different companies can work together to design new products or offer new services, and businesses can share information with customers to improve customer relations.

The Internet can simplify communication, change business relationships, and offer new opportunities to both consumers and businesses. As e-business matures and more companies conduct business online, consumers can engage in comparison shopping more easily, for example. Even though direct buyer–seller communication has increased, there are still new opportunities for

Global Information Systems at Rohm & Haas

Rohm & Haas, part of Dow Chemical, has production units in many countries. In the past, each country site operated independently and had its own inventory system. A major problem with this setup was that a country site might not to able to supply customers with the products they wanted. For example, if the site in France reported to customers that it was out of a certain product, it didn't have an easy way to check whether the site in Germany, only 20 miles away, had a supply of this product. To solve these problems, Rohm & Haas overhauled its global information system by upgrading the order entry system and installing a company-wide materials management system. These systems were tied in with a global demand planning system. Rohm & Haas can now provide better service to its customers and ship products from other sites as quickly as needed. These improvements have given Rohm & Haas more of a competitive advantage in the global marketplace.[4]

> With a GIS in place, an international company can increase its control over and enhance the coordination of its subsidiaries, and be able to access new global markets.

intermediaries. For example, some businesses can become intermediaries or brokers to track special markets, notify clients of bargains, change market conditions, locate hard-to-find items, and even conduct searches for special products on clients' behalf.

The Internet, of course, is what makes e-business possible. Small companies have discovered that they can conduct business online just as large companies do, and that they can use the Internet as a communication medium or as a way of replacing internal networks in order to lower costs. The following section discusses the Internet's growth, which has contributed to the increase in e-business.

1.2 Growth of the Internet

Chapter 7 covered the Internet and its phenomenal growth, and in this section, you explore its worldwide growth in greater depth. Today, the Internet is a part of daily life in most parts of the world. Minwatts Marketing Group presents the worldwide Internet growth from 2000 to 2010.[5] According to this source, growth has been highest in the Middle East and lowest in North America. As of June 30, 2010, there were approximately 2 billion worldwide Internet users, with Asia having over 825 million, Europe having 475 million, North America having 266 million, Latin America/Caribbean having 204 million, Africa having 110 million, the Middle East having 63 million, and Oceana/Australia having 21 million.[5]

2 Global Information Systems: An Overview

global information system (GIS) is an information system that works across national borders, facilitates communication between headquarters and subsidiaries in other countries, and incorporates all the technologies and applications found in a typical

information system to store, manipulate, and transmit data across cultural and geographic boundaries.[2] In other words, a GIS is an information system for managing global operations, supporting an international company's decision-making processes, and dealing with complex variables in global operations and decision making.

With a GIS in place, an international company can increase its control over and enhance its coordination of its subsidiaries, and be able to access new global markets.[1] Strategic planning is also a core function of a GIS. By being able to share information between subsidiaries more efficiently, international companies can track performance, production schedules, shipping alternatives, and accounting items. Tracking all this information with a GIS enables management to coordinate business objectives on an international scale.

A GIS can be defined along two dimensions: control and coordination. Control consists of using managerial power to ensure adherence to the organization's goals. Coordination is the process of managing the interaction between activities in different, specialized parts of an organization. Control requires a centralized architecture for data, standardized definitions used across the organization, standard formats for reports, defined behaviors for different processes (such as how to respond when a customer has a complaint), and performance-tracking systems. Coordination requires a decentralized architecture for data, standardization within departments, and the ability to communicate these standards to other departments, collaboration systems, and technologies that support informal communication and socialization. The trade-off between the amount of control and coordination needed define the organization's globalization strategy. Global organizations might use a combination of high control and

> A **global information system (GIS)** is an information system that works across national borders, facilitates communication between headquarters and subsidiaries in other countries, and incorporates all the technologies and applications found in a typical information system to store, manipulate, and transmit data across cultural and geographic boundaries.

high coordination, high control and low coordination, low control and high coordination, and low control and low coordination.[6]

Having high coordination, for example, has the following advantages:[7]

- Flexibility in responding to competitors in different countries and markets
- Ability to respond in one country to a change in another country
- Ability to maintain control of market needs around the world
- Ability to share and transfer knowledge between departments and international branches
- Increased efficiency and effectiveness in meeting customers' needs
- Reduced operational costs

2.1 Components of a Global Information System

Although a GIS can vary quite a bit depending on a company's size and business needs, most GISs have these basic components:

- A network capable of global communication, including transmission equipment and communication media
- A global database
- Information-sharing technologies

International companies can use a variety of technologies for an integrated GIS. Small companies might outsource for expertise that's not available inside the company. On the other hand, large companies with the resources and technical expertise might develop custom applications to be shared across borders. Depending on the system's use, a GIS might consist of a network for e-mail, remote data entry, video and computer conferencing, and distributed databases. However, small companies might take advantage of existing public network providers, such as the Internet or value-added networks, for multicountry communication.[8] Value-added networks are private multipoint networks managed by a third party and used by organizations on a subscription basis. They offer electronic data interchange standards, encryption, secure e-mail, data synchronization, and other services. However, with the Internet, they're not used as much now; businesses of all sizes typically use the Internet to conduct international business. No matter the organization's size or scope, an integrated network for global control over the organization's resources is the foundation of any GIS.

An information system manager faces design and implementation issues when developing a global network. In addition to the usual components of a domestic network, a global network requires bridges, routers, and gateways that allow several networks to connect worldwide. In addition, a global network must have switching nodes to guide packets to their destinations.[9] (These components were discussed in Chapter 6.)

An information system manager must also determine the best communication media to meet global performance and traffic needs, such as fiber optics, satellites, microwaves, or conventional phone lines. Factors to consider include bandwidth, range, noise, and cost. You learned about bandwidth and range in Chapter 6. Global providers such as SprintLink, AT&T, and MCI can supply information on the range specifications for companies. The noise factor involves how immune a medium is to outside electronic interference. As always, component, installation, and leasing costs must be balanced with these other factors.

Additionally, an information system manager must choose the best transmission technology for the global network's needs. Without reliable transmission, a network has no value. Current transmission technologies are synchronous, asynchronous, multiplexing, digital (baseband), and analog (broadband). Synchronous transmission requires both parties to be connected, as with phone calls; with asynchronous transmission, both parties don't need to be connected, which is true of e-mail. However, an international company is restricted to transmission technologies supported by the telecommunication infrastructures of the countries where subsidiaries are located. Information system managers must also select the right network and protocol to manage connections and minimize error rates.

Next, information system managers must consider the company's objectives when determining the network architecture. For example, if the company's international communication requirements mean only simple file sharing is needed and response time isn't a critical factor, half-duplex transmission (one direction at a time) used with a value-added network is probably adequate. However, if the company uses multimedia applications, such as video conferencing or electronic meeting systems, in addition to normal file and database sharing, full-duplex transmission (both directions simultaneously) is more efficient for the processing requirements of multimedia transfers. Further, a private network or dedicated leased lines provide stability in

Global information systems require substantial resource commitments, usually years in advance, so decision makers must justify the investment in a GIS.

transmission protocols when there are inadequate telecommunication infrastructures in developing countries, such as in Africa, the Middle East, and Latin America.

Designing and implementing global databases, whether they're centralized or distributed, are technical challenges in GIS design, mainly because of the longer last names or the different character sets required for names and the different formats required for phone numbers and postal codes. Currency conversion is also a challenge in database development, although some software is available for this task. For example, SunSystems' Systems Union offers multicurrency software that converts local currencies into U.S. dollars. As you see later in the Industry Connection, SAP (originally called Systems Applications and Products in Data Processing) also offers valuable features and capabilities for GISs. Because of these challenges, having a global database that's standardized across the organization is critical.

Although a network infrastructure is necessary for a GIS, the major requirement of a network is making communication and information-sharing possible. After a global network is in place, therefore, an international company must decide on the type of information-sharing technologies to use, such as electronic meeting systems or video conferencing, group support systems, FTP, data synchronization, and application sharing.

With all of these decisions, information system managers should keep in mind that standardized software and hardware are always ideal but not always feasible. For example, hardware seems easy enough to duplicate in other countries, but it's not as simple as shipping the same kind of system to another country and plugging it in. Vendors might not offer technical support in that country, or electrical standards might differ. Using the same software in other countries becomes more complicated because of differences in language, business methods, and **transborder data flow (TDF)**, which restricts what type of data can be captured and transmitted. TDF includes national

laws and international agreements on privacy protection and data security. However, with cooperation and coordination between countries improving, these problems are becoming more manageable.

2.2 Requirements of Global Information Systems

What makes an information system global? A GIS must be capable of supporting complex global decisions. This complexity stems from the global environment in which **multinational corporations (MNCs)** operate. A global environment includes many factors, such as legal (intellectual property laws, patent and trademark laws, TDF regulations, and so forth), cultural (languages, ethical issues, religious beliefs), economic (currency, tax structure, interest rates, monetary and fiscal policies), and political (government type and stability, policies toward MNCs, and so on).[2]

Transborder data flow (TDF) restricts what type of data can be captured and transmitted in foreign countries.

A **multinational corporation** or enterprise refers to a corporation that has assets and operations in at least one country other than its home country. This corporation delivers products and services across its national borders and is usually managed centrally from its headquarters.

> A GIS, like any information system, can be classified according to the different managerial support that it provides: operational, tactical, and strategic.

In international business planning, it is critical to understand the global risks unique to operating an MNC—specifically, the political, foreign exchange, and market risks. Political risks include the problems caused by an unstable government, which is an important consideration, given the many political uprisings of recent years. An unstable government can result in currency rates fluctuating, power changing hands rapidly and unpredictably, and other issues that affect company operations. A company considering political risks might not want to set up offices in India or Ireland, for example, despite the low costs and high technological capabilities available in those countries, because of their unstable political situations. In addition, managing global operations requires considering potential conflicts of interest between the government in the company's headquarters (the parent government) and the government in the subsidiary's location (the host government).

A GIS, like any information system, can be classified according to different managerial support that it provides: operational, tactical, and strategic. The complexities of global decision making mean that a GIS has some functional requirements that differ from a domestic information system's requirements. In addition, the line between tactical and operational management has blurred. The first four requirements that follow are classified as operational, the remaining ones as strategic:

- *Global data access*—Online access to information from locations around the world allows management to monitor global operations from the company headquarters. Ideally, global networks provide a real-time communication link with global subsidiaries by integrating voice, data, and video. Several MNCs, such as Hewlett-Packard, General Electric, Texas Instruments, and IBM, have corporate databases linked worldwide.

- *Consolidated global reporting*—This is a crucial tool for managing overseas subsidiaries. These reports should include accounting and financial data, manufacturing updates, inventory, and so forth, and they enable management to compare financial information in all the subsidiaries. Because of differences in accounting procedures and regulatory standards, these comparisons can be difficult; however, consolidated global reporting can help lessen these difficulties.

- *Communication between headquarters and subsidiaries*—To facilitate decision-making and planning processes, a GIS should provide an effective means of communication between the MNC headquarters and subsidiaries.

- *Management of short-term foreign exchange risks*—A mix of free-floating (no government intervention), managed-floating, and fixed-exchange rates characterizes today's international monetary systems. Currency rates can change daily, so management must minimize the impact of currency fluctuations in countries where the parent company and the subsidiaries are located. To manage foreign exchange risks, many companies have developed expert systems and decision support systems (discussed in Chapters 12 and 13). For example, Morgan Stanley's Trade Analysis and Processing Systems (TAPS) tracks financial data, such as bonds, equity, interest rates, and currency conversions.

- *Strategic planning support*—This is the core of any GIS, a focus on regionalizing resources more effectively and responding to rapid environmental changes, such as increased political and foreign exchange risks and global competition.

- *Management of conflicts and political risks*—A major difference between decision making at an MNC and at a domestic company is that an MNC's decisions must always take into account the conflicting objectives of the MNC, national governments, and multinational organizations. Conflict management in MNCs, particularly with host governments and multinational organizations, is critical for survival in the global marketplace.

- *Management of long-term foreign exchange risks*—When formulating a worldwide financial strategy, MNCs must resolve several relevant issues, including risks in long-term foreign exchange. For example, a company should consider the risks involved if Japanese goods flooded Asian markets and crowded out many products made in China after the yen was allowed to devalue.

- *Management of global tax risks*—Designing tax-risk management systems requires detailed knowledge of international finance, international monetary systems, and international tax law.

2.3 Goals of Global Information Systems

Implementing a GIS can be difficult because of differences between countries in culture, politics, social and economic infrastructures, and business methods. International policies can vary, too, which affect communication and standardization processes. Furthermore, some argue that a truly global corporation doesn't exist, much less a GIS to support its operations. A GIS can't simply be added to an existing organization. Several issues must be addressed before adding a GIS, including the following:[10],[11]

- The organization's business opportunities in the global marketplace must be identified.
- Substantial resource commitments must be made, usually years in advance; so decision makers must justify the organization's investment in a GIS.
- Implementing a GIS is more challenging than implementing a domestic information system. Consequently, organization personnel need to be screened for technical and business expertise.
- Migration to the GIS needs to be coordinated carefully to help personnel move from the old familiar system to the new one.

Using information systems on a global scale is more challenging than doing so on a local scale. The challenges, discussed in more detail later in this chapter, include such factors as infrastructure, languages, time zones, and cultures. To address some basic issues with GISs, management must first determine the kind of information that international companies need to share. In addition, management can't assume that the company's products or services will continue selling the same way because of possible changes in customers' needs and preferences and global competition.[10] Considering the entire organization's operational efficiency is critical in coordinating international business activities, so international companies need to change their production and marketing strategies in an effort to respond to the global market.

GISs can be categorized in different ways, depending on their function or application. Global marketing information systems, strategic intelligent systems, transnational management support systems, and global competitive intelligent systems are some different names for GISs. Regardless of its name, a GIS must enable resource sharing across borders. However, the complexity and financial investment required for a GIS might not be feasible from an economic standpoint.

3 Organizational Structures and Global Information Systems

t he most important factor for the effective operation of an MNC is coordination, and a global information system can provide essential information for this task. Here are four common types of global organizations:

- Multinational
- Global
- International
- Transnational

The organizational structure usually determines the architecture of a GIS, as you see in the following sections.

3.1 Multinational Structure

In a **multinational structure,** as shown in Exhibit 9.1, production, sales, and marketing are decentralized,

> In a **multinational structure**, production, sales, and marketing are decentralized, and financial management remains the parent's responsibility.

Exhibit 9.1 *A multinational structure*

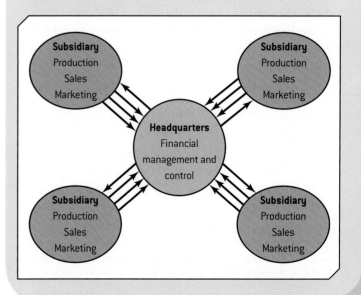

McDonald's, in order to appeal to local tastes, changed the burgers it serves in India to a 100% vegetarian product consisting of potatoes, peas, carrots, and some Indian spices.

and financial management remains the parent's responsibility. For example, Tyco Corporation has a multinational structure.[12] The focus is on local responsiveness—responding to customers' needs in a subsidiary's location—so subsidiaries operate autonomously but report to the parent company regularly. Nestle, for example, has 140 financial systems used at different locations around the world. However, this structure reduces the need for communication between subsidiaries and headquarters, because subsidiaries can make many decisions without consulting the parent first.[13]

Local hardware and software vendors influence choices in applications. Inevitably, each subsidiary operates on a different platform, and uniform connections are economically impractical. However, new methods are altering the definition of MNCs by decentralizing all other functions except information systems.

3.2 Global Structure

An organization with a **global structure**, sometimes called a "franchiser," manages highly centralized information systems.[1] Subsidiaries have little autonomy and rely on headquarters for all process and control decisions as well as system design and implementation. Consequently, an extensive communication network is necessary to manage this type of organization, and a GIS fits well into this structure.

Unfortunately, the integration needed to manage factors such as production, marketing, and human resources is difficult and impractical because of the heavy reliance on headquarters for new products and ideas. To manage the organization as efficiently as possible, duplicate information systems are developed.[14] Products are usually created, financed, and produced

in the headquarters country, and subsidiaries have the responsibility of sales and marketing for their countries and must tailor products to local requirements and tastes. For example, McDonald's, in order to appeal to local tastes, changed the burgers it serves in India to a 100% vegetarian product consisting of potatoes, peas, carrots, and some Indian spices.

McDonald's, Mrs. Field's Cookies, and Kentucky Fried Chicken are examples of corporations that have a global structure.[15] Another example is General Motors, which uses a GIS to integrate inventory information from all over the world. Exhibit 9.2 shows this structure, with a one-way flow of services, goods, information, and other resources.

A **global structure** (also known as a headquarters-driven structure) manages highly centralized information systems.[1] Subsidiaries have little autonomy and rely on headquarters for all process and control decisions as well as system design and implementation.

© Jonathan Larsen/Shutterstock.com

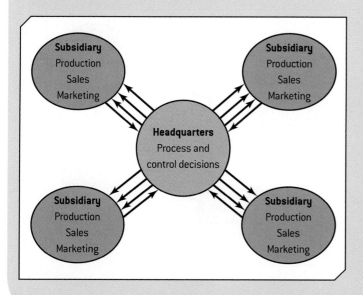

Exhibit 9.2 *A global structure*

Subsidiary
Production
Sales
Marketing

Subsidiary
Production
Sales
Marketing

Headquarters
Process and
control decisions

Subsidiary
Production
Sales
Marketing

Subsidiary
Production
Sales
Marketing

3.3 International Structure

An organization with an **international structure** operates much like a multinational corporation, but subsidiaries depend on headquarters more for process and production decisions. Domestic exporters are examples of this structure. Information systems personnel are regularly exchanged among locations to encourage joint development of applications for marketing, finance, and production. This exchange encourages a cooperative culture in geographically dispersed personnel, and using a GIS to support an international structure is more feasible because of this cooperative nature. Subsidiaries' GISs can be centralized or decentralized, depending on the extent to which they cooperate.

Exhibit 9.3 shows an international structure that uses two-way communication; for example, expertise information flows from headquarters to subsidiaries, and financial information flows from subsidiaries to headquarters. Caterpillar Corporation and other heavy equipment manufacturers usually have this structure.

3.4 Transnational Structure

In an organization with a **transnational structure**, the parent and all subsidiaries work together in designing policies, procedures, and logistics for

delivering products and services to the right market. This type of organization might have several regional divisions that share authority and responsibility, but in general, it does not have its headquarters in a particular country. A transnational organization usually focuses on optimizing supply sources and using advantages available in subsidiary locations. Many companies do this when they look for manufacturing facilities in countries where labor is less expensive than it is in the parent country. For example, China, India, Vietnam, and other countries have cheaper labor costs than the United States does. Again, a GIS fits into this structure well by integrating global activities through cooperation and information sharing between headquarters and subsidiaries.

The architecture of the GIS in a transnational structure requires a higher level of standardization and uniformity for global efficiency but must maintain local responsiveness. Universal data dictionaries and standard databases, for example, enhance the integration of GISs.

An organization with an **international structure** operates much like a multinational corporation, but subsidiaries depend on headquarters more for process and production decisions.

In an organization with a **transnational structure**, the parent and all subsidiaries work together in designing policies, procedures, and logistics for delivering products and services to the right market.

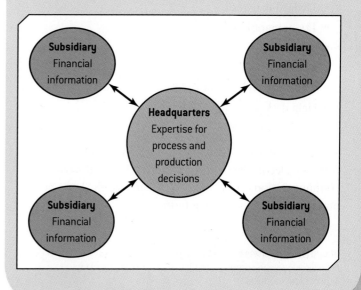

Exhibit 9.3 *An international structure*

Subsidiary
Financial
information

Subsidiary
Financial
information

Headquarters
Expertise for
process and
production
decisions

Subsidiary
Financial
information

Subsidiary
Financial
information

The level of cooperation and worldwide coordination needed for a transnational structure doesn't fully exist in today's global environment. However, with increasing cooperation between nations, this structure is becoming more feasible. Citigroup, Sony, and Ford, to name a few, have been striving to adopt a transnational structure. In fact, the globalizing trend is forcing many organizations to gravitate toward this structure because of competition with companies who already share innovations across borders and maintain responsiveness to local needs. These companies have increased efficiency in production costs because production can be spread across more locations.[13] Exhibit 9.4 shows this structure, with cooperation between headquarters and subsidiaries as well as between subsidiaries. Foreign exchange systems that allow traders and brokers from around the world to interact are an example of information systems that support this structure.

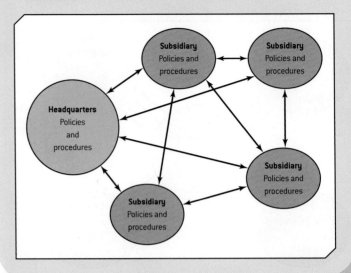

Exhibit 9.4 *A transnational structure*

3.5 Global Information Systems Supporting Offshore Outsourcing

Offshore outsourcing is an alternative for developing information systems. With this approach, an organization chooses an outsourcing firm in another country that can provide needed services and products. Initially, offshore outsourcing was used mostly in manufacturing to find cheap labor, but now it's used for many information technology tasks, including the following:

- Medical diagnosis
- Tax preparation
- Programming
- Application development
- Web site development
- Help desk/user support
- Quality assurance/software testing

The widespread availability of the Internet, improved telecommunication systems, reduced cost of communication, and increased bandwidth have all made offshore outsourcing more attractive for all types

of organizations. A GIS plays an important role in supporting offshore outsourcing by providing a global network that all participants can use for coordinating development activities, such as product design and global marketing campaigns. Table 9.1 lists the top offshore locations for outsourcing in 2008.[16] Some of the criteria used to rate countries include language proficiency, local government's support of offshore business, potential labor pool, existing infrastructure

Table 9.1

Top offshore locations for outsourcing in 2008

Americas	Asia/Pacific	Europe, Middle East, and Africa
Argentina	Australia	Czech Republic
Brazil	China	Hungary
Canada	India	Ireland
Chile	Malaysia	Israel
Costa Rica	New Zealand	Northern Ireland
Mexico	Pakistan	Poland
Uruguay	Philippines	Romania
	Singapore	Russia
	Sri Lanka	Slovakia
	Vietnam	South Africa
		Spain
		Turkey
		Ukraine

With **offshore outsourcing**, an organization chooses an outsourcing firm in another country that can provide needed services and products.

> To achieve true integration on an international scale, organizations must empower key personnel in other countries and rely on feedback and information-sharing technologies to maintain a global perspective.

(roads, rail service, and airports), and quality of the educational system.

The information box on this page highlights the Internet's role in globalization.

4 Obstacles to Using Global Information Systems

lobal information systems offer advantages in improving global coordination, managing the factors that promote globalization, and maintaining a competitive edge by supporting strategic planning. However, like any information system project, there are problems to overcome in implementing and maintaining a GIS. Companies planning to use GISs should analyze these problems and try to address them. Taking a proactive approach can increase the chance of success in using this technology. The following factors, some of which are discussed in more detail in the following sections, can hinder the success of a GIS[18,19]:

- Lack of standardization (including differences in time zones, taxes, language, and work habits)
- Cultural differences
- Diverse regulatory practices
- Poor telecommunication infrastructures
- Lack of skilled analysts and programmers

In addition, a more subtle obstacle to GIS development is the unwillingness to delegate control of information systems to host countries. To achieve true integration on an international scale, organizations must empower key personnel in other countries and rely on feedback and information-sharing technologies to maintain a global perspective.

4.1 Lack of Standardization

Lack of standardization can affect the development of a GIS on two fronts. First, the lack of international standards impedes the development of a cohesive system that's capable of sharing information resources across borders. Electronic data interchange, e-mail, and telecommunication standards vary throughout the world, so considering all standards is impractical. Although open-source systems are increasing in popularity, and the technology to link diverse systems is available, few organizations can afford the costs of integrating different platforms. Consequently, the lack of international system development standards prevents many organizations from using a GIS.

On the other hand, too much standardization can prevent flexibility in responding to local preferences and even time differences. For example, a global production system shouldn't dictate using the metric system for all

The Internet and Globalization in Action

The Internet allows entrepreneurs in developing countries to start and expand businesses without making large investments. Muhammad Hassaan Khan, a young entrepreneur, established a design and consulting business called Zuha Innovation that is based in Faisalabad, Pakistan. The Internet has lowered the barriers for small businesses all over the world, and entrepreneurs have responded, including ones from India, China, Mexico, and Brazil. Online businesses can be created regardless of where the creator is located; all that is needed is Internet access. These businesses vary widely in the activities they're engaged in, from e-commerce sites selling local products to bloggers making an income from advertising revenue. As Adam Toren, president of YoungEntrepreneur.com, puts it, "The Internet brings all continents, races, cities, and villages together into a global network of trade and communications."[17]

In some cultures, using technology is considered a boring, low-level task, but in others, being technologically knowledgeable is seen as a sign of social importance.

subsidiaries. It should allow subsidiaries the flexibility to use the measuring system used in their locations and be capable of handling conversions from one system to the other.

Additionally, information systems personnel managing a centralized GIS under international standards and sharing information resources across time zones might have difficulties finding the right time to take the system offline for backup and maintenance.[20] A balance between international system development standards, allowing ease of integration, modularization, custom tailoring of systems, and applications for local responsiveness, is needed.

Sharing software is difficult and impractical when these factors are considered. Only 5–15% of a company's applications are truly global in nature. Most applications are local in nature and can't be integrated into a GIS infrastructure. Even if the software can be integrated globally, support and maintenance problems might result. If the network goes down, who's responsible for

© tonobalaguer/Shutterstock.com

bringing the system back online? Moreover, employees calling the help desk might not speak the same language as the help desk personnel. Therefore, coordination and planning for variations in local needs are critical for using a GIS.[21]

4.2 Cultural Differences

Cultural differences include differences in values, attitudes, and behaviors from country to country, and these play an important role in using GISs. For example, in some cultures, using technology is considered a boring, low-level task, but in others, being technologically knowledgeable is seen as a sign of social importance. Some countries only allow purchasing local hardware and software for a GIS; others are more open.

For example, a travel Web site aimed at customers who make last-minute travel reservations at lower prices worked well in the United Kingdom but didn't translate well to other countries, such as Germany, where advance planning is expected and last-minute reservations are not "rewarded" with lower prices.[22]

Organizations might also need to look at changing content or images on their Web sites; photos of women dressed in a certain way might be acceptable in the Western world but unacceptable in the Middle East, for instance. Cultural issues are best addressed with education and training.

4.3 Diverse Regulatory Practices

Diverse regulatory practices also impede the integration process. This obstacle doesn't necessarily apply to TDF regulations; it applies to policies on business practices and technological use. Many countries also restrict the type of hardware and software that can be imported or used, and the vendors that an organization normally deals with might not service certain countries.[20] For example, in August 2010, the United Arab Emirates, citing security concerns, announced that BlackBerry phones will not be allowed to access e-mail or the Web. Saudi Arabia also plans to join the ban.[23]

Adopting open-source systems could eliminate part of this problem. However, as mentioned, few organizations are capable of adopting these systems.

Jurisdiction issues on contents of a GIS could also be challenging. ISPs, content providers, servers, and organizations owning these entities might be scattered throughout the world and operating under different rules and regulations. For example, Yahoo! was sued in French courts because Nazi memorabilia were being sold on its auction site, which is an illegal activity in France. To date, French and U.S. courts haven't agreed on the resolution or even on which court has jurisdiction.[24] Determining jurisdiction in cases involving cyberspace is still difficult.

The nature of intellectual property laws and how they're enforced in different countries also varies. Software piracy is a problem in all countries, but several have piracy rates higher than 90 piracy. This problem has resulted in an estimated loss of $40 billion worldwide.[25] Other legal issues include privacy and cybercrime laws as well as censorship and government control, which vary widely from country to country.

4.4 Poor Telecommunication Infrastructures

As mentioned, before adding a GIS, international companies must take into consideration the telecommunication infrastructures of the countries where subsidiaries are located. An organization might have the resources and skills to implement a worldwide integrated system, but it can't change an existing telecommunication infrastructure. Furthermore, the differences in telecommunication systems make consolidating them difficult. Implementing a GIS that encompasses 25 countries, for instance, is expensive and cumbersome when each country might have different service offerings, price schedules, and policies.

In countries where Internet access is slower or more costly, Web pages shouldn't have content with lots of graphics and animation that require more bandwidth. However, people in countries such as South Korea, where high-speed access is common, expect sophisticated Web sites with many graphics features.

Even when the telecommunication structure in two countries is comparable, differences in standards can cause problems. For example, a company with branches in the United States and Egypt might face the problems of different Internet protocols, higher costs, slower speed, and less reliability in Egypt.

4.5 Lack of Skilled Analysts and Programmers

Having skilled analysts and consultants with the knowledge to implement a GIS is critical, particularly with the severe shortage of qualified information systems professionals in the United States and Western Europe. When forming integrated teams, companies must consider the nature of each culture and differences in skills in other countries.[20] For example, experts from Singapore and Korea have been regarded as the best consultants in Asia because of their work ethic and their broad skill base. Germans are recognized for their project management skills, and Japanese are known for their quality process controls and total quality management. Ideally, an organization would link the skills of people from different countries to form a sort of "dream team." However, cultural and political differences can affect the cooperative environment needed for global integration. Training and certification programs, many of which are offered through the Internet, are one possible solution for narrowing this skills gap in developing nations.

The Industry Connection highlights the SAP Corporation as a leader in enterprise computing and global information systems.

5 Chapter Summary

In this chapter, you've learned about factors contributing to the globalization trend and the roles that e-business and the Internet have played in this trend. To meet the needs of this trend, global information systems are becoming more widespread, and you've learned about their components, the requirements for using them, how they're used in a variety of multinational companies' structures, and different applications of GISs, including offshore outsourcing. This chapter concluded with an overview of the obstacles to using global information systems.

SAP Corporation*

SAP, founded by five former IBM employees in 1972, is one of the leading providers of business software. Its applications can be used to manage finances, assets, production operations, and human resources, for example. The latest version of its enterprise resource planning software, SAP ERP 7.0, includes comprehensive Web-enabled products. SAP offers a Web interface for customers, called mySAP.com, and e-business applications, including customer relationship management (CRM) and supply chain management (SCM) systems. From the beginning, SAP products have been designed to be used with multiple languages and currencies, which is particularly useful for companies that need to support global operations. In addition, SAP includes upgrades for global aspects of information systems, such as making the transition from European currencies to Euros. In addition to products for enterprise resource planning and supply chain management (both discussed in Chapter 11), SAP offers the following:

- *SAP Supplier Relationship Management*—Helps manage a company's relationships with its suppliers by automating procurement processes, managing the supply chain, and creating a collaborative environment between the company and its suppliers.

- *SAP Product Lifecycle Management (PLM)*—Includes services for coordinating manufacturing processes, from developing product prototypes to producing the final product to ensuring compliance with industry standards and regulations.

*This information has been gathered from the company Web site (*www.sap.com*) and other promotional materials. For more detailed information and updates, visit the Web site.

Key Terms

global information system (GIS) (161)

global structure (166)

international structure (167)

multinational corporations (MNCs) (163)

multinational structure (165)

offshore outsourcing (168)

transborder data flow (TDF) (163)

transnational structure (167)

Problems, Activities, and Discussions

1. What are some advantages of a GIS?

2. What are some requirements of a GIS for operations management? For tactical management? For strategic management?

3. How does a GIS support offshore outsourcing?

4. Visit the following Web page, consult other sources, and write a one-page paper that discusses the importance of language proficiency in the success of a global company. What are some software tools and services that can increase the possibility of success?

   www.internetretailer.com/dailyNews.asp?id=20102

5. Visit the following Web page, consult other sources, and write a one-page paper that discusses the importance of international search capabilities offered by search engines for the success of a global company.

   www.btobonline.com/apps/pbcs.dll/ article?AID=/20091116/FREE/311109998/1156/ ISSUENETMARKETING

6. Visit the following Web page, consult other sources, and write a one-page paper that discusses the importance of understanding culture in order to be successful as a global company. Why is cultural understanding as important as technology and operations?

   www.internetretailer.com/article.asp?id=18102

7. Visit the following Web page, consult other sources, and write a one-page paper that discusses the importance of international marketing on the success of a global company. Why are emerging markets so important to a global company's success?

www.btobonline.com/apps/pbcs.dll/
article?AID=/20090406/FREE/304069975/1109/
FREE

8. Visit the following Web page, consult other sources, and write a one-page paper that discusses how global companies such as PepsiCo, Hyatt, and the Mayo Clinic are using Twitter to promote their products and services.

www.informationweek.com/news/global-cio/inter-views/showArticle.jhtml?articleID=218900376

9. Which of the following is an obstacle to developing and using a GIS? (Choose all that apply.)

a. Lack of standardization

b. Too many analysts and programmers

c. Diverse regulatory practices

d. Poor telecommunication infrastructures

10. Even in countries where Internet access is slower or more costly, Web pages should contain sophisticated features, such as animation and graphics, to maintain a good image. True or False?

casestudy

MULTINATIONAL COMPANIES COULD BREAK THE LANGUAGE BARRIERS ON THE WEB

Although English is the universal language of the Internet, billions of potential customers do not speak English. Therefore, proficiency at various languages, not to mention being able to translate the Web into various languages, could bring huge benefits to global companies. The Pop!Tech (*http://poptech.org*) conference is an annual gathering of leaders and scientists from various disciplines who explore together the effect that new technologies have on society. In 2007, the presentations from the conference were made available online, with subtitled versions in eight languages. To achieve this, Pop!Tech teamed up with dotSUB (*http://dotsub.com*), a service that offers Wiki-style translations. Traditional translation methods are usually expensive and time consuming, and embedding subtitles on a video is a complex task. With the dotSUB service, however, users from all over the world can upload videos, films, and TV programs. For example, ArcelorMittal, the world's largest steel company, translated a corporate video into

© Toria/Shutterstock.com

the 14 languages that the company does business in. dotSUB offers two different services: closed and open. With the closed option, which is available for a fee, dotSUB's team of freelancers does the translating. With the open method, volunteers using dotSUB's software tools post their own translated videos on dotSUB.com.[26]

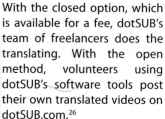

Answer the following questions:

1. What type of conference is the Pop!Tech? Who participates in this conference? What is its goal?

2. Why could language translation on the Web be beneficial to a global company?

3. How could dotSUB services be used? What are the differences between the closed and open options using dotSUB?

BUILDING SUCCESSFUL INFORMATION SYSTEMS

t his chapter explains the systems development life cycle (SDLC), a model for developing a system or project. The cycle is usually divided into five phases, and you learn about the tasks involved in each phase. For example, in the first phase—planning—a feasibility study is typically conducted, and the SDLC task force is formed. You also learn about two alternatives to the SDLC model: self-sourcing and outsourcing. Finally, you review new trends in systems analysis and design, such as service-oriented architecture, rapid application development, extreme programming, and agile methodology.

learning outcomes

After studying this chapter, you should be able to:

LO1 Describe the systems development life cycle (SDLC) as a method for developing information systems.

LO2 Explain the tasks involved in the planning phase.

LO3 Explain the tasks involved in the requirements-gathering and analysis phase.

LO4 Explain the tasks involved in the design phase.

LO5 Explain the tasks involved in the implementation phase.

LO6 Explain the tasks involved in the maintenance phase.

LO7 Describe new trends in systems analysis and design, including service-oriented architecture, rapid application development, extreme programming, and agile methodology.

1 Systems Development Life Cycle: An Overview

I n the information systems field, system failure can happen for several reasons, including missed deadlines, users' needs that aren't met, dissatisfied customers, lack of support from top management, and going over the budget. Using a system development method can help prevent these failures. Designing a successful information system requires integrating people, software, and hardware. To achieve this integration, designers often follow the **systems development life cycle (SDLC)**, also known as the "waterfall model." It's a series of well-defined phases performed in sequence that serves as a framework for developing a system or project. Exhibit 10.1 shows the phases of the SDLC, which are explained throughout this chapter. In this model, each phase's output (results) becomes the input for the next phase. When following this model, keep in mind that the main goal of an information system is delivering useful information in a timely manner to the right decision maker.

> **Systems development life cycle (SDLC)**, also known as the "waterfall model," is a series of well-defined phases performed in sequence that serves as a framework for developing a system or project.

Systems planning today is about evaluating all potential systems that need to be implemented. A preliminary analysis of requirements for each is done first, and a feasibility study is conducted for each system. Then the organization decides which ones are a "go" and proceeds to the next phase.

Information system projects are often an extension of existing systems or involve replacing an old technology with a new one. However, sometimes an information system needs to be designed from scratch, and the SDLC model is particularly suitable in these situations. For existing information systems, some phases might not be applicable, although the SDLC model can still be used. In addition, when designing information systems, projecting the organization's growth rate is important; otherwise, the system could become inefficient shortly after it's designed.

> During the **planning phase**, which is one of the most crucial phases of the SDLC model, the systems designer must understand and define the problem the organization faces, taking care not to define symptoms rather than the underlying problem.

 ## 2 Phase 1: Planning

during the **planning phase**, which is one of the most crucial phases of the SDLC model, the systems designer must understand and define the problem the organization faces, taking care not to define symptoms rather than the underlying problem. The problem can be identified in-

Exhibit 10.1 *Phases of the SDLC*

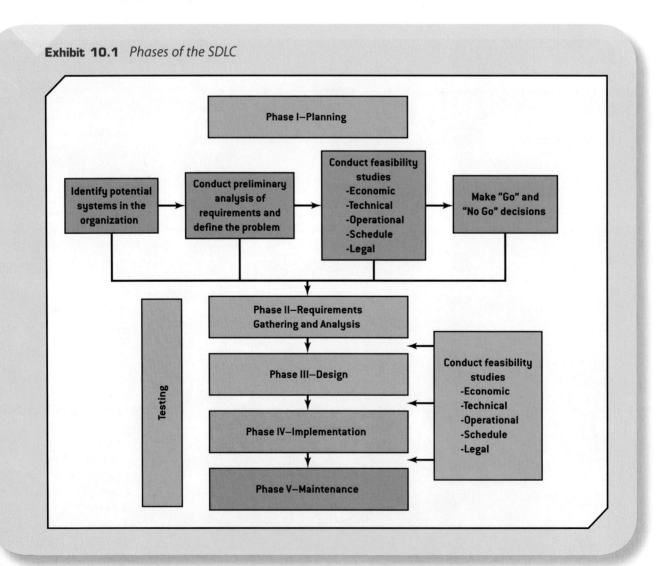

ternally or externally, such as from customers or suppliers. For example, a problem identified internally would be management's concern about the organization's lack of a competitive edge in the marketplace; a problem identified externally would be suppliers noting inefficiency in the inventory control procedure.

After identifying the problem, an analyst or team of analysts assesses the current and future needs of the organization or a specific group of users by answering the following questions:

- Why is this information system being developed?
- Who are the system's current and future users?
- Is the system new or an upgrade or extension of an existing system?
- Which functional areas (departments) will be using the system?

As part of this assessment, analysts must examine the organization's strategic goals, how the proposed system can support these goals, which factors are critical to the proposed system's success, and the criteria for evaluating the proposed system's performance. Establishing evaluation criteria ensures objectivity throughout the SDLC process.

In addition, analysts must get feedback from users on the problem and the need for an information system. During this phase, they need to make sure users understand the four *W*s:

- *Why*—Why is the system being designed? Which decisions will be affected?
- *Who*—Who is going to use the system? Is it going to be used by one decision maker or a group of decision makers? This question is also about types of users. For example, will the marketing department be using the system? Will manufacturing also be using the

system as suppliers or consumers of information?
- *When*—When will the system be operational? When in the business process (in what stages) will the system be used?
- *What*—What kind of capabilities will the system provide? How will these capabilities be used?

The end result of this phase should give users and top management a clear view of what the problem is and how the information system will solve the problem. As an example, here's a look at how ABC Furniture is planning for an information system to solve the problem of inaccurate inventory forecasts. Currently, ABC Furniture buys wood from New England Wood (NEW).

- *Why*—ABC Furniture needs an information system to track inventory, generate a more accurate forecast of product demand, and track requirements for wood to be ordered from NEW. Clearly, a more accurate inventory will help reduce inventory costs, improve ABC Furniture's relationship with NEW and with distributors, ensure that the company's products are available for retailers, and improve ABC's image in the marketplace.

- *Who*—The main users of the information system will be the procurement group responsible for placing orders with NEW, the manufacturing division responsible for tracking inventory and ensuring that demand for finished goods is met, the sales personnel who take orders from distributors, and possibly distributors who take orders from retailers.

- *When*—The system must become operational within the next four months because the company's main competitor is planning to open a new store in six months. Further, the system must support the materials-ordering stage, the production-planning stage, and the shipping stage of the manufacturing process. It must also supply information for the marketing campaign that ABC Furniture is planning to run in five months and support ABC's expansion into a new region.

- *What*—On the inbound side, the system must track pending and received deliveries, quantities of raw materials, orders placed for raw materials, and raw material levels from all of ABC's suppliers, including NEW. On the operations side, the system must provide information on inventory levels of all products, raw materials, work in progress at each stage of manufacturing, quality of raw materials received, quality of finished goods inspected, and rejects. On the outbound side, the system must track placed orders, unfulfilled orders, and fulfilled orders for each finished product as well as the order history for each distributor and retailer demands.

© Supertrooper/Shutterstock.com

Feasibility is the measure of how beneficial or practical an information system will be to an organization and should be measured continuously throughout the SDLC process.

2.1 Formation of the Task Force

To ensure an information system's success, users must have input in the planning, requirements-gathering and analysis, design, and implementation phases. For this reason, a task force is formed, consisting of representatives from different departments (including IT), systems analysts, technical advisors, and top management. This team collects user feedback and works toward getting users involved from the beginning.

The system designers and analysts should explain the goals and benefits of the new system so that the task force knows what to look for in user input. Generally, an information system has two groups of users the task force should gather feedback from: internal and external. **Internal users** are employees who will use the system regularly, and they can offer important feedback on the system's strengths and weaknesses. **External users** aren't employees but do use the system; they include customers, contractors, suppliers, and other business partners. Although they aren't normally part of the task force, their input is essential.

Using a task force for designing an information system is similar to using the joint application design approach. **Joint application design (JAD)** is a collective activity involving users, top management, and IT professionals. It centers on a structured workshop (called

a JAD session) where users and system professionals come together to develop an application. It involves a detailed agenda, visual aids, a leader who moderates the session, and a scribe who records the specifications. It results in a final document containing definitions for data elements, workflows, screens, reports, and general system specifications. An advantage of the JAD approach is that it incorporates varying viewpoints from different functional areas of an organization to help ensure that collected requirements for the application aren't too narrow and one-dimensional in focus.[1]

2.2 Feasibility Study

Feasibility is the measure of how beneficial or practical an information system will be to an organization and should be measured continuously throughout the SDLC process (see the WestJet Airlines Feasible Project information box). Upper management is often frustrated by information systems that are unrelated to the organization's strategic goals or have an inadequate payoff, by poor communication between system users and designers, and by designers' lack of consideration for users' preferences and work habits. A detailed feasibility study that focuses on these factors can help ease management's frustration with investing in information systems.[2]

During the planning phase, analysts investigate a proposed solution's feasibility and determine how best

Internal users are employees who will use the system regularly, and they can offer important feedback on the system's strengths and weaknesses.

External users aren't employees but do use the system; they include customers, contractors, suppliers, and other business partners. Although they aren't normally part of the task force, their input is essential.

Joint application design (JAD) is a collective activity involving users, top management, and IT professionals. It centers on a structured workshop (called a JAD session) where users and system professionals come together to develop an application. It involves a detailed agenda, visual aids, a leader who moderates the session, and a scribe who records the specifications. It results in a final document containing definitions for data elements, workflows, screens, reports, and general system specifications.

© Yuri Arcurs/Shutterstock.com

Part 3: IS Development, Enterprise Systems, MSS, and Emerging Trends

178

A Feasible Project Becomes Unfeasible

WestJet Airlines, a Canadian discount airline in Calgary, announced in July 2007 that it was stopping development on a new reservation system called AiRES, even though it had invested $30 million in the project. The problem wasn't with Travelport, the company developing the system. WestJet simply grew faster than anticipated, and the original specifications for the reservation system didn't address this fast growth. Management wanted to add features, such as the capability to partner with international carriers, but the system had been planned to fit a small discount airline, not a large international airline. WestJet suspended the work of 150 internal IT specialists and about 50 outside consultants. This example shows the need for conducting feasibility studies through a project's life cycle. If WestJet had continued the project, the potential losses would have been more than $30 million.[3]

to present the solution to management to get funding. The tool used for this purpose is a **feasibility study**, and it usually has five major dimensions, discussed in the following sections: economic, technical, operational, scheduling, and legal.

2.2.1 Economic Feasibility

Economic feasibility assesses a system's costs and benefits. Simply put, if implementing the system results in a net gain of $250,000 but the system cost $500,000, the system isn't economically feasible. To conduct an economic feasibility study, the systems analyst team must identify all costs and benefits—tangible and intangible—of the proposed system. The team must also be aware of opportunity costs associated with the information system. Opportunity costs measure what you would miss by not having a system or feature. For example, if your competitor has a Web site and you don't, what's the cost of not having a site, even if you don't really need one? What market share are you likely to lose if you don't have a Web site?

To assess economic feasibility, the team tallies tangible development and operating costs for the system and compares them with expected financial benefits of the system. Development costs include the following:

- Hardware and software
- Software leases or licenses
- Computer time for programming, testing, and prototyping
- Maintenance costs for monitoring equipment and software
- Personnel costs—salaries for consultants, systems analysts, network specialists, programmers, data entry clerks, computer operators, secretaries, and technicians
- Supplies and other equipment
- Training employees who will be using the system

Operating costs for running the system are typically estimated, although some vendors and suppliers can supply costs. These costs can be fixed or variable (depending on rate of use). After itemizing these costs, the team creates a budget. Many budgets don't allow enough for development costs, especially technical expertise (programmers, designers, and managers), and for this reason, many information system projects go over budget.

An information system's scope and complexity can change after the analysis or design phases, so the team should keep in mind that an information system project that is feasible at the outset could become unfeasible later. Integrating feasibility checkpoints into the SDLC process is a good idea to ensure the system's success. Projects can always be canceled or revised at a feasibility checkpoint, if needed.

To complete the economic feasibility study, the team must identify benefits of the information system, both tangible and intangible. Tangible benefits can be quantified in terms of monthly or annual savings, such as the new system allowing an organization to operate with three employees rather than five or the new system resulting in increased profits. The real challenge is assessing intangible costs and benefits accurately; attaching a realistic monetary value to these factors can be difficult.

Intangible benefits are difficult to quantify in terms of dollar amounts, but if they aren't at least identified, many information system projects can't be justified. Some examples of intangible benefits include improved employee morale, better customer satisfaction, more

> A **feasibility study** analyzes a proposed solution's feasibility and determines how best to present the solution to management. It usually has five major dimensions: economic, technical, operational, scheduling, and legal.
>
> **Economic feasibility** assesses a system's costs and benefits.

efficient use of human resources, increased flexibility in business operations, and improved communication. For example, you could quantify customer service as maintaining current total sales and increasing them by 10 percent to improve net profit. Other measures have been developed to assess intangibles, such as quantifying employee morale with rates of on-time arrival to work or working overtime. Customer satisfaction, though intangible, can be measured by using satisfaction surveys, and the Internet has made this method easier.

After collecting information on costs and benefits, the team can do a cost-effectiveness analysis. This analysis is based on the concept that a dollar today is worth more than a dollar one year from now. If the system doesn't produce enough return on the investment, the money can be better spent elsewhere. The most common analysis methods are payback, net present value (NPV), return on investment (ROI), and internal rate of return (IRR). The final result of this task is the cost-benefit analysis (CBA) report, used to sell the system to top management. This report can vary in format but should include the following sections: executive summary, introduction, scope and purpose, analysis method, recommendations, justifications, implementation plans, summary, and appendix items, which can include supporting documentation. Some examples of useful supporting documentation are organizational charts, workflow plans, floor plans, statistical information, project sequence diagrams, and timelines or milestone charts.

2.2.2 Technical Feasibility

Technical feasibility is concerned with the technology that will be used in the system. The team needs to assess whether technology to support the new system is available or feasible to implement. For example, a full-featured voice-activated monitoring system isn't technically feasible at this point. However, given the pace of technological development, many of these problems will eventually have solutions. Lack of technical

feasibility can also stem from an organization lacking the expertise, time, or personnel to implement the new system. This problem is also called "a lack of organizational readiness." In this case, the organization can take steps to address its shortcomings and then consider the new system. Extensive training is one solution to this problem.

2.2.3 Operational Feasibility

Operational feasibility is the measure of how well the proposed solution will work in the organization and how internal and external customers will react to it. The major question to answer is "Is the information system worth implementing?" To assess operational feasibility, the team should address the following questions:

- Is the system doing what it's supposed to do? For example, will the information system for ABC reduce orders for raw materials by tracking inventory more accurately?
- Will the information system be used?
- Will there be resistance from users?
- Will top management support the information system?
- Will the proposed information system benefit the organization?
- Will the proposed information system affect customers (both internal and external) in a positive way?

2.2.4 Scheduling Feasibility

Scheduling feasibility is concerned with whether the new system can be completed on time. For example, an organization might need a wireless network immediately because of a disaster that destroyed the existing network. However, if the new system can't be delivered in time, the loss of customers could force the organization

Technical feasibility is concerned with technology to be used in the system. The team needs to assess whether technology to support the new system is available or feasible to implement.

Operational feasibility is the measure of how well the proposed solution will work in the organization and how internal and external customers will react to it.

Scheduling feasibility is concerned with whether the new system can be completed on time.

© emily2k/Shutterstock.com

out of business. In this case, the proposed system isn't feasible from a schedule viewpoint. The problem of missing deadlines is common in the information systems field, but designers can often minimize this problem by using project management tools.

2.2.5 Legal Feasibility

Legal feasibility is concerned with legal issues and typically addresses questions such as the following:

- Will the system violate any legal issues in the country where it will be used?
- Are there any political repercussions of using the system?
- Is there any conflict between the proposed system and legal requirements? For example, does the system take the Information Privacy Act into account?

3 Phase 2: Requirements Gathering and Analysis

I n the **requirements-gathering and analysis phase**, analysts define the problem and generate alternatives for solving it. During this phase, the team attempts to understand the requirements for the system, analyzes these requirements to determine the main problem with the current system or processes, and looks for ways to solve problems by designing the new system.

The first step in this phase is gathering requirements. Several techniques are available for this step, including interviews, surveys, observations, and the JAD approach described earlier in the chapter. The intent is to find out what users do, how they do it, what problems they face in performing their jobs, how the new system would address these problems, what users expect from the system, what decisions are made, what data is needed to make decisions, where data comes from, how data should be presented, and what tools are needed to examine data for the decisions users make. All this information can be recorded, and the team uses this information to determine what the new system should do (process analysis) and what data is needed for this process to be performed (data analysis).

The team uses the information collected in requirements-gathering to understand the main problems: Define the project's scope, including what it should and shouldn't do, and create a document called the "system specifications." This document is then sent to all key users and task force members for approval. The creation of this document indicates the end of the analysis phase and the start of the design phase.

There are two major approaches for analysis and design of information systems: the structured systems analysis and design (SSAD) approach and the object-oriented approach. (The object-oriented approach was introduced in Chapter 3 with the discussion of object-oriented databases.) The onset of the Web and Java, an object-oriented language, created the push for a different approach than SSAD. To understand the difference between the two approaches, first realize that any system has three parts: process, data, and user interface. Analyzing requirements in the analysis phase is done from the perspective of the process and data. The SSAD approach treats process and data independently and is a sequential approach that requires completing the analysis before beginning the design. The object-oriented approach combines process and data analysis, and the line between analysis and design is so thin that analysis and design seem to be a single phase instead of the two distinct phases shown previously in Exhibit 10.1.

These two approaches use different tools for creating analysis models. Table 10.1 shows some examples of tools used in the SSAD approach.

Exhibit 10.2 shows an example of a data flow diagram for ABC's inventory management system, and Exhibit 10.3 shows a context diagram.

Legal feasibility is concerned with legal issues, including political repercussions and meeting the requirements of the Information Privacy Act.

In the **requirements-gathering and analysis phase**, analysts define the problem and generate alternatives for solving it.

Table 10.1

Examples of tools used in SSAD analysis models

Modeling tool	What's analyzed	What it's used for
Data flow diagram (DFD)	Process analysis and design	Helps break down a complex process into simpler, more manageable, and more understandable subprocesses; shows how data needed by each process flows between processes and what data is stored in the system; it also helps define the system's scope
Flowchart	Process analysis	Illustrates the logical steps in a process but doesn't show data elements and associations; it can supplement a DFD and help analysts understand and document how a process works
Context diagram	Process analysis and design	Shows a process at a more general level and is helpful for showing top management and the task force how a process works
Conceptual data model (such as an entity relationship model)	Data analysis	Helps analysts understand the data requirements a system must meet by defining data elements and showing the associations between them

Notice in Exhibit 10.2 that processes are indicated with a circle. Anything that interacts with the system but isn't part of it is considered an "external entity" and is shown as a blue rectangle. Data stores (databases, file systems, even file cabinets) are shown as a grey rectangle.

In Exhibit 10.3, the DFD has been simplified into a context diagram, also called a "Level 0 diagram." Each process in this context diagram could be broken down into a separate diagram called "Level 1."

Both modeling tools show data flows between processes and external entities, and the DFD also shows data flows between processes and data stores. These data flows are general, so they don't show specific data elements. For example, "Purchase order" is shown in Exhibit 10.2 instead of all the pieces of data making up a purchase order, such as order number, order date, item number, and item quantity.

The models created during the analysis phase constitute the design specifications. After confirming these specifications with users, analysts start designing the system.

Exhibit 10.2 *A data flow diagram for ABC's inventory management system*

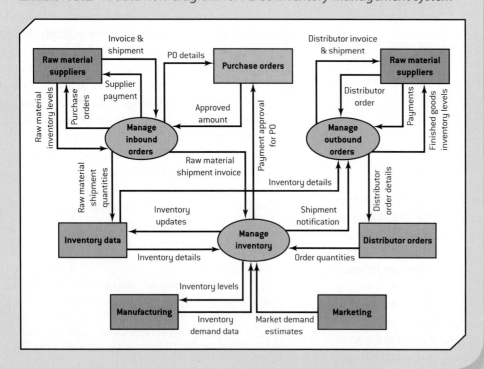

During the design phase, analysts choose the solution that's the most realistic and offers the highest payoff for the organization.

4 Phase 3: Design

during the **design phase**, analysts choose the solution that's the most realistic and offers the highest payoff for the organization. Details of the proposed solution are outlined, and the output of this phase is a document with exact specifications for implementing the system, including files and databases, forms and reports, documentation, procedures, hardware and software, networking components, and general system specifications. For large projects in particular, CASE tools (discussed in the next section) are helpful in the analysis and design phases.

The design phase consists of three parts: conceptual design, logical design, and physical design. The conceptual design is an overview of the system and doesn't include hardware or software choices. The logical design makes the conceptual design more specific by indicating hardware and software, such as specifying Linux servers, Windows clients, an object-oriented programming language, and a relational DBMS. These choices usually require changing the conceptual design to fit the platforms and programming languages chosen. Finally, the physical design is created for a specific platform, such as choosing Dell servers running Ubuntu Linux, Dell laptops running Windows 7 and Internet Explorer, Java for the programming language, and SQL Server 2008 for the relational DBMS.

During the **design phase**, analysts choose the solution that's the most realistic and offers the highest payoff for the organization. Details of the proposed solution are outlined, and the output of this phase is a document with exact specifications for implementing the system, including files and databases, forms and reports, documentation, procedures, hardware and software, networking components, and general system specifications.

Computer-aided systems engineering (CASE) tools automate parts of the application development process. These tools are particularly helpful for investigation and analysis in large-scale projects because they automate parts of the design phase.

4.1 Computer-Aided Systems Engineering

Systems analysts use **computer-aided systems engineering (CASE)** tools to automate parts of the application development process. These tools are particularly helpful for investigation and analysis in large-scale projects because they automate parts of the design phase. Analysts can use them to modify and update several design versions in an effort to choose the best version. CASE tools support the design phase by helping analysts do the following:

- Keep models consistent with each other.
- Document models with explanations and annotations.
- Ensure that models are created according to specific rules.
- Create a single repository of all models related to a single system, which ensures consistency in analysis and design specifications.

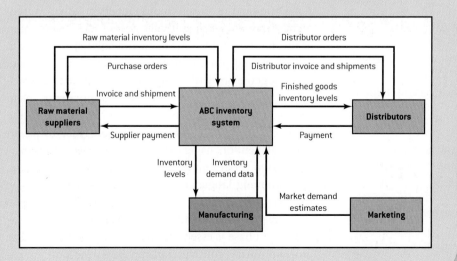

Exhibit 10.3 *A context diagram for ABC's inventory management system*

- Track and manage changes to the design.
- Create multiple versions of the design.

CASE tools are similar to computer-aided design (CAD) tools used by architects and engineers. Their capabilities vary, depending on the product, but generally include the following:

- Graphics tools, such as data flow diagrams, to illustrate a system's operation
- Dictionary tools designed to record the system's operation in detail
- Prototyping tools for designing input and output formats, forms, and screens
- Code generators to minimize or eliminate programming efforts
- Project management tools to help control the system's schedule and budget

Several CASE tools are available, including CA ERwin Process Modeler (*www.ca.com/us/products/product. aspx?id=1001*), Oracle Designer (*www.oracle.com/ technology/products/designer/index.html*), and Visible System's Visible Analyst (*www.visible.com/Products/ Analyst/vacorporate.htm*). CASE tools usually include the following output:

- Specifications documents
- Documentation of the analysis, including models and explanations
- Design specifications with related documentation
- Logical and physical design documents based on the conceptual design
- Code modules that can be incorporated into the system

4.2 Prototyping

Prototyping has been around for many years in physical science because building a small working model first is easier and less expensive than building the entire system. Prototypes can also be tested to detect potential problems and devise solutions.

In **prototyping**, a small-scale version of the system is developed, but one that's large enough to illustrate the system's benefits and allow users to offer feedback.

A **proof-of-concept prototype** shows users how a particular task that wasn't technically feasible can be done.

A **selling prototype** is used to sell a proposed system to users or management by showing some of its features.

Prototyping has gained popularity in designing information systems because needs can change quickly and lack of specifications for the system can be a problem. Typically, a small-scale version of the system is developed, but one that's large enough to illustrate the system's benefits and allow users to offer feedback. Prototyping is also the fastest way to put an information system into operation. Prototypes are usually used for the following purposes:

- *Gathering system requirements*—During the planning phase, designing a prototype and showing it to users is a good way to gather additional information and refine requirements for the proposed system.
- *Helping to determine system requirements*—If users aren't sure about the type of information system they want, a prototype can serve as a valuable tool for demonstrating the system's functional capabilities and then asking for users' reactions.
- *Determining a system's technical feasibility*—If a system isn't technically feasible or appears to be unfeasible, a prototype can be used to show users that a particular task can be done. This type of prototype is called a **proof-of-concept prototype**.
- *Selling the proposed system to users and management*—Prototypes are sometimes used to sell a proposed system to users and management by showing some of its features and demonstrating how beneficial it could be to the organization. This type of prototype is called a **selling prototype**.

Prototyping is usually done in four steps:[4]

1. Define the initial requirements.
2. Develop the prototype.
3. Review and evaluate the prototype.
4. Revise the prototype.

Defining the initial requirements involves agreement between users and designers that prototyping is the most suitable approach for solving a problem. After agreeing on the approach, users and designers work together to gather information about the prototype's components and how these components relate to one another. The team might decide on one of these approaches for constructing the prototype: using an external vendor, using software packages or fourth-generation programming languages, or using high-level programming languages and developing the prototype from scratch.

Including users and top management in the construction phase is essential because some problems that crop up during construction can be solved only by users or top management. For example, top management typically must solve problems of financing a system, and lack of specifications is a problem better suited for users to solve. In addition, during this phase, users and top management can learn more about the problems the information system will solve, and the team of users and designers can learn a lot about decision making in the organization.

After completing the prototype, users begin using it and evaluating its performance. Depending on the outcome, one of the following decisions is made: Revise the prototype, cancel the information system project, develop a new prototype, or build a complete system based on the prototype. Regardless of the decision, the prototype has provided useful information to the team of users and designers. At this point, the problem is better defined, and the system's operations are understood more clearly.

4.2.1 Prototyping Development Tools

Numerous tools can be used for constructing a prototype of a system. Some widely used tools include spreadsheet packages, such as Microsoft Excel, and database management packages, such as Microsoft Access. In addition, Visual Basic is commonly used to code the logic required for processes. CASE tools and third- and fourth-generation programming languages can also be used to develop prototypes quickly. In addition, prototyping tools for user interface design include GUImagnets (*www.guimagnets.com*), Designer Vista (*http://designervista.com*), and GUI Design Studio (*www.carettasoftware.com/guidesignstudio*).

4.2.2 Advantages and Disadvantages of Prototyping

As mentioned, prototyping offers several advantages:

- They provide a method for investigating an environment in which the problem is poorly defined and information is difficult to gather.
- They reduce the need for training information system users because the users are involved in developing the system.
- They reduce costs because building a model is less expensive than building the complete system. If users and top management decide the system shouldn't be developed, the organization hasn't lost all the money that would have been spent on building a complete system.
- They increase the chance of the system's success by encouraging users' involvement.
- They are easier to modify than a complete system.

- They improve documentation because users and designers can walk through several versions of the system.
- They improve communication among users, top management, and information systems personnel because seeing a concrete model often prompts potential users of the system to ask questions, express opinions, point out shortcomings and strengths, and so forth.

Even with all these advantages, prototyping has some disadvantages:

- Developing them might require more support and assistance from users and top management than they're willing to offer.
- They might not reflect the final system's actual operation and, therefore, be misleading.
- Developing them might lead analysts and designers to forego comprehensive testing and documentation. If the prototype works, the team might be convinced that the final system will work, too, and this assumption can be misleading.

5 Phase 4: Implementation

during the **implementation phase**, the solution is transferred from paper to action, and the team configures the system and procures components for it. A variety of tasks takes place in the implementation phase, including the following:

- Acquiring new equipment
- Hiring new employees
- Training employees
- Planning and designing the system's physical layout
- Coding
- Testing
- Designing security measures and safeguards
- Creating a disaster recovery plan

> During the **implementation phase**, the solution is transferred from paper to action, and the team configures the system and procures components for it.

With the help of development tools, such as query languages, report generators, and fourth-generation programming languages, self-sourcing has become an important part of information system resources.

When an information system is ready to be converted, designers have several options, including the following:

- *Parallel conversion*—In **parallel conversion**, the old and new systems run simultaneously for a short time to ensure that the new system works correctly. However, this approach is costly and can be used only if an operational system is already in place.
- *Phased-in-phased-out conversion*—In **phased-in-phased-out conversion**, as each module of the new system is converted, the corresponding part of the old system is retired. This process continues until the entire system is operational. Although this approach isn't suitable in all situations, it can be effective in accounting and finance.
- *Plunge (direct cutover) conversion*—In **plunge (direct cutover) conversion**, the old system is stopped and the new system is implemented. This approach is risky if there are problems with the new system, but the organization can save on costs by not running the old and new systems concurrently.
- *Pilot conversion*—In **pilot conversion**, the analyst introduces the system in only a limited area of the organization, such as a division or department. If the system works correctly, it's implemented in the rest of the organization in stages or all at once.

In **parallel conversion**, the old and new systems run simultaneously for a short time to ensure that the new system works correctly.

In **phased-in-phased-out conversion**, as each module of the new system is converted, the corresponding part of the old system is retired. This process continues until the entire system is operational.

In **plunge (direct cutover) conversion**, the old system is stopped and the new system is implemented.

In **pilot conversion**, the analyst introduces the system in only a limited area of the organization, such as a division or department. If the system works correctly, it's implemented in the rest of the organization in stages or all at once.

A **request for proposal (RFP)** is a written document with detailed specifications used to request bids for equipment, supplies, or services from vendors.

5.1 Request for Proposal

A **request for proposal (RFP)** is a written document with detailed specifications used to request bids for equipment, supplies, or services from vendors. It's usually prepared during the implementation phase and contains detailed information about functional, technical, and business requirements of the proposed information system. Drafting an RFP can take 6 to 12 months, but with software, the Internet, and other online technologies, time and costs can be reduced.

A crucial part of this process is comparing bids from single and multiple vendors. Using a single vendor to provide all the information system's components is convenient, but the vendor might not have expertise in all areas of the information system's operations.

The main advantage of an RFP is that all vendors get the same information and requirements, so bids can

Exhibit 10.4 *Main components of an RFP for ABC Furniture's inventory system*

1. **Introduction**
    a. Background and organizational goals of ABC
2. **System Requirements**
    a. Problems to be solved
    b. Details of preliminary analyses
    c. Key insights gained
3. **Additional Information**
    a. Hardware available
    b. Software preferences
    c. Other existing systems and integration requirements
    d. Understanding of benefits to be gained
    e. Business understanding necessary (for the bidder to work with ABC)
    f. Technology and technical know-how necessary
4. **Project Time Frame**
5. **Contact Information and Submission Procedures**

be evaluated more fairly. Furthermore, all vendors have the same deadline for submitting bids, so no vendor has the advantage of having more time to prepare an offer. RFPs are also useful in narrowing down a long list of prospective vendors.

A major disadvantage of an RFP is the time involved in writing and evaluating proposals. With the rapid changes in information technologies, a lengthy time frame makes RFPs less appealing. Many companies can't wait 6 to 12 months to decide on a vendor for an information system. Exhibit 10.4 shows the main components of an RFP. You can find free templates for RFPs by searching at Business-in-a-Box (*www.biztree.com*), TEC (*http://technologyevaluation.com*), and Klariti (*www.klariti.com*).

With the typical need to complete information system projects quickly, shortening the time needed to write and evaluate proposals is often necessary. One alternative to an RFP is a **request for information (RFI)**, a screening document for gathering vendor information and narrowing the list of potential vendors. An RFI can help manage the selection of vendors by focusing on the project requirements that are crucial to selecting vendors. However, an RFI has its limitations. It's not suitable for complex projects because it can be used only for selecting three or four finalists from a list of candidates.

5.2 Implementation Alternatives

The SDLC approach is sometimes called **insourcing**, meaning an organization's team develops the system internally. However, two other approaches used for developing information systems are self-sourcing and outsourcing, discussed in the following sections.

5.2.1 Self-Sourcing

The increasing demand for timely information has put pressure on information systems teams, who are already overloaded with maintaining and modifying existing systems. In many organizations, the task of keeping existing systems running takes up much of the available computing resources and personnel, leaving few resources for developing new systems. The resulting inability to respond to users' needs has increased employee dissatisfaction and caused a backlog in systems development in both well-managed and poorly managed organizations. So in recent years, more end users have been developing their own information systems with little or no formal assistance from the information systems team. These users might not

know how to write programming code but are typically skilled enough to use off-the-shelf software, such as spreadsheet and database packages, to produce custom-built applications.[5,6] This trend, called **self-sourcing** (or end-user development), has resulted from long backlogs in developing information systems, the availability of affordable hardware and software, and organizations' increasing dependency on timely information.

With the help of development tools, such as query languages, report generators, and fourth-generation programming languages, self-sourcing has become an important part of information system resources. It's also useful in creating one-of-a-kind applications and reports. Self-sourcing can help reduce the backlog in producing information systems and improve flexibility in responding to users' information needs. Backlogs, however, are just the tip of the iceberg. When the backlog list is long, end users often stop making new requests for many of the applications they need because they believe these requests would just make the list longer. The list of applications that aren't requested is often longer than the backlog, and it's called the "invisible" backlog.

Although self-sourcing can solve many current problems, managers are concerned about end users' lack of adequate systems analysis and design background and loosening of system development standards. Other disadvantages of self-sourcing include the following:

- Possible misuse of computing resources
- Lack of access to crucial data
- Lack of documentation for the applications and systems that end users develop
- Inadequate security for the applications and systems that end users develop
- Applications developed by end users not up to information systems standards
- Lack of support from top management
- Lack of training for prospective users

A **request for information (RFI)** is a screening document for gathering vendor information and narrowing the list of potential vendors. It can help manage the selection of vendors by focusing on the project requirements that are crucial to selecting vendors.

Insourcing happens when an organization's team develops the system internally.

Self-sourcing is when end users develop their own information systems with little or no formal assistance from the information systems team. These users might not know how to write programming code but are typically skilled enough to use off-the-shelf software, such as spreadsheet and database packages, to produce custom-built applications.

Self-sourcing gives end users the power to build their own applications in a short time and create, access, and modify data. This power can be destructive, however, if the organization doesn't apply control and security measures. For example, end users' access to computing resources must be controlled to prevent interfering with the efficiency of the organization's information-processing functions.

To prevent the proliferation of information systems and applications that aren't based on adequate systems development principles, organizations should develop guidelines for end users and establish criteria for evaluating, approving or rejecting, and prioritizing projects. Criteria could include asking questions such as "Can any existing application generate the proposed report?" or "Can the requirements of multiple users be met by developing a single application?"

Classifying and cataloging existing applications are necessary to prevent end users from developing applications that basically handle the same functions as an existing application; this redundancy can be costly. In addition, data administration should be enforced to ensure the integrity and reliability of information. Creating private data should be minimized, if not eliminated. Sometimes, for the sake of efficient data processing, redundant data can exist; however, it should be monitored closely. This task is becoming more difficult, however, because the number of end users using diverse data is growing. The best approach to control the proliferation of invalid and inconsistent data in corporate databases is controlling the flow of data, such as with rigorous data entry procedures the database administrator establishes.

5.2.2 Outsourcing

With the **outsourcing** approach, an organization hires an external vendor or consultant who specializes in providing development services. This approach can save the cost of hiring additional staff and meet the demands for more timely development of information systems projects. Companies offering outsourcing services include IBM Global Services, Accenture, Infosys Technologies, and Computer Sciences Corporation.

With the development of Web 2.0, another form of outsourcing has become popular: **crowdsourcing**. This refers to the process of outsourcing tasks that are traditionally performed by employees or contractors to a large group of people (a crowd) through an open call. Let's say your town's City Hall is developing a Web site in order to be able to better serve the community. Using, it would invite everybody to participate in the design process. Crowdsourcing has become popular with publishers, journalists, editors, and businesses that want to take advantage of the collaborative capabilities offered by Web 2.0. InnoCentive (*www2.innocentive.com/*) is a company that is very active in crowdsourcing. It works with organizations to solve their problems, taking advantage of the power of diverse thinking inside and outside the organization.

An outsourcing company can employ the SDLC approach to develop the requested system by using the following options:

- *Onshore outsourcing*—The organization chooses an outsourcing company in the same country.

- *Nearshore outsourcing*—The organization chooses an outsourcing company in a neighboring country, such as Canada or Mexico for an organization in the United States.

- *Offshore outsourcing*—The organization chooses an outsourcing company in any part of the world (usually in a country farther away than a neighboring country), as long as it can provide the needed services.

Although outsourcing has the advantages of being less expensive, delivering information systems more quickly, and giving an organization the flexibility to concentrate on its core functions as well as other projects, it does have some disadvantages. They include the following:

- *Loss of control*—Relying on the outsourcing company to control information system functions could result in the system not fully meeting the organization's information requirements.

- *Dependency*—If the organization becomes too dependent on the outsourcing company, changes in the outsourcing company's financial status or managerial structure could have a major impact on the organization's information system.

- *Vulnerability of strategic information*—Because third parties are involved in outsourcing, the risk of leaking confidential information to competitors increases.

With the **outsourcing** approach, an organization hires an external vendor or consultant who specializes in providing development services.

Crowdsourcing is the process of outsourcing tasks that are traditionally performed by employees or contractors to a large group of people (a crowd) through an open call.

6 Phase 5: Maintenance

In the **maintenance phase**, the information system is operating, enhancements and modifications to the system have been developed and tested, and hardware and software components have been added or replaced. The maintenance team assesses how the system is working and takes steps to keep the system up and running. As part of this phase, the team collects performance data and gathers information on whether the system is meeting its objectives by talking with users, customers, and other people affected by the new system. If the system's objectives aren't being met, the team must take corrective action. Creating a help desk to support users is another important task in this phase. With the ongoing nature of the SDLC approach, maintenance can lead to starting the cycle over at the planning phase if the team discovers the system isn't working correctly.

7 New Trends in Systems Analysis and Design

The SDLC model might not be appropriate in the following situations:

- Lack of specifications—that is, the problem under investigation isn't well defined.
- The input–output process can't be identified completely.
- The problem is "ad hoc," meaning it's a one-time problem that's not likely to reoccur.
- Users' needs change constantly, meaning the system has to undergo several changes until it satisfies their needs. The SDLC model might work in the short term, but in the long term, it's not suitable in this situation.

For these situations, other approaches, described in the following sections, are more suitable.

In the **maintenance phase**, the information system is operating, enhancements and modifications to the system have been developed and tested, and hardware and software components have been added or replaced.

Service-oriented architecture (SOA) is a philosophy and a software and system development methodology that focuses on the development, use, and reuse of small, self-contained blocks of codes (called services) to meet the software needs of an organization.

7.1 Service-Oriented Architecture

Service-oriented architecture (SOA) is a philosophy and a software and system development methodology that focuses on the development, use, and reuse of small, self-contained blocks of codes (called services) to meet the software needs of an organization. SOA attempts to solve software development issues by recognizing, accepting, and leveraging the existing services. Checking shipping status, customer credit, or the inventory status are a few examples of such services. More specifically, a service could be a database table, a set of related database tables, one or more data files in any format, or data obtained from another service.

The fundamental principle behind SOA is that the "blocks of codes" can be reused in a variety of different applications, allowing new business processes to be created from a pool of existing services. These services should be organized so that they can be accessed when needed via a network. SOA offers many potential benefits to organizations, including reduced application development time, greater flexibility, and an improved return on investment.

In any business organization, there are things that do not change very often, such as an order processing system. These often represent a major part of a business and are therefore called core business functions. At the same time, there are functions and activities that change on a regular basis, such as taxes and contents of a marketing campaign. SOA advocates that core business functions and the dynamic functions that change all the time should be decoupled. SOA allows an organization to pick and choose those services that respond most effectively to the customer's needs and market demands. Services or "blocks of codes" can be replaced, changed, or even combined.

Many organizations use SOA as a philosophy and methodology. For example, Starwood Hotels and Resorts Worldwide is replacing its legacy room-reservation system with an SOA-based system. By using SOA, they can offer as many as 150 service-based applications built on Web standards. T-Mobile is also employing SOA, both for internal integration and reuse and for external, revenue-generating services. This enables T-Mobile to work effectively with third-party content providers, such as Time Warner and the Bertelsmann Group, to deliver services to customers.[7]

7.2 Rapid Application Development

Rapid application development (RAD) concentrates on user involvement and continuous interaction between users and designers. It combines the planning and analysis phases into one phase and develops a prototype of the system. RAD uses an iterative process (also called "incremental development") that repeats the design, development, and testing steps as needed, based on feedback from users. After the initial prototype, the software library is reviewed, reusable components are selected from the library and integrated with the prototype, and testing is conducted. After these steps, the remaining phases are similar to the SDLC approach. One shortcoming of RAD is a narrow focus, which might limit future development. In addition, because these applications are built quickly, the quality might be lower.

© iofoto/Shutterstock.com

Rapid application development (RAD) concentrates on user involvement and continuous interaction between users and designers. It combines the planning and analysis phases into one phase and develops a prototype of the system. It uses an iterative process (also called "incremental development") that repeats the design, development, and testing steps as needed, based on feedback from users.

Extreme programming (XP) divides a project into smaller functions, and developers can't go on to the next phase until the current phase is finished. Each function of the overall project is developed in a step-by-step fashion.

Pair programming is where two programmers participate in one development effort at one workstation. Each programmer performs the action the other is not currently doing.

7.3 Extreme Programming

Extreme programming (XP) is a recent method for developing software applications and information system projects. Kent Beck, during his work on the Chrysler Comprehensive Compensation System, created this method as a way to establish specific goals and meet them in a timely manner. XP divides a project into smaller functions, and developers can't go on to the next phase until the current phase is finished. Analysts write down features the proposed system should have—called the "story"—on index cards. The cards include the time and effort needed to develop these features, and then the organization decides which features should be implemented and in what order, based on current needs.[8] Each function of the overall project is developed in a step-by-step fashion. At the beginning, it's similar to a jigsaw puzzle; individually, the pieces make no sense, but when they're combined, a complete picture can be seen. The XP method delivers the system to users as early as possible and then makes changes that the user suggests. In the XP environment, programmers usually work on the same code in teams of two (often referred to as "sharing a keyboard"). This practice is also called **pair programming**, where two programmers participate in one development effort at one workstation. Each programmer performs the action the other is not currently doing. In this way, they can detect and correct programming mistakes as they go, which is faster than correcting them after the entire program has been written. Another advantage is more communication between programmers during code development.

This approach is a major departure from traditional software development, such as the SDLC model, which looks at the project as a whole. XP uses incremental steps to improve a system's quality by addressing major issues that haven't been examined before. SDLC develops the entire system at once; XP develops the system in incremental phases. Like RAD, XP uses a software library for reusable pieces that can be integrated into the new system. IBM, Chrysler, and Microsoft, among others, have used this method successfully. Its key features are as follows:

- Customer and user satisfaction
- Simplicity
- Incremental process
- Responsiveness to changing requirements and changing technology
- Teamwork
- Continuous communication among key players
- Immediate feedback from users

The Extreme Programming information box highlights the use of XP at Sabre Holdings Corp. and other companies.

7.4 Agile Methodology

Agile methodology is similar to XP in focusing on an incremental development process and timely delivery of working software. However, there's less emphasis on team coding and more emphasis on limiting the project's scope. Agile methodology focuses on setting a minimum number of requirements and turning them into a working product. The Agile Alliance organization (*www.agilealliance.org*) has developed guidelines for this method, which emphasizes collaboration between programmers and business experts, preferably with face-to-face communication, and working in teams. Goals of this step-by-step approach include responding to changing needs instead of sticking to a set plan and developing working, high-quality software. The agile methodology also strives to deliver software quickly to better meet customers' needs.

Agile Alliance has written a manifesto that includes the following principles:[9]

- Satisfy the customer through early and continuous delivery of valuable software.
- Welcome changing requirements, even late in development.
- Have business people and developers work together daily throughout the project.
- Build projects around motivated individuals. Give them the environment and support they need, and trust them to get the job done.
- Always attend to technical excellence. Good design enhances agility.
- At regular intervals, the team should reflect on how to become more effective, then tune and adjust its behavior accordingly.

The information box below covers agile methodology used in Overstock.Com.

The Industry Connection highlights CA Technologies, which offers several systems development tools.

> **Agile methodology** is similar to XP in focusing on an incremental development process and timely delivery of working software. However, there's less emphasis on team coding and more emphasis on limiting the project's scope.

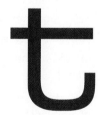

8 Chapter Summary

t his chapter explained the systems development life cycle (SDLC), which consists of five phases: planning, requirements-gathering and analysis, design, implementation, and maintenance.

The chapter discussed CASE tools for automating some activities in the SDLC, the prototyping process, and using a request for proposal for screening vendors. You also learned about self-sourcing and outsourcing, two alternatives for developing an information system. Finally, you were given an overview of new trends in systems development, such as service-oriented architecture, rapid application development, extreme programming, and agile methodology.

Industry Connection

CA Technologies*

CA Technologies (originally Computer Associates International, Inc.) offers several products and services for use in enterprise IT management, which includes managing information systems, networks, security platforms, storage, applications, and databases. The company offers a variety of hardware, software, and services for business as well as home use in the following categories: application development and databases; application performance management; database management; infrastructure and operations management; mainframe applications; project, portfolio, and financial management; security management; service management; and storage and recovery management. It is also involved in the cloud computing environment.

One product from CA Technologies is the ERwin Process Modeler, a CASE tool used for a variety of systems analysis and design activities. It's a modeling tool that can help the systems analyst visualize complex systems with many inputs, processes, and outputs and create workflow and data flow modeling, such as DFDs. ERwin can also be used to create databases and design, share, and reuse physical and logical models.

*This information has been gathered from the company Web site (*www.ca.com*) and other promotional materials. For more information and updates, visit the Web site.

Key Terms

agile methodology (191)

computer-aided systems engineering (CASE) (183)

crowdsourcing (188)

design phase (183)

economic feasibility (179)

external users (178)

extreme programming (XP) (190)

feasibility study (179)

implementation phase (185)

insourcing (187)

internal users (178)

joint application design (JAD) (178)

legal feasibility (181)

maintenance phase (189)

operational feasibility (180)

outsourcing (188)

pair programming (190)

parallel conversion (186)

phased-in-phased-out conversion (186)

pilot conversion (186)

planning phase (176)

plunge (direct cutover) conversion (186)

proof-of-concept prototype (184)

prototyping (184)

rapid application development (RAD) (190)

request for information (RFI) (187)

request for proposal (RFP) (186)

requirements-gathering and analysis phase (181)

scheduling feasibility (180)

self-sourcing (187)

selling prototype (184)

service-oriented architecture (SOA) (189)

systems development life cycle (SDLC) (175)

technical feasibility (180)

Problems, Activities, and Discussions

1. What are some of the activities that must be performed during the planning phase of SDLC? What are some of the questions that must be answered?

2. What are the five major dimensions of a feasibility study?

3. What are the two major alternatives to SDLC for the construction of an information system?

4. Watch the two videos posted on the following Web sites, consult other sources, and write a one-page paper on SOA. How might this methodology save organizations money when they're developing information systems?

   www.youtube.com/watch?v=sbd_1G8Kqjs

   www.zdnet.com/videos/whiteboard/what-is-soa/155698

5. What is crowdsourcing? Why and when is it used? Which companies are involved in this approach?

6. Visit the following Web page, consult other sources, and write a one-page paper that discusses why media outsourcing can be risky.

   www.useit.com/alertbox/social-mega-ia.html

7. Visit the following Web sites, consult other sources, and write a one-page paper that discusses the advantages and disadvantages of outsourcing.

   www.socialtext.net/ism4300/index.cgi?the_future_of_outsourcing_charles_elliott management. about.com/cs/people/a/offshoring104.htm

8. Tele-Talk, a multinational textile company is developing an online customer service for better serving its customers. List three tangible and three intangible benefits of this system. What are some of the costs associated with this information system development?

9. Which of the following activities isn't done during the implementation phase?

   a. Acquiring new equipment

   b. Creating a help desk

   c. Training employees

   d. Designing the system's physical layout

10. Lack of technical feasibility is sometimes referred to as "a lack of organizational readiness." True or False?

casestudy

SYSTEMS DEVELOPMENT AT SEB LATVIA

Skandinaviska Enskilda Banken (SEB), one of northern Europe's largest banking groups, operates in several countries, including Germany, Poland, and Russia. In the past, many of SEB's decisions were made as a result of an approval process that involved circulating paper documents. In addition, important data was stored on decentralized systems that were accessible to only certain decision makers. Delays and bottlenecks resulted, with a consequent effect on the speed and effectiveness of customer service. To solve these problems, SEB Latvia chose the IBM Lotus Domino messaging and collaboration platform. After the system was installed, Exigen (an IBM business partner) and the SEB Latvia's IT team started working with

© 616975807/6/Shutterstock.com

users to gather requirements and automate processes to increase efficiency. The Lotus system now provides centralized information access via a Web interface that all decision makers can use.[12]

Answer the following questions:

1. What processes did the Lotus platform automate and streamline?

2. What are some benefits of the new Web interface?

3. Which of the development methods discussed in this chapter would be most suitable at SEB Latvia?

ENTERPRISE SYSTEMS

a n **enterprise system** is an application used in all the functions of a business that supports decision making throughout the organization. For example, an enterprise resource planning system is used to coordinate operations, resources, and decision making among manufacturing, production, marketing, and human resources. As you've learned in previous chapters, intranets and Web portals are used by many organizations to improve communication among departments as well as increase overall efficiency, and enterprise systems are another way to make important information readily available to decision makers throughout an organization.

In this chapter, you learn about supply chain management (SCM), customer relationship management (CRM), knowledge management systems, and enterprise resource planning (ERP) enterprise systems. With each type of enterprise system, you review the system's goals, the information technologies used for it, and any applicable issues.

An **enterprise system** is an application used in all the functions of a business that supports decision making throughout the organization.

learning outcomes

After studying this chapter, you should be able to:

LO1 Explain how supply chain management is used.

LO2 Describe customer relationship management systems.

LO3 Explain knowledge management systems.

LO4 Describe enterprise resource planning systems.

1 Supply Chain Management

a **supply chain** is an integrated network consisting of an organization, its suppliers, transportation companies, and brokers used to deliver goods and services to customers. As Exhibit 11.1 shows, in a manufacturing firm's supply chain, raw materials flow from suppliers to manufacturers (where they're transformed into finished goods), to distributors, and finally to consumers. Supply chains exist in both service and manufacturing organizations, although the chain's complexity can vary widely in different organizations and industries. In manufacturing, the major links in the supply chain are suppliers, manufacturing facilities, distribution centers, and customers. In service organizations, such as real estate, the travel industry, temporary labor, and advertising, these links include suppliers (service providers), distribution centers (such as travel agencies), and customers.

A **supply chain** is an integrated network consisting of an organization, its suppliers, transportation companies, and brokers used to deliver goods and services to customers.

SIX

© Jetta Productions/Getty images

Exhibit 11.1 *A supply chain configuration*

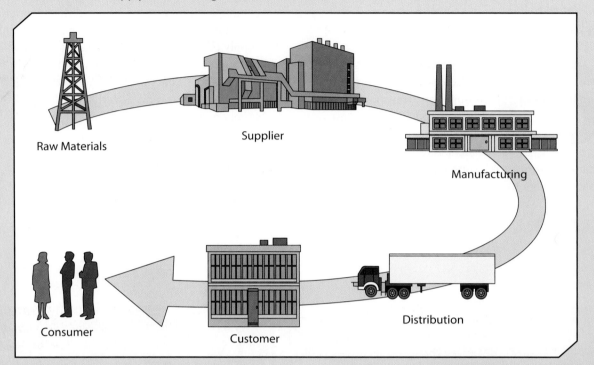

Raw Materials

Supplier

Manufacturing

Distribution

Customer

Consumer

Supply chain management (SCM) is the process of working with suppliers and other partners in the supply chain to improve procedures for delivering products and services. An SCM system coordinates the following functions:

- Procuring materials (in service organizations, this can include resources and information)
- Transforming materials into intermediate and finished products or services
- Distributing finished products or services to customers

In a manufacturing firm's SCM system, communication takes place among the following areas:

- *Product flow*—Managing the movement of goods all the way from suppliers to customers; customer service and support are included in this flow.
- *Information flow*—Overseeing order transmissions and delivery status updates throughout the order-processing cycle

Supply chain management (SCM) is the process of working with suppliers and other partners in the supply chain to improve procedures for delivering products and services.

© Rade Kovac/Shutterstock.com

- *Finances flow*—Handling credit terms, payment schedules, and consignment and title ownership arrangements

An SCM system must manage communication in all these areas as part of overseeing the manufacturing process and managing inventory and delivery. Four key decisions in supply chain management related to manufacturing are the following:

- *Location*—Where should manufacturing facilities be placed?
- *Inventory*—When should an order be placed? How much should be ordered?
- *Production*—What should be produced? How much should be produced?
- *Transportation*—Which transportation systems will reduce costs and expedite the delivery process?

For organizations that don't have in-house resources to develop an SCM system, several vendors offer comprehensive solutions for this purpose, such as SAP (*www.sap.com/solutions/business-suite/scm/index.epx*), Oracle (*www.oracle.com/us/solutions/scm/index.htm*), JDA Software (*www.jda.com*), Ariba (*www.ariba.com*), and Manhattan Associates (*www.manh.com*). In addition, hosting services are now available for SCM systems; this trend is called "software as a service" (SaaS, discussed in Chapter 14). The information box below highlights the supply chain at Dell Computer.

1.1 SCM Technologies

Information technologies and the Internet play a major role in implementing an SCM system. These tools are explained in the following sections.

1.1.1 Electronic Data Interchange

Electronic data interchange (EDI) enables business partners to send and receive information on business transactions.

Many companies substitute EDI for printing, mailing, and faxing paper documents, such as purchase orders, invoices, and shipping notices. By using the Internet and established Web protocols for electronic exchange of information, companies can improve the efficiency and effectiveness of the supply chain process. EDI expedites the delivery of accurate information in the following processes, among others:

- Transaction acknowledgments
- Financial reporting
- Invoice and payment processing
- Order status
- Purchasing
- Shipping and receiving
- Inventory management and sales forecasting

In addition, using the Internet and Web protocols for EDI lowers the cost of transmitting documents. This method is called "Web-based EDI" or "open EDI." It also has the advantage of being platform independent and easy to use, but transmitting across the Internet does involve more security risks than traditional EDI, which uses proprietary protocols and networks.

However, using EDI has some drawbacks. For instance, EDI uses proprietary standards. An EDI provider sets up an EDI network (as a VPN), and organizations enroll in the network. EDI is more beneficial when there are more companies in the EDI network, because when the number of partners is small, the cost per partner is higher. For this reason, large companies tend to insist on their suppliers and distributors becoming part of the same EDI network, which many small suppliers and distributors can't afford. With the advent of XML, organizations can use the Internet and Open EDI to perform the same function that EDI performs, so traditional EDI has declined in popularity.

Dell Computer's Supply Chain

Dell Computer has modified its supply chain from a "push" to a "pull" manufacturing process to give the company a competitive advantage. (Push and pull technologies are discussed in Chapter 14.) This strategy is also known as "built to order" (BTO). The main sales channel is direct sales to customers, eliminating extra steps between the manufacturer and the customer. Customers place orders by phone or at Dell's Web site. Dell assembles computers according to customers' specifications and then ships them directly to customers. With this model, Dell has been able to reduce costs by eliminating intermediaries and shorten the delivery time, which has resulted in increased customer satisfaction. Also, the company needs to keep only the minimum amount of inventory on hand, which has reduced the total inventory cost.

> E-distributors offer fast delivery of a wide selection of products and services, usually at lower prices, and they help companies reduce the time and expense of searching for goods.

1.1.2 Internet-Enabled SCM

Internet-enabled SCM improves information sharing throughout the supply chain, which helps reduce costs for information transmission and improves customer service. For instance, many companies use point-of-sale (POS) systems that scan what's being sold and collect this data in real time. This information helps organizations decide what to reorder to replenish stock, and it is sent via the Internet to suppliers so they can synchronize production with actual sales. Internet-enabled SCM can improve the following SCM activities:

- *Purchasing/procurement*—Purchasing and paying for goods and services online, bargaining and renegotiating prices and term agreements, using global procurement strategies

- *Inventory management*—Providing real-time stock information, replenishing stock quickly and efficiently, tracking out-of-stock items

- *Transportation*—Customers using the Internet for shipping and delivery information

- *Order processing*—Checking order placement and order status, improving the speed and quality of order processing, handling returned goods and out-of-stock notifications to customers

- *Customer service*—Responding to customers' complaints, issuing notifications (such as product recalls), providing around-the-clock customer service

- *Production scheduling*—Coordinating just-in-time (JIT) inventory programs with vendors and suppliers, coordinating production schedules between companies and their vendors and suppliers, conducting customer demand analysis

1.1.3 E-marketplaces

An **e-marketplace** is a third-party exchange (a B2B business model) that provides a platform for buyers and sellers to interact with each other and trade more efficiently online. E-marketplaces help businesses maintain

© Indeed/Getty Images

a competitive edge in the supply chain in the following ways:

- Providing opportunities for sellers and buyers to establish new trading partnerships

- Providing a single platform for prices, availability, and stock levels that's accessible to all participants

- Solving time-constraint problems for international trade, and making it possible to conduct business around the clock

- Making it easy to compare prices and products from a single source instead of spending time contacting each seller

- Reducing marketing costs more than traditional sales channels can

E-distributors are common examples of e-marketplaces. An e-distributor is a marketplace owned and operated by a third party that provides an electronic catalog of products. For example, an e-distributor might offer a catalog containing a variety of hardware and

An **e-marketplace** is a third-party exchange (B2B model) that provides a platform for buyers and sellers to interact with each other and trade more efficiently online.

software products so that a network administrator can order all the equipment and applications needed for an organization's network instead of purchasing components from several different vendors. Another common offering from e-distributors is maintenance, repair, and operations (MROs) services; a company can purchase an MRO package that might include services from different vendors, but the e-distributor coordinates them into one package for customers. This packaging is an example of a horizontal market, which concentrates on coordinating a business process or function involving multiple vendors. E-distributors offer fast delivery of a wide selection of products and services, usually at lower prices, and they help companies reduce the time and expense of searching for goods.

As you learned in Chapter 8, third-party exchanges bring together buyers and sellers in vertical as well as horizontal markets. Buyers can gather information on products and sellers, and sellers have access to more potential buyers. Examples of third-party exchanges include PowerSource Online, FoodTrader.com, and Farms.com.

© Jason Reed/Ryan McVay/Getty Images

1.1.4 Online Auctions

Auctions help determine the price of goods and services when there is no set price in the marketplace. An **online auction** is a straightforward yet revolutionary business concept. By using the Internet, it brings traditional auctions to customers around the globe and makes it possible to sell far more goods and services than at a traditional auction. It's based on the brokerage business model discussed in Chapter 8, which brings buyers and sellers together in a virtual marketplace. Typically, the organization hosting the auction collects transaction fees for the service. Online auctions are particularly cost-effective for selling excessive inventory. Some companies use **reverse auctions**, which invite sellers to submit bids for products and services. In other words, there's one buyer and many sellers: a one-to-many relationship. The buyer can choose the seller that offers the service or product at the lowest price.

1.1.5 Collaborative Planning, Forecasting, and Replenishment

Collaborative planning, forecasting, and replenishment (CPFR) is used to coordinate supply chain members through point-of-sale (POS) data sharing and joint planning (see Exhibit 11.2). In other words, any data collected with POS systems is shared with all members of the supply chain, which is useful in coordinating production and planning for inventory needs. The goal is to improve operational efficiency and manage inventory. With a structured process of sharing information among supply chain members, retailers can compare customer demands or sales forecasts with a manufacturer's order forecast, for example. If there's a discrepancy

By using the Internet, an **online auction** brings traditional auctions to customers around the globe and makes it possible to sell far more goods and services than at a traditional auction.

A **reverse auction** invites sellers to submit bids for products and services. In other words, there's one buyer and many sellers: a one-to-many relationship. The buyer can choose the seller that offers the service or product at the lowest price.

Collaborative planning, forecasting, and replenishment (CPFR) is used to coordinate supply chain members through point-of-sale (POS) data sharing and joint planning.

Many companies substitute EDI for printing, mailing, and faxing paper documents, such as purchase orders, invoices, and shipping notices.

between forecasts, members can get together and decide on the correct quantity to order. One main obstacle to improving supply chain performance is companies not knowing enough about what customers want, which can lead to lost sales and unsold inventory for retailers and manufacturers. CPFR has the advantage of decreasing merchandising, inventory, and logistics costs for all supply chain members.

Coordinating the supply chain can be difficult. To understand these problems, let's return to the example of ABC Furniture Company, which was used in Chapter 10. Recall that ABC Furniture buys wood from New England Wood and hardware (nuts, bolts, and so forth) from Vermont Hardware. ABC Furniture also uses a distributor, Furniture Distribution Company (FDC), to send the final products to retailers. Therefore, ABC Furniture's supply chain includes New England Wood and Vermont Hardware on what's called the "upstream"

side of the supply chain, and FDC, retailers, and customers on the "downstream" side of the supply chain.

The only partner in this supply chain that knows exactly how many pieces of furniture have been sold and how many are available in inventory is the retailer, and without agreements in place, most retailers don't share this information. The distributor, FDC, knows only how many pieces retailers order to replenish their stock, which doesn't indicate exact numbers sold. So, FDC places an order with ABC Furniture based on its forecasts of orders from retailers. ABC Furniture doesn't know exactly how much inventory FDC has on hand, only the quantity of orders FDC has placed. So, ABC Furniture orders supplies from New England Wood and Vermont Hardware based on the distributor's forecasts.

If a retailer gets more products than it can sell or stock, it returns the products to FDC, which absorbs part of the cost. FDC, in turn, sends some products

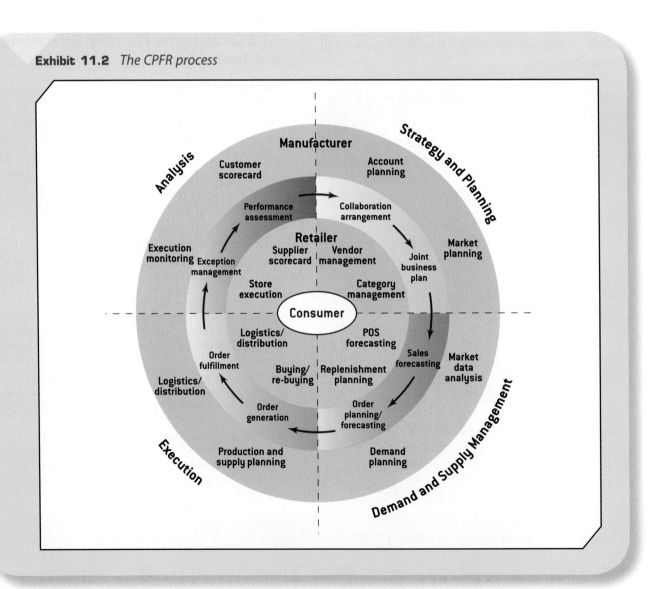

Exhibit 11.2 *The CPFR process*

back to ABC Furniture, which absorbs this cost. ABC Furniture now has too many products in its inventory and must slow down manufacturing. However, it can't return lumber to New England Wood, so it ends up paying for carrying extra raw materials in inventory. As a result, ABC Furniture might suffer from having the highest costs in the supply chain, and retailers suffer the least.

CPFR ensures that inventory and sales data is shared across the supply chain so that everyone knows the exact sales and inventory levels. The collaboration part of this process is the agreement between all supply chain partners that establishes how data is shared, how problems with overstock are solved, and how to ensure that costs for each partner are shared or minimized. The agreement also encourages retailers to share important data with the distributor and manufacturer, often by offering them better discounts. Retailers are also motivated to sell more to give themselves more leverage with ABC Furniture, which works for ABC Furniture because the more products it sells, the better its bottom line.

Even with an agreement in place, unforeseen problems can crop up, so planning for these "exceptions" is important. Handling unforeseen problems is called "exception management," and lessons learned during this process can be used in future planning.

2 Customer Relationship Management

Customer relationship management (CRM) consists of the processes a company uses to track and organize its contacts with customers. The main goal of a CRM system is to improve services offered to customers and use customer contact information for targeted marketing. Businesses know that keeping and maintaining current customers is less expensive than attracting new customers, and an effective CRM system is useful in meeting this goal.

Marketing strategies in a CRM system focus on long-term relationships with customers instead of transactions. These strategies include identifying customer segments, improving products and services to meet customers' needs, improving customer retention, and identifying a company's most profitable (and loyal) customers. To get the most out of these strategies, a CRM system helps organizations make better use of data, information, and knowledge to understand their customers.[1] A CRM system captures information about customer interactions for sales personnel and customer service representatives so they can perform their jobs more effectively and efficiently. This information can include customers' preferences, background, income, gender, and education.

CRM is more than just tracking and organizing contacts with customers. It gives organizations more complete pictures of their customers. CRM systems include tools for conducting complex analyses on customer data, such as a data warehouse and data-mining tools, discussed in Chapter 3. With these systems, organizations can integrate demographic and other external data with customers' transaction data to better understand customer behavior. Based on this analysis, organizations can better target products to customers and manage customer issues, which increases customer satisfaction and retention. In addition, organizations can classify customers based on how valuable they are to the organization and manage them accordingly.

Grocery stores offering loyalty cards with discounts to customers are an example of how transaction data can be used in a CRM system. Knowing that a customer bought 4 gallons of milk the previous week, for example, doesn't give a grocery store much information, but with loyalty cards, a grocery store can track all sorts of information on specific customers. For example, when customers apply for loyalty cards, they must give demographic information, such as name, age, marital status, and address. So, instead of knowing that "Customer 49 bought 4 gallons of milk last week," a store

Customer relationship management (CRM) consists of the processes a company uses to track and organize its contacts with customers. It improves services offered to customers and uses customer contact information for targeted marketing.

can learn that "James Smith, 35 years old, married and residing in zip code 11223, bought four gallons of milk last week." With this information, the store can assume James Smith has young children (or clearly isn't lactose intolerant!). In addition, if James Smith purchases no cereal that same week, the store can assume he's buying cereal from another store (because with the purchase of that amount of milk and the assumption that he has young children, it's likely his children are eating cereal). Therefore, the store decides to send coupons for discounts on cereal to James Smith. This is referred to as "cross-selling"—getting the customer to buy additional products. The store might also send James Smith coupons for a more expensive brand of milk, in the hope that his family will decide it prefers that brand. This practice is called "upselling."

Organizations can also pay external agencies for additional data about their potential customers. This data might be public or semiprivate, such as whether they own their homes, the value of their homes, and their estimated mortgage or rent payments. This gives organizations more information to analyze.

With a CRM system, an organization can do the following:[2]

- Provide services and products that meet customers' needs.
- Offer better customer service through multiple channels (traditional as well as the Internet).
- Increase cross-selling and upselling of products to increase revenue from existing customers.
- Help sales personnel close deals faster by offering data on customers' backgrounds.
- Retain existing customers and attract new ones.

Several IT tools discussed throughout this book are used to improve customer service. For example, e-mail, the Internet, Web portals, and automated call centers have played a major role in CRM systems. E-commerce sites use e-mail to confirm items purchased, confirm shipping arrangements, and send notifications on new products and services. Web portals and extranets, such as FedEx.com, allow customers to perform tasks, such as checking the status of shipments and arranging a package pickup. Database systems, data warehouses, and data-mining tools are effective in tracking and analyzing customers' buying patterns, which helps businesses meet customers' needs. A CRM system includes the following activities:

- Sales automation
- Order processing
- Marketing automation
- Customer support
- Knowledge management
- Personalization technology

These activities, performed by CRM software, are discussed in more detail in the following sections. The information box below highlights a real-life application of a CRM system.

CRM in Action

Time Warner Cable Business Class adopted a CRM system from Salesforce.com to analyze business data, improve the accuracy of forecasts, improve problem solving, and monitor sales and business activities. Some important features of the system include dashboards that display analyses of key business variables, features for "drilling down" through information, a Web-based knowledge base that employees and customers can use, and a Web log for sales personnel communication. Overall, the system has increased productivity by 10 percent and decreased the time needed to perform business processes, such as reducing the time for survey completion from seven days to two days.[3]

Customization, which is somewhat different from personalization, allows customers to modify the standard offering, such as selecting a different home page to be displayed each time you open your Web browser.

© Wavebreakmedia Ltd/Shutterstock.com

2.1 CRM Applications

Typically, CRM applications are implemented with one of two approaches: on-premise CRM or Web-based CRM. Organizations with an established IT infrastructure often choose an on-premise CRM, which is implemented much like any other IT system. With Web-based CRM, the company accesses the application via a Web interface instead of running the application on its own computers and pays to use CRM software as a service (SaaS), which is similar to Web-hosting services. The SaaS vendor also handles technical issues. (SaaS is covered in more detail in Chapter 14.) Several software packages are available for setting up a CRM system, including Amdocs CRM (*www.amdocs.com/ os/home/home.htm*), Optima Technologies ExSellence (*www.optima-tech.com*), Infor CRM (*www.infor. com/solutions/crm/*), SAP mySAP, (*www.sap.com/solutions/business-suite/crm/index.epx*), Oracle PeopleSoft CRM (*www.oracle.com/applications/peoplesoft/crm/ ent/index.html*), and Oracle Siebel (*www.oracle.com/ us/products/applications/siebel/index.htm*). Although these packages vary in capabilities, they share the following features:

- *Salesforce automation*—Assists with such tasks as controlling inventory, processing orders, tracking

customer interactions, and analyzing sales forecasts and performance. It also assists with collecting, storing, and managing sales contacts and leads.

- *eCRM or Web-based CRM*—Allows Web-based customer interaction and is used to automate e-mail, call logs, Web site analytics, and campaign management. Companies use campaign management to customize marketing campaigns, such as designing a marketing campaign tailored to customers in southern California or customers in the 18–35 age bracket.

- *Survey management*—Automates electronic surveys, polls, and questionnaires, which is useful for gathering information on customers' preferences.

- *Automated customer service*—Used to manage call centers and help desks and can sometimes answer customers' queries automatically.

2.2 Personalization Technology

Personalization is the process of satisfying customers' needs, building customer relationships, and increasing profits by designing goods and services that meet customers' preferences better. It involves not only customers' requests, but also the interaction between customers and the company. You're probably familiar with Web sites that tailor content based on your interests and preferences. Amazon.com, for example, suggests products you might enjoy, based on your past browsing and purchasing habits.

Customization, which is somewhat different from personalization, allows customers to modify the standard offering, such as selecting a different home page to be displayed each time you open your Web browser.

Personalization is the process of satisfying customers' needs, building customer relationships, and increasing profits by designing goods and services that meet customers' preferences better. It involves not only customers' requests, but also the interaction between customers and the company.

Customization allows customers to modify the standard offering, such as selecting a different home page to be displayed each time you open your Web browser.

As another example, after registering with Yahoo!, you can customize the start page by choosing your preferred layout, content, and colors. You can find many examples of customization in retail, too, such as Build-A-Bear Workshops, where children can design their own teddy bears, or Nike, which allows customers to create their own shoes by selecting styles and colors.[4]

Because personalization and customization help companies meet customers' preferences and needs, customers often experience a more efficient shopping process and, as a result, are less likely to switch to competitors to get similar products or services. However, using personalization requires gathering a lot of information about customers' preferences and shopping patterns, and some customers get impatient with answering long surveys about their preferences. In addition, collecting this information might affect customers' sense of privacy. For example, drugstore customers might be concerned that the drugstore has their prescription histories, that the information might be misused and even affect their insurance coverage. To ease these concerns, companies should include clear privacy policies on their Web sites stating how personal information is collected and used.

Amazon is known for using personalization to recommend products to customers. You're probably familiar with the message "Customers who bought this item also bought" followed by a list of suggestions. Amazon's recommendation system is made up of a huge database containing customers' previous purchases and a recommendation algorithm. When a customer logs on to Amazon.com, the recommendation system first checks the customer's purchase history and that of other similar customers. Using this information, a list of recommended products is displayed, based on the customer's shopping history and choices by other customers who have similar purchase histories. In addition, Amazon gives customers an opportunity to rate the recommendations. The more items the customer purchases and the more recommendations the customer rates, the better recommendations are tailored to the customer.[5]

Many other companies use personalization technology to improve customer service. For example, if you buy a suit from Nordstrom.com, the site might suggest shoes or a tie that goes with the suit or a similar suit in the same category. If you buy a song from Apple iTunes, other songs that listeners like you purchased are suggested.

Google also provides personalized services for Google account holders. Users can get personalized search results that are reordered based on their searching histories. For example, Avni Shah, Google product manager, explains that if a user has "fly fishing" in his or her search history and then searches on "bass," more weight is given to Web pages about fish than pages about musical instruments in the search results. Google also has a bookmark feature so that users can save useful search results for later use. Unlike Yahoo!'s MyWeb feature, which saves the text of Web pages, this feature simply saves the link to the page.[6]

To implement a personalization system, several IT tools are needed, including the Internet, databases, data warehouse/data marts, data-mining tools, mobile networks, and collaborative filtering. **Collaborative filtering (CF)** is a search for specific information or patterns using input from multiple business partners and data sources. It identifies groups of people based on common interests and recommends products or services based on what members of the group purchased or didn't purchase. It works well for a single product category, such as books, computers, and so forth. One drawback of CF is that it needs a large sample of users and content to work well. In addition, it isn't useful for making recommendations across unrelated categories, such as predicting that customers who liked a certain CD would also like a particular computer.[7]

One application of collaborative filtering is making automatic predictions about customers' preferences and interests based on similar users. For example, if a user rates several movies and is then added to a database that contains other users' ratings, a CF system can predict the user's ratings for movies he or she hasn't evaluated. You may have seen this feature used on Netflix.com, where lists of other movies you might like are displayed. Recently, Netflix paid $1 million to the team that won a contest to come up with the best algorithm for improving the accuracy of the Netflix recommendation system. Other Web sites that use CF systems to improve customer services are Amazon, Barnes and Noble, and Half.com.

3 Knowledge Management

Knowledge management draws on concepts of organizational learning, organizational culture, and best practices to convert tacit knowledge into explicit knowledge, create a knowledge-sharing culture in an organization, and eliminate obstacles to sharing knowledge.

Knowledge management (KM) is a technique used to improve CRM systems (and many other systems) by identifying, storing, and disseminating "know-how"—facts about how to perform tasks. Know-how can be explicit knowledge (formal, written procedures) or tacit knowledge (personal or informal knowledge). Knowledge is an asset that should be shared throughout an organization to generate business intelligence and maintain a competitive advantage in the marketplace. Knowledge management, therefore, draws on concepts of organizational learning, organizational culture, and best practices to convert tacit knowledge into explicit knowledge, create a knowledge-sharing culture in an organization, and eliminate obstacles to sharing knowledge. In this respect, knowledge management shares many of the goals of information management but is broader in scope because information management tends to focus on just explicit knowledge.

Knowledge is more than information and data. It's also contextual. Explicit knowledge, such as how to close a sale, can be captured in data repositories and shared. Expert salespeople can document how they close sales successfully, and this documentation can be used to train new salespeople or those who are struggling with closing sales. Tacit knowledge, however, can't be captured as easily. Knowledge someone has gained through experience might vary depending on the situation in which it was used—the context. Typically, the best way to gather this information is interactively, such as asking the employee specific questions about how he or she would handle an issue. Because interaction is a key part of managing tacit knowledge, a knowledge management system must encourage open communication and the exchange of ideas, typically via e-mails, instant messaging, internal company wikis, videoconferencing, and tools such as WebEx, that create virtual instructional environments.

By storing knowledge captured from experts, a knowledge repository can be created for employees to refer to when needed. The most common example is creating a knowledge base of typical customer complaints and solutions. Dell Computer uses this type of knowledge base, so when customers call about a problem with their computers, the steps for solving the problem are documented and readily accessible, which shortens response times.

Knowledge bases can also be used when new products are being designed. A company can store past experiences with similar designs, mistakes made in testing, and so forth to help speed up the delivery timetable and avoid making the same mistakes. This use of knowledge bases is particularly helpful in designing software products and services.

Employees might be reluctant to share their expertise because they think it will diminish their value in the organization. To motivate them to share knowledge, rewards must be offered. A knowledge management system can track how often an employee participates in knowledge-sharing interactions with other employees and track any resulting improvements in performance. This information can be used to reward employees for sharing tacit knowledge. Reward systems can be set up for sharing explicit knowledge, too, by tracking how often an employee contributes to a company's internal wiki, for example.

A simple knowledge management system might consist of using groupware (discussed in Chapter 12), such as IBM Lotus Notes or Microsoft Sharepoint Server, to create, manage, and distribute documents in an organization. These documents include the kind of information discussed previously, such as outlines of procedures for customer service representatives or reports of past design efforts. Other tools and technologies might include DBMSs, data-mining tools, and decision support systems (discussed in Chapter 12). Knowledge management plays a key role in the success of a CRM system because it helps businesses use their knowledge assets to improve customer service and productivity, reduce costs, and generate more revenue. A knowledge management system should help an organization do one or more of the following:[8]

- Promote innovation by encouraging the free exchange of ideas.
- Improve customer service by reducing response time.
- Increase revenue by reducing the delivery time for products and services.
- Improve employee retention rates by rewarding employees for their knowledge.

4 Enterprise Resource Planning

nterprise resource planning (ERP) is an integrated system that collects and processes data and manages and coordinates resources, information, and functions throughout an organization. A typical ERP system has many components, including hardware, software, procedures, and input from all functional areas. To integrate information for the entire organization, most ERP systems use a unified database to store data for various functions in an organization (see Exhibit 11.3).

Table 11.1 summarizes the functions of these components.

A well-designed ERP system offers the following benefits:

- Increased availability and timeliness of integrated information
- Increased data accuracy and improved response time
- Improved customer satisfaction
- Improved employee satisfaction
- Improved planning and scheduling
- Improved supplier relationship
- Improved reliability of information
- Reduction in inventory costs
- Reduction in labor costs
- Reduction in order-to-fulfillment time

With all the advantages listed previously, an ERP system also has drawbacks, such as high cost, difficulties

Enterprise resource planning (ERP) is an integrated system that collects and processes data and manages and coordinates resources, information, and functions throughout an organization.

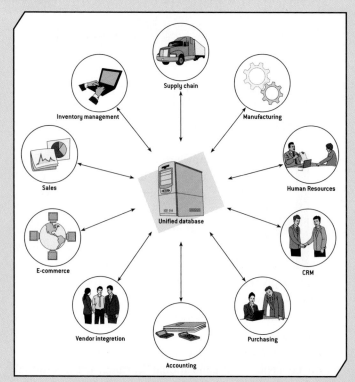

Exhibit 11.3 *An ERP configuration*

Supply chain

Inventory management

Manufacturing

Sales

Human Resources

E-commerce

Unified database

CRM

Vendor integration

Accounting

Purchasing

ERP Streamlines Operations at Naghi Group

Naghi Group, based in Jeddah, Saudi Arabia, operates several companies that together offer a wide range of products and services throughout the Middle East. Unfortunately, the legacy software that the firm was using wasn't able to communicate and integrate with the distribution and manufacturing software it was using. This lack of integration cost Naghi Group many hours each month as it tried to reconcile the data from various sources and generate critical financial-management reports. It needed an ERP system to integrate its major functional areas, including finance, sales, and supply chain management. It also needed to monitor inventory status and purchasing activity. The solution was ERP software—specifically, VAI's (Vormittag Associates, Inc.'s) S2K Distribution Suite. Now, the firm's managers are able to view financial data, keep track of inventory status, and analyze its customers' purchasing activities in real time. The ERP system has streamlined operations and improved customer service; it has also contributed to more timely business decisions.[9]

Table 11.1

ERP components

Component	Function
Unified database	Collects and analyzes relevant internal and external data and information needed by other functions
Inventory management	Provides inventory status and inventory forecasts
Supply chain	Provides information on supply chain members, including suppliers, manufacturing, distribution, and customers
Manufacturing	Supplies information on production costs and pricing
Human resources	Provides information on assessing job candidates, scheduling and assigning employees, and predicting future personnel needs
CRM	Supplies information on customers and their needs and preferences
Purchasing	Provides information related to the purchasing function, including e-procurement
Accounting	Tracks financial information, such as budget allocations and debits and credits
Vendor integration	Integrates information for vendors, such as offering automated downloads of data on product pricing, specifications, and availability
E-commerce	Provides B2C information related to order status and B2B information related to suppliers and business partners
Sales	Supplies information on sales and marketing

in installation, extensive training, and compatibility problems with legacy systems.

The previous information box summarizes some of the benefits in operational efficiency that Naghi Group gained from an ERP system.

Most ERP systems are available as modules, so an organization can purchase only the components it needs and add others later, if needed. Having modular components is a major factor in the success of ERP systems because it keeps costs down. More than 40 vendors, such as SAP, Oracle, Sage Group, and Microsoft, offer ERP software, with varying capabilities. If an organization decides to use a full-featured ERP system, the system development life cycle (SDLC) method introduced in Chapter 10 can be used.

The information box below describes a real-life application of a global ERP system.

The Industry Connection highlights Salesforce.com as a leader in enterprise systems.

5 Chapter Summary

In this chapter, you've learned what enterprise systems are used for and seen how supply chain management (SCM) plays a role in improving an organization's efficiency and effectiveness. You also learned about the technologies used in SCM systems, including electronic data interchange (EDI), Internet-based SCM, online auctions, e-marketplaces, and collaborative planning, forecasting, and replenishment. Next, you learned about customer relationship management (CRM) systems and how personalization and customization are used to improve customer service. Finally, you learned how knowledge management systems and enterprise resource planning (ERP) systems are used.

Global ERP in Action

Jabil Circuit is one of the top five electronic manufacturing service providers worldwide. Its main business is manufacturing network interface cards, and the company has nine manufacturing plants on three continents. To generate reports on the company's overall performance, Jabil needed to integrate information from all its branch plants. Jabil teamed with IBM Business Consulting Services to migrate all branch sites to a global ERP system so that it can link information from all its plants more easily. This system has reduced the time needed to integrate information and generate reports by 60 percent. In addition, Jabil has reduced overall costs and can now share resources among all its plants more easily, which streamlines operations.[10]

Industry Connection

SALESFORCE.COM*

Salesforce.com, a leader in CRM services, offers enterprise applications that can be customized to meet companies' needs. Its products and services include the following:

- *CRM applications*—Products such as Cloud Platform for CRM and Cloud Infrastructure for CRM are used for salesforce automation, sales management, and contact management.

- *Sales analytics*—Sales Cloud enables management to discover which salespeople are closing the most deals and how long tasks take. Customizable dashboards offer instant access to real-time information, allow monitoring of critical factors on sales, marketing, service, and other departments, as well as produce consolidated analyses from a variety of data sources.

- *Chatter*—A social networking and collaborative application that works with Sales Cloud. All users of Salesforce can access Chatter for no additional cost. Similar to Facebook Pages, Chatter enables groups to collaborate on projects, share information and documents, and control the privacy so that information is only shown to appropriate team members.

- *Service and support*—Service Cloud offers a customer portal, call center, and knowledge base. With information from these features, users can analyze who's asking for support and how long responses take, examine employee performance, and determine which reps handle most of the customer inquiries.

- *Marketing automation*—Includes Google Adwords, campaign management, marketing analytics, and marketing dashboards. Users can track multichannel campaigns, from generating sales leads to closing sales.

- *Force.com Builder*—Allows developers to create add-on applications that can be integrated into Salesforce.com applications and hosted by Salesforce.com.

* This information has been gathered from the company Web site (*www.salesforce.com*) and other promotional materials. For more information and updates, visit the Web site.

Key Terms

collaborative filtering (CF) (204)

collaborative planning, forecasting, and replenishment (CPFR) (199)

customer relationship management (CRM) (201)

customization (203)

electronic data interchange (EDI) (197)

e-marketplace (198)

enterprise resource planning (ERP) (206)

enterprise system (194)

knowledge management (KM) (205)

online auction (199)

personalization (203)

reverse auction (199)

supply chain (195)

supply chain management (SCM) (196)

Problems, Activities, and Discussions

1. List some functions that an SCM system can coordinate.

2. What is a reverse auction?

3. What is an enterprise resource planning system?

4. According to the source below, the World Bank, Southern Co., Dow Jones, Shuffle Master, and Pratt & Whitney have all implemented knowledge management systems, and each has achieved significant operational efficiency. Summarize these gains in a one-page paper. Do you see any similarities among these companies in their use of knowledge management systems?

   www.cioinsight.com/c/a/Case-Studies/5-Big-Companies-That-Got-Knowledge-Management-Right/1/

5. According to the source below, a logistics and procurement exchange connects Chinese original equipment manufacturers with their parts suppliers and with international customers. Summarize the advantages of this system in a one-page paper.

   www.informationweek.com/news/software/enterpriseapps/showArticle.jhtml?articleID=18400995

6. The source below presents several case studies of companies using CRM systems. Read three of the

case studies and write a two-page paper that summarizes the advantages of these systems.

www.concentrix.co.uk/software/crm/case-studies/

7. After visiting the following Web pages, write a one-page paper that describes the various ways RFID can help optimize supply chain management.

www.ameinfo.com/66090.html

www.eweek.com/c/a/Mobile-and-Wireless/ RFID-Reshapes-Supply-Chain-Management/

8. Visit the following Web site, consult other sources, and write a one-page paper that summarizes the ways organizations can enhance online-customer loyalty.

www.tmcnet.com/call-center/oeA0201.htm

9. Which of the following is a key decision in supply chain management for manufacturing? (Choose all that apply.)

a. Location

b. Inventory

c. Production

d. Financing

10. Salesforce automation is a common feature of CRM software packages. True or False?

casestudy

ERP AT JOHNS HOPKINS INSTITUTIONS

Maryland's largest private employer, Johns Hopkins Institutions (JHI) has more than 45,000 full-time staff members at Johns Hopkins Hospital and Health System, Johns Hopkins University, and several other hospitals and institutions. To improve data quality and reporting, JHI decided to use a centralized ERP system instead of several different business applications that weren't fully integrated. Getting an overall view of JHI's operations and performance was difficult with these different applications.

© Richard Thornton/Shutterstock.com

JHI faced technical challenges in adopting an ERP system. To overcome these challenges, it chose BearingPoint's SAP applications with IBM Power

570 servers as the hardware platform. JHI also decided to use server clusters to make the system more fault tolerant and installed a storage area network (SAN) for high performance. The ERP system now provides a centralized method of gathering information for the entire organization.[11]

Answer the following questions:

1. What was the main reason for using an ERP system at JHI?

2. How did the ERP system that JHI adopted help address technical challenges?

3. What did the ERP system achieve at JHI?

MANAGEMENT SUPPORT SYSTEMS

t his chapter begins by summarizing types of decisions and phases in the decision-making process. Next, you learn about a decision support system (DSS), its components and its capabilities, and see how it can benefit an organization. In addition, you learn about other management support systems used in decision making: executive information systems (EISs), group support systems (GSSs), and geographic information systems (GISs). This chapter concludes with an overview of guidelines for designing a management support system.

learning outcomes

After studying this chapter, you should be able to:

LO1 Define types of decisions and phases of the decision-making process in a typical organization.

LO2 Describe a decision support system.

LO3 Explain an executive information system's importance in decision making.

LO4 Describe group support systems, including groupware and electronic meeting systems.

LO5 Summarize uses for a geographic information system.

LO6 Describe guidelines for designing a management support system.

1 Types of Decisions in an Organization

In a typical organization, decisions usually fall into one of these categories:

- *Structured decisions*—**Structured decisions**, or programmable tasks, can be automated because a well-defined standard operating procedure exists for these types of decisions. Record keeping, payroll, and simple inventory problems are examples of structured tasks. Information technologies are a major support tool for making structured decisions.

- *Semistructured decisions*—**Semistructured decisions** aren't quite as well defined by standard operating procedures, but they include a structured aspect that benefits from information retrieval, analytical models, and information systems technology. For example, preparing budgets has a structured aspect in calculating percentages of available funds for each department. Semistructured decisions are often used in sales forecasting, budget preparation, capital acquisition analysis, and computer configuration.

Structured decisions, or programmable tasks, can be automated because a well-defined standard operating procedure exists for these types of decisions.

Semistructured decisions include a structured aspect that benefits from information retrieval, analytical models, and information systems technology.

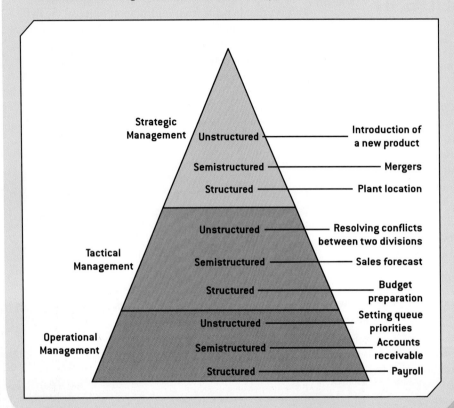

Exhibit 12.1 *Organizational levels and types of decisions*

Strategic Management
- Unstructured — Introduction of a new product
- Semistructured — Mergers
- Structured — Plant location

Tactical Management
- Unstructured — Resolving conflicts between two divisions
- Semistructured — Sales forecast
- Structured — Budget preparation

Operational Management
- Unstructured — Setting queue priorities
- Semistructured — Accounts receivable
- Structured — Payroll

- *Unstructured decisions*—**Unstructured decisions** are typically one-time decisions, with no standard operating procedure pertaining to them. The decision maker's intuition plays the most important role, as information technology offers little support for these decisions. Areas involving unstructured decisions include research and development, hiring and firing, and introducing a new product.

Semistructured and unstructured decisions are challenging because they involve multiple criteria, and often users have to choose between conflicting objectives. For example, a manager might want to give raises

Unstructured decisions are typically one-time decisions, with no standard operating procedure pertaining to them.

Management support systems (MSSs) are the different types of information systems that have been developed to support certain aspects and types of decisions. Each type is designed with unique goals and objectives.

In the **intelligence phase**, a decision maker examines the organization's environment for conditions that need decisions. Data is collected from a variety of sources (internal and external) and processed. From this information, the decision maker can discover ways to approach the problem.

to employees to boost morale and increase employee retention but has been asked to reduce the total cost of production. These two objectives conflict, at least in the short run. Artificial intelligence applications (discussed in Chapter 13) might be helpful in the future for handling qualitative decisions. Exhibit 12.1 shows organizational levels (operational, tactical, and strategic) and types of decisions.

Different types of information systems have been developed to support certain aspects and types of decisions. Collectively, these systems are called **management support systems (MSSs)**, and each type is designed with its own goals and objectives, as discussed in this chapter.

1.1 Phases of the Decision-Making Process

Herbert Simon, winner of the 1978 Nobel Prize in economics, defines three phases in the decision-making process: intelligence, design, and choice.[1] A fourth phase, implementation, can be added. The following sections explain these phases.

1.1.1 The Intelligence Phase

In the **intelligence phase**, a decision maker (a marketing manager, for example) examines the organization's environment for conditions that need decisions. Data is collected from a variety of sources (internal and external) and processed. From this information, the decision maker can discover ways to approach the problem. This phase has three parts: First, determine what the reality is—identify what's really going in order to help define the problem. Second, get a better understanding of the problem by collecting data and information about it. Third, gather data and information needed to define alternatives for solving the problem.

For example, an organization has noticed a decrease in total sales over the past six months. To pinpoint the cause of the problem, the organization can collect data from customers, the marketplace, and the competition.

Semistructured and unstructured decisions are challenging because they involve multiple criteria, and often users have to choose between conflicting objectives.

After the data has been processed, analysis can suggest possible remedies. Information technologies, particularly database management systems, can help in this analysis. In addition, many third-party vendors, such as Nielsen and Dow Jones, specialize in collecting data about the marketplace, competition, and the general status of the economy. The information they collect can support the intelligence phase of decision making.

1.1.2 The Design Phase

In the **design phase**, the objective is to define criteria for the decision, generate alternatives for meeting the criteria, and define associations between the criteria and the alternatives. Criteria are goals and objectives that decision makers establish in order to achieve certain performance. For example, the criterion in the previous example of decreased sales might simply be to increase sales. To make this criterion more specific, you can state it as "Increase sales by 3 percent each month for the next three months." Next, the following alternatives could be generated:

- Assign more salespeople to the target market.
- Retrain and motivate current salespeople.
- Reassign current salespeople.
- Revamp the product to adjust to consumers' changing tastes and needs.
- Develop a new advertising campaign.
- Reallocate existing advertising to other media.

In the **design phase**, the objective is to define criteria for the decision, generate alternatives for meeting the criteria, and define associations between the criteria and the alternatives.

© Sai Yeung Chan/Shutterstock.com

Defining associations between alternatives and criteria involves understanding how each alternative affects the criteria. For example, how would increasing the salesforce increase sales? By how much does the salesforce need to be increased to achieve a 3 percent increase in sales? Generally, information technology doesn't support this phase of decision making much, but group support systems and electronic meeting systems, discussed later in this chapter, can be useful. Expert systems (covered in Chapter 13) are helpful in generating alternatives, too.

1.1.3 The Choice Phase

The **choice phase** is usually straightforward. From the practical alternatives, the best and most effective course of action is chosen. It starts with analyzing each alternative and its relationship to the criteria to determine whether it's feasible. For instance, for each salesperson added, how are sales expected to increase? Will this result be economically beneficial? After a thorough analysis, the choice phase ends with decision makers recommending the best alternative. For the problem of decreased sales, the organization decided to use the first alternative, assigning more salespeople to the target market. A decision support system (DSS) can be particularly useful in this phase. DSSs are discussed later in the chapter, but these systems help sort through possible solutions to choose the best one for the organization. Typically, they include tools for calculating cost–benefit ratios, among others. For example, an organization is trying to decide which of three transportation systems to use for shipping its products to retail outlets. A DSS can assess cost factors and determine which transportation system minimizes costs and maximizes profits. Generally, information technologies are more useful in the intelligence and choice phases than in the design phase.

During the **choice phase**, the best and most effective course of action is chosen.

In the **implementation phase**, the organization devises a plan for carrying out the alternative selected in the choice phase and obtains the resources to implement the plan.

A **decision support system (DSS)** is an interactive information system consisting of hardware, software, data, and models (mathematical and statistical) designed to assist decision makers in an organization. Its three major components are a database, a model base, and a user interface.

The **model base** component includes mathematical and statistical models that, along with the database, enable a DSS to analyze information.

1.1.4 The Implementation Phase

In the **implementation phase**, the organization devises a plan for carrying out the alternative selected in the choice phase and obtains the resources to implement the plan. In others words, ideas are converted into actions. Information technologies, particularly DSSs, can also be useful in this phase. A DSS can do a follow-up assessment on how well a solution is performing. In the previous example of selecting a transportation system, a DSS might reveal that the system the organization chose isn't performing as well as expected and suggest an alternative.

2 Decision Support Systems

for the purposes of this book, a **decision support system (DSS)** is an interactive information system consisting of hardware, software, data, and models (mathematical and statistical) designed to assist decision makers in an organization. The emphasis is on semistructured and unstructured tasks. A DSS should meet the following requirements:

- Be interactive.
- Incorporate the human element as well as hardware and software.
- Use both internal and external data.
- Include mathematical and statistical models.
- Support decision makers at all organizational levels.
- Emphasize semistructured and unstructured tasks.

2.1 Components of a Decision Support System

A DSS, shown in Exhibit 12.2, includes three major components: a database, a model base, and a user interface. In addition, a fourth component, the DSS engine, manages and coordinates these major components. The database component includes both internal and external data, and a database management system (DBMS) is used for creating, modifying, and maintaining the database. This component enables a DSS to perform data analysis operations.

The **model base** component includes mathematical and statistical models that, along with the database, enable a DSS to analyze information. A model base management system (MBMS) performs tasks similar to a DBMS in accessing, maintaining, and updating models in the model base. For example, an MBMS might include tools for conducting what-if analysis so that a forecasting model can generate reports showing how forecasts vary, depending on certain factors.

Finally, the user interface component is how users access the DSS, such as when querying the database or model base, for help in making decisions. From the end user's point of view, the interface is the most important part of a DSS and must be as flexible and user friendly as possible. Because most DSS users are senior executives with little computer training, user friendliness is essential in these systems.[2]

2.2 DSS Capabilities

DSSs include the following types of features to support decision making:

- *What-if analysis*—This shows the effect of a change in one variable, answering questions such as "If labor costs increase by 4 percent, how is the final cost of a product affected?" and "If the advertising budget increases by 2 percent, what's the effect on total sales?"

> A **managerial designer** defines the management issues in designing and using a DSS. These issues don't involve the technological aspects of the system; they're related to management's goals and needs.

- *Goal seeking*—This is the reverse of what-if analysis. It asks what has to be done to achieve a particular goal—for example, how much to charge for a product in order to generate $200,000 profit, or how much to advertise a product to increase total sales to $50,000,000.
- *Sensitivity analysis*—This enables you to apply different variables, such as determining the maximum price you could pay for raw materials and still make a profit or determining how much the interest rate has to go down for you to be able to afford a $100,000 house with a monthly payment of $700.
- *Exception reporting analysis*—This monitors the performance of variables that are outside a defined range, such as pinpointing the region that generated the highest total sales or the production center that went over budget.

A typical DSS has many more capabilities, such as graphical analysis, forecasting, simulation, statistical analysis, and modeling analysis.

2.3 Roles in the DSS Environment

To design, implement, and use a DSS, several roles are involved. These include the user, managerial designer, technical designer, and model builder.[2]

Users are the most important category because they're the ones using the DSS; therefore, the system's success depends on how well it meets their needs. Users can include department or organizational units in addition to people.

A **managerial designer** defines the management issues in designing and using a DSS. These issues don't involve the technological aspects of

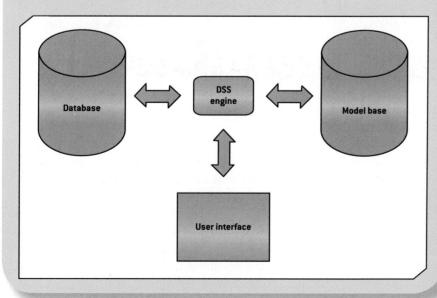

Exhibit 12.2 *Components of a DSS*

Database ⟷ DSS engine ⟷ Model base

User interface

the system; they're related to management's goals and needs. This person specifies data requirements, what models are needed, how these models might be used, and how users want to view the results (graphics, text, and so forth). This role often includes questions such as the following:

- What type of data should be collected, and from what sources?
- How recent should the collected data be?
- How should the data be organized?
- How should the data be updated?
- What should the balance between aggregated (lump sum) and disaggregated (itemized) data be?

The **technical designer** focuses on how the DSS is implemented and usually addresses the following questions:

- How should the data be stored (centralized, decentralized, or distributed)?
- What type of file structure should be used (sequential, random, or indexed sequential)?
- What type of user access should be used? Menu driven, such as QBE? Or command line, such as SQL?
- What type of response time is required?
- What types of security measures should be installed?

The technical designer might be a computer specialist or a consultant from outside the company and may use a commercial DSS package or write the system's code from scratch.

A **model builder** is the liaison between users and designers. For example, during the design phase, the model builder might explain users' needs to the managerial designer or technical designer. Later, during the implementation phase, this person might explain the output of a regression analysis, for example, to users, describing the assumptions underlying the model, its limitations, and its strengths. He or she is responsible for supplying information on what the model does, what data inputs it accepts, how the model's output should be interpreted, and what assumptions go into creating and using the model. Typically, requirements

for what the model should do come from the managerial designer, implementation of the model is carried out by the technical designer, and specifications for the model come from the model builder. The model builder can also suggest new or different applications of a DSS.

2.4 Costs and Benefits of Decision Support Systems

Some DSSs can be developed from resources already available in the organization, which can reduce costs, but many require new hardware and software. Before making this investment, organizations should weigh the costs and benefits of using a DSS. Costs and benefits can be difficult to assess, however, because these systems are focused on effectiveness rather than efficiency. In addition, a DSS facilitates improvements but doesn't necessarily cause them. How do you assign a monetary value to facilitating communication or expediting problem solving, for instance?

Peter G. Keen, a former MIT professor, conducted an interesting study on how organizations use DSSs and concluded that the decision to build a DSS seems to be based on value rather than cost. He outlined the benefits of a DSS as follows:[3]

- Increase in the number of alternatives examined
- Fast response to unexpected situations
- Ability to make one-of-a-kind decisions
- New insights and learning
- Improved communication
- Improved control over operations, such as controlling the cost of production
- Cost savings from being able to make better decisions and analyze several scenarios (what-ifs) in a short period
- Better decisions
- More effective teamwork
- Time savings
- Better use of data resources

As this study indicates, most of the benefits are intangible and difficult to assess. However, they can be quantified to a degree, although the quantification might vary, depending on the person doing the calculations. You can quantify the benefit of saving time, for instance, by measuring the two hours a manager wasted looking for information that a DSS could have made available immediately. Of course, you'd probably also

> You can quantify the benefit of saving time, for instance, by measuring the two hours a manager wasted looking for information that a DSS could have made available immediately.

notice that a manager who didn't have to waste this much time is less frustrated and more productive, but quantifying these results is harder, or at least requires more work, such as conducting interviews or surveys.

The benefit of improving communication and interactions between management and employees is perhaps the most difficult to quantify, but it is one of the most important.[4] DSSs can, and are, improving how decision makers view themselves, their jobs, and the way they spend time. Therefore, improving communication and expediting learning are among the main objectives of a DSS.

A DSS is said to have achieved its goals if employees find it useful in doing their jobs. For example, a portfolio manager who uses a financial DSS to analyze different scenarios would certainly find the ease of analyzing a variety of variables, such as the interest rate and economic forecasts, to be useful. The manager can try different values for these variables quickly and easily to determine which variable has the greatest effect and decide which portfolio will be the most profitable. In addition, some DSSs result in saving on clerical costs, and others improve the decision-making process.

3 Executive Information Systems

xecutive information systems (EISs), a branch of DSSs, are interactive information systems that give executives easy access to internal and external data and typically include "drill-down" features (explained in Chapter 3) and a digital dashboard for examining and analyzing information. (Although some experts consider executive support systems and executive management systems variations of EISs, this book considers them to fall under the term EIS.)

Ease of use plays an important role in the success of an EIS. Because most EIS users aren't computer experts, simplicity of the system is crucial, and EIS designers should focus on simplicity when developing a user interface. Typically, GUIs are used, but adding features such as multimedia, virtual reality, and voice input and output can increase ease of use.

Another important factor in an effective EIS is access to both internal and external data so that executives can spot trends, make forecasts, and conduct different types of analyses. For an EIS to be useful, it should also collect data related to an organization's "critical success factors"—issues that make or break a business. In banks, interest rates are considered a critical success factor; for car manufacturers, location of dealerships might be a critical success factor. An EIS should be designed to provide information related to an organization's critical success factors.

Most EISs include a **digital dashboard**, which integrates information from multiple sources and presents it in a unified, understandable format, often charts and graphs. Digital dashboards and scorecards offer up-to-the minute snapshots of information and assist decision makers in identifying trends and potential problems. Many digital dashboards are Web-based, such as the one included in Microsoft SharePoint. Exhibit 12.3 shows an example of a digital dashboard.

The following are some important characteristics of an EIS:[5]

- Tailored to meet management's information needs
- Can extract, compress, filter, and track critical data
- Provides online status access, trend analysis, and exception reporting
- Offers information in graphical, tabular, and text formats

> **Executive information systems (EISs),** branches of DSSs, are interactive information systems that give executives easy access to internal and external data and typically include "drill-down" features and a digital dashboard for examining and analyzing information.
>
> A **digital dashboard** integrates information from multiple sources and presents it in a unified, understandable format, often charts and graphs. It offers up-to-the minute snapshots of information and assists decision makers in identifying trends and potential problems.

Exhibit 12.3 *A digital dashboard*

Used with permission from Microsoft.

- Includes statistical analysis techniques for summarizing and structuring data
- Retrieves data in a wide range of platforms and data formats
- Contains customized application-development tools
- Supports electronic communication, such as e-mail and video conferencing

3.1 Reasons for Using EISs

An EIS can put a wealth of analytical and decision-making tools at managers' fingertips and includes graphical representations of data that helps them make critical decisions. In addition, executives can use EISs to share information with others more quickly and easily. Managers can use these tools to improve the efficiency and effectiveness of decision making in the following ways:

- Increase managers' productivity by providing fast and easy access to relevant information.
- Convert information into other formats, such as bar charts or graphs, to help managers analyze different business scenarios and see the effect of certain decisions on the organization.
- Spot trends and report exceptions. For example, a manager might use an EIS to gather data on profitability and production costs at a manufacturing plant and determine whether closing the plant is more beneficial than keeping it open.

3.2 Avoiding Failure in Design and Use of EISs

As with other management support systems, effective design and implementation of an EIS requires top-management support, user involvement, and the right technologies. The following are factors that can lead to a failed EIS:[6, 7]

- The corporate culture isn't ready, there's organizational resistance to the project, or the project is viewed as unimportant.
- Management loses interest or isn't committed to the project.
- Objectives and information requirements can't be defined clearly, or the system doesn't meet its objectives.
- The system's objectives aren't linked to factors critical to the organization's success.
- The project's costs can't be justified.
- Developing applications takes too much time, or the system is too complicated.
- Vendor support has been discontinued.
- Some of today's senior executives missed the computer revolution and might feel uncomfortable using computers. Ongoing education and increasing computer awareness should solve this problem.
- Executives' busy schedules and frequent travel make long training sessions difficult, don't allow much uninterrupted time for system use, and often prevent daily use of an EIS. The result is that senior executives are unlikely to use systems that need considerable training

© FERNANDO BLANCO CALZADA/Shutterstock.com

> The intervention aspect of a GSS reduces communication barriers and introduces order and efficiency into situations that are inherently unsystematic and inefficient, such as group meetings and brainstorming sessions.

and regular use to learn. A user-friendly interface can encourage executives to use an EIS more often, however.

- Some EISs don't contain the information that senior executives need because there's a lack of understanding about what executives' work involves. Designers must determine what types of information executives need before designing a system.

3.3 EIS Packages and Tools

EISs are generally designed with two or three components: an administrative module for managing data access, a builder module for developers to configure data mapping and screen sequencing, and a runtime module for using the system. Sometimes, administrative and builder modules are combined into one module. Some EIS packages provide a data storage system, and some simply package data and route it to a database, usually on a LAN. Most EIS packages come with a standard graphical user interface (GUI). Exhibit 12.4 shows a screen from one EIS package, Business Intelligence (SAS Institute).

Generally, managers perform six tasks for which an EIS is useful: tracking performance, flagging exceptions, ranking, comparing, spotting trends, and investigating/exploring. Most EIS packages provide tools for these tasks, such as displaying summaries of data in report or chart format and sequencing screens to produce slide shows. Exception or variance reporting is another useful technique managers use to flag data that's unusual or out of normal boundaries. Both unusual and periodic events can be defined to trigger visual cues or activate intelligent agents to perform a specific task. Intelligent agents, covered in Chapter 13, are "smart" programs that carry out repetitive tasks and can be programmed to make decisions based on specified conditions. Some widely used EIS packages include SAS Business Intelligence (*www.sas.com/technologies/bi/*), Monarch Datawatch (*www.datawatch.com*), and Cognos PowerPlay (*www-01.ibm.com/software/data/cognos/products/series7/powerplay/*).

> **Group support systems (GSSs)** assist decision makers working in groups. These systems use computer and communication technologies to formulate, process, and implement a decision-making task and can be considered a kind of intervention technology that helps overcome the limitations of group interactions.

4 Group Support Systems

In today's business environment, decision makers often work in groups, so you hear the terms *group computing* or *collaborative computing* used often. All major software vendors are competing to enter this market or increase their market share in this fast-growing field. In this collaborative environment, there has been an increase in **group support systems (GSSs)**, which are intended to assist decision makers working in groups. DSSs are usually designed to be used by a particular decision maker; a GSS is designed to be used by more than one decision maker. These systems use computer and communication technologies to formulate, process, and implement a decision-making task and can

Exhibit 12.4 *The Business Intelligence interface*

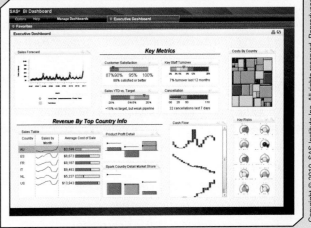

> **Groupware** assists groups in communicating, collaborating, and coordinating their activities. It is a collection of applications that supports decision makers by providing access to a shared environment and information.

be considered a kind of intervention technology that helps overcome the limitations of group interactions. The intervention aspect of a GSS reduces communication barriers and introduces order and efficiency into situations that are inherently unsystematic and inefficient, such as group meetings and brainstorming sessions. A GSS, with the help of a human facilitator, enhances decision making by providing a clear focus for group discussion, minimizing politicking, and focusing attention on critical issues. The success of a GSS depends on the following:

- Matching the GSS's level and sophistication to the group's size and the scope of the task
- Providing supportive management (especially at the CEO level) that is willing to "champion" using a GSS in the organization

Related technologies for group support, such as electronic meeting systems (EMSs), groupware, computer-mediated communication (CMC), computer-supported cooperative work (CSCW), and e-collaboration, aren't considered full-function GSSs because they don't have decision-making tools, but they're less expensive and include communication and problem-solving mechanisms for effective team management.[8, 9]

GSSs are useful for committees, review panels, board meetings, task forces, and decision-making sessions that require input from several decision-makers. They can be used to find a new plant location, introduce a new product or advertising campaign, participate in an international bid, brainstorm alternatives, and other tasks. In addition to all the capabilities of a DSS, a GSS should include communication features so that decision makers in many different locations can still work together to participate in the decision-making process.

4.1 Groupware

The goal of **groupware** is to assist groups in communicating, collaborating, and coordinating their activities. It's intended more for teamwork than for decision support. For the purposes of this book, groupware is a collection of applications that supports decision makers by providing access to a shared environment and information. A shared environment can consist of an e-mail message, a memo, a single file, or even an entire database.

Groupware is software that helps a group of decision makers work with the same application, regardless of their locations. Groupware tools include e-mail, chat applications, video conferencing, and database sharing. IBM Lotus Notes, Microsoft Sharepoint, and Novell GroupWise are common examples of groupware, the capabilities of which include the following:

- Audio and video conferencing
- Automated appointment books
- Brainstorming
- Database access
- E-mail
- Online chat
- Scheduling
- To-do lists
- Workflow automation

Groupware and Health IT

Hospital personnel have gotten used to electronic health records (EHRs) over the last 25 years, but the technology, which presumes a client-server, physician-centric model, has never been very popular. Today, there's an alternative: IT systems that integrate and coordinate the data provided by every member of the health care team, from the doctor to the patient. Under the general heading of "clinical groupware," there are now software products available—RMDNetworks and Shared Health are two examples—that are Web-based, networkable, and interactive, thereby improving health care through better communication and coordination. Clinical groupware is used by groups, not individuals, and the patient is considered a vital member of that group. The overall aim is to provide a unified view of the patient by collecting and analyzing data and information from a variety of sources. Clinical groupware has the following additional benefits:[10]

- *Inexpensive to acquire and use*
- *Offers evidence-based guidance*
- *Routinely collects quality and performance measures*
- *Provides a collaborative workflow platform*

LANs, WANs, and MANs, discussed in Chapter 6, are the network foundations for groupware. Although e-mail isn't the same thing as groupware, it provides the main functions of groupware: transmitting text messages across a network. The information box on the previous page introduces an application of groupware in the health care industry.

The Internet has become an important part of groupware. The most important advantage of a Web-based GSS is being able to use open network standards, meaning the GSS can be used on any operating system or type of workstation. The most notable disadvantages are speed limitations (because the Internet is often slower than a company's proprietary network) and security issues. Some examples of Web-based GSS tools are Microsoft Office SharePoint Server (see the following information box) and IBM Lotus Domino. Another type of software used for e-collaboration is an electronic meeting system, such as Microsoft Live Meeting, Metastorm, and IBM FileNet.

4.2 Electronic Meeting Systems

Electronic meeting systems enable decision makers in different locations to participate in a group decision-making process. There are various types of electronic meeting systems, but they all perform the following tasks:

- Real-time computer conferencing allows a group of people to interact via their workstations and share files, such as documents and images. This conference often includes an audio link but has no video capabilities.
- Video teleconferencing, the closest thing to a face-to-face meeting, requires special equipment and sometimes trained operators. Video cameras are used to transmit live pictures and sounds, and this is more effective than phone conferencing but also more expensive. The main drawback is that participants can't share text and graphics.

- Desktop conferencing combines the advantages of video teleconferencing and real-time computer conferencing. Using these systems, participants can have multiple video windows open at one time. Participants also have interfaces to a conference installed on their workstations, so these systems are easier for employees to use.

4.3 Advantages and Disadvantages of GSSs

Advantages of GSSs include the following:

- Because decision makers don't have to travel as much and pay for planes, hotels, and meals, costs as well as stress levels are reduced.
- Because decision makers aren't traveling long distances, they have more time to talk with one another and solve problems.
- Shyness isn't as much of an issue in GSS sessions as it is in face-to-face meetings.
- Increasing collaboration improves the effectiveness of decision makers.

Disadvantages of GSSs include the following:

- *Lack of the human touch*—Gestures, handshakes, eye contact, and other nonverbal cues can be lost, which can hinder the effectiveness of meetings. New developments in virtual reality technologies (discussed in Chapter 14) could solve this problem, however.
- *Unnecessary meetings*—Because arranging a GSS session is easy, there's a tendency to schedule more meetings than are necessary, which wastes time and energy.

> **Electronic meeting systems** enable decision makers in different locations to participate in a group decision-making process.

Microsoft Office SharePoint Server: A New Type of Groupware

Microsoft Office SharePoint Server 2010, part of the Microsoft Office 2010 suite, is used to improve collaboration, provide content management features, carry out business processes, and provide access to information that's essential to organizational goals (http://sharepoint.microsoft.com/en-us/Pages/default.aspx). You can create SharePoint sites that support content publishing, content management, records management, and business intelligence needs. You can also conduct searches for people, documents, and data as well as access and analyze large amounts of business data. SharePoint Server 2010 provides a single, integrated location where employees can collaborate with team members, find organizational resources, search for experts and information, manage content and workflow, and use the information they've found to make better decisions.

- *Security problems*—GSS sessions have the same security problems as other data communication systems, so there's the possibility of private organizational information falling into the hands of unauthorized people. Tight security measures for accessing GSS sessions and transferring data are essential.

- The costs of GSS implementation are high because the system includes many features.

5 Geographic Information Systems

executives often need to answer questions such as the following:

- Where should a new store be located to attract the most customers?

- Where should a new airport be located to keep the environmental impact to a minimum?

- What route should delivery trucks use to reduce driving time?

- How should law enforcement resources be allocated?

A well-designed **geographic information system (GIS)** can answer these questions and more. This system captures, stores, processes, and displays geographic information or information in a geographic context, such as showing the location of all city streetlights on a map. A GIS uses spatial and nonspatial data and specialized techniques for storing coordinates of complex geographic objects, including networks of lines (roads, rivers, streets) and reporting zones (zip codes, cities, counties, states). Most GISs can superimpose the results of an analysis on a map, too. Typically, a GIS uses three geographic objects:

- *Points*—The intersections of lines on a map, such as the location of an airport or a restaurant

> A **geographic information system (GIS)** captures, stores, processes, and displays geographic information or information in a geographic context, such as showing the location of all city streetlights on a map.

- *Lines*—Usually a series of points on a map (a street or a river, for example)

- *Areas*—Usually a section of a map, such as a particular zip code or a large tourist attraction

Digitized maps and spatially oriented databases are two major components of a GIS. For example, say you want to open a new store in southwest Portland, Oregon, and would like to find out how many people live within walking distance of the planned location. With a GIS, you can start with the map of the United States and zoom in repeatedly until you get to the street map level. You can mark the planned store location on the map and draw a circle around it to represent a reasonable walking distance. Next, you can request a summary of U.S. census data on everyone living inside the circle who meet certain conditions, such as a particular income level, age, marital status, and so forth. A GIS can provide all kinds of information that enables you to zero in on specific customers. A GIS can perform the following tasks:

- Associate spatial attributes, such as a manufacturing plant's square footage, with points, lines, and polygons on maps.

- Integrate maps and database data with queries, such as finding zip codes with a high population of senior citizens with relatively high income. A GIS can support some sophisticated data management operations, such as the following:

  - Show the customers who live within a 5-mile radius of the Super Grocery at the corner of 34th and Lexington. A database can't answer this question because it can't determine the latitude/longitude coordinates of the store, compute distances using the specified location as the center, identify all zip codes within this circle, and pull out the customers living in these zip codes.

  - Show the customers whose driving route from work to home and back takes them through the intersection of 34th and Lexington. A database can't address this query, either. In this case, the GIS maps customers' home and work locations and determines all possible routes. It can then narrow down the customer list by picking only those whose shortest route takes them through the specified intersection.

Exhibit 12.5 *Locations of child care centers: an example of a pin map*

Courtesy of the Community Connection for Child Care

A GIS with analytical capabilities evaluates the impact of decisions by interpreting spatial data. Modeling tools and statistical functions are used for forecasting purposes, including trend analysis and simulations. Many GISs offer multiple windows so that you can view a mapped area and related nonspatial data simultaneously, and points, lines, and polygons can be color coded to represent nonspatial attributes. A zoom feature is common for viewing geographic areas in varying levels of detail, and map overlays can be useful for viewing such things as gas lines, public schools, or fast-food restaurants in a specified region. A buffering feature creates pin maps that highlight locations meeting certain criteria, such as finding a new store location based on population density. Exhibit 12.5 shows an example of output from Environmental Systems Research Institute (ESRI), a major vendor of GIS software. To see examples of different types of maps, visit *www.esri.com.*

A common example of a GIS—and one you've probably used often—is getting driving directions from Google Maps. It's an interactive GIS that identifies routes from start to destination, overlays routes on a map, shows locations of nearby landmarks, and estimates distances and driving times. It's also considered a DSS because you can change routes by dragging different points and have what-if analysis performed on alternative routes (such as taking back roads instead of the highway), including estimates of driving time to help you decide which route is best. Google Maps has a user-friendly interface that helps you visualize the route, and after you make a decision, you can print driving directions and a map.

5.1 GIS Applications

GISs integrate and analyze spatial data from a variety of sources. Although they're used mainly in government and utility companies, more businesses are using them, particularly in marketing, manufacturing, insurance, and real estate. No matter what category a GIS falls into, most applications require a GIS to handle converting data to information, integrating data with maps, and conducting different types of analysis. GIS applications can be classified in the following categories, among several others:

- *Education planning*—Analyzing demographic data toward changing school district boundaries or deciding where to build new schools.

- *Urban planning*—Tracking changes in ridership on mass-transit systems and analyzing demographic data to determine new voting districts, among many other uses.

- *Government*—Making the best use of personnel and equipment while dealing with tight budgets, dispatching personnel and equipment to crime and fire locations, maintaining crime statistics.

- *Insurance*—Combining data on community boundaries, street addresses, and zip codes with search capabilities to find information on potential hazards, such as natural disasters, auto-rating variables, and crime rate indexes. (Some of this information comes from federal and state agencies.)[11, 12]

- *Marketing*—Pinpointing areas with the greatest concentration of potential customers, displaying sales statistics in a geographic context, evaluating demographic and lifestyle data to identify new markets, targeting new products to specific populations, analyzing market share and population growth in relation to new store locations, and evaluating a company's market position based on geographic location.[13, 14] For example, Pepsico uses a GIS to find the best locations for new Pizza Hut and Taco Bell outlets.

- *Real estate*—Finding properties that meet buyers' preferences and price ranges, using a combination of census data, multiple listing files, mortgage information, and buyer profiles. GISs also help establish selling prices for homes by surveying an entire city to identify comparable neighborhoods and average sales prices. GISs can be used for appraisal purposes to determine relationships between national, regional, and local economic trends and the demand for local real estate.[15, 16]

- *Transportation and logistics*—Managing vehicle fleets, coding delivery addresses, creating street networks for predicting driving times, and developing maps for scheduling routing and deliveries.[17]

The information box on the next page discusses how GIS can be used to monitor and reduce the spread of disease.

© Susan Law Cain/Shutterstock.com

6 Guidelines for Designing a Management Support System

before designing any management support system, the system's objectives should be defined clearly, and then the system development methods discussed in Chapter 10 can be followed. Because MSSs have a somewhat different purpose than other information systems, the important factors in designing one are summarized in the following list:

- *Get support from the top*—Support and commitment must come from the top; without a full commitment from top management, the system's chances of success are low.

- *Define objectives and benefits clearly*—Because many benefits of an MSS are intangible, this step is challenging. Costs are always in dollars, but benefits are qualitative. The design team should spend time identifying all costs and benefits in order to present a convincing case to top management. When benefits are intangible, such as improving customer service, the design team should associate the benefit with a measurable factor, such as increased sales.

- *Identify executives' information needs*—Examine the decision-making process executives use to find out what kinds of decisions they're making—structured, semistructured, or unstructured—and what kind of information they need to make these decisions.

- *Keep the lines of communication open*—This is important to ensure that key decision makers are involved in designing the MSS.

- *Hide the system's complexity and keep the interface simple*—Avoid using technical jargon when explaining the system to executives because they might lose interest if they think the system is too technical. Executives aren't interested in the choice of platform or software, for example. Their main concern is getting the information they need in the simplest way possible. In addition, the system must be easy for executives to learn with little or no training. To most executives, the interface is the system, so its ease of use is a crucial factor in the system's success.

GISs for Fighting Disease

Public health officials and government agencies around the globe use GISs, demographic information, remote-sensing data, and even Google Maps to help fight diseases such as avian flu, malaria, H1N1 (swine flu), SARS, and more. With a GIS, health officials can map the spread of epidemics and identify high-risk population areas before the epidemic reaches them, which can help decrease the death rate and reduce the spread of infectious diseases by tracking their origins. GISs can also be useful in the following tasks:[18,19]

- *Locating contaminated water sources for waterborne diseases*
- *Plotting confirmed and suspected cases of a disease on a map*
- *Identifying existing medical infrastructures*
- *Determining how far people have to travel to reach a health care center and whether public transportation is available*
- *Monitoring virus mutations and their locations during an outbreak and assisting in early identification of infected people or animals*
- *Determining the geographic distribution of diseases*
- *Planning and targeting interventions*
- *Routing health care workers, equipment, and supplies to remote locations*
- *Locating the nearest health care facility*

dynamic business environment. A flexible system can incorporate changes quickly.

- *Make sure response time is fast*—MSS designers must monitor the system's response time at regular intervals, as executives rarely tolerate slow response times. In addition, when a system function takes more than a few seconds, make sure a message is displayed stating that the system is processing the request. Using a progress bar can help reduce frustration, too.

The Industry Connection highlights the software and services available from SAS Corporation, one of the leaders in decision support systems.

- *Keep the "look and feel" consistent*—Designers should use standard layouts, formats, and colors in windows, menus, and dialog boxes for consistency and ease of use. That way, a user who has learned the database portion of the system, for example, should be able to switch to the report-generating portion with little trouble because the interface is familiar. You can see this in Microsoft Office, which uses similar features, such as formatting toolbars, in all its applications. Users accustomed to Word, for example, can learn how to use Excel quickly because the interface is familiar.
- *Design a flexible system*—Almost all aspects of an MSS, including the user interface, change over time because of rapid developments in technology and the

7 Chapter Summary

In this chapter, you've learned about different types of decisions and of decision-making stages in a typical organization. You reviewed the components of a decision support system, along with its capabilities, key players, and costs and benefits. In addition, you learned how executive information systems are used and what factors affect their success, and the advantages and disadvantages of group support systems. Finally, you learned about the capabilities and possible uses of geographic information systems and reviewed guidelines for designing management support systems.

Industry Connection

SAS, INC. *

SAS was founded in the early 1970s to analyze agricultural research data. Eventually, it developed into a vendor of software for conducting comprehensive analysis and generating business intelligence information for a variety of decision-making needs. Products and services offered by SAS include the following:

- *SAS Business Intelligence*—Analyzes past data to predict the future. It provides features for reporting, queries, analysis, OLAP, and integrated analytics.

- *Data Integration*—Gives organizations a flexible, reliable way to respond quickly to data integration requirements at a reduced cost.
- *SAS Analytics*—Provides an integrated environment for modeling analyses, including data mining, text mining, forecasting, optimization, and simulation. Includes several techniques and processes for collecting, classifying, analyzing, and interpreting data to reveal patterns and relationships that can help with decision-making.

* This information has been gathered from the company Web site (*www.sas.com*) and other promotional materials. For more information and updates, visit the Web site.

Key Terms

choice phase (214)

decision support system (DSS) (214)

design phase (213)

digital dashboard (217)

electronic meeting systems (221)

executive information systems (EISs) (217)

geographic information system (GIS) (222)

group support systems (GSSs) (219)

groupware (220)

implementation phase (214)

intelligence phase (212)

management support systems (MSSs) (212)

managerial designer (215)

model base (214)

model builder (216)

semistructured decisions (211)

structured decisions (211)

technical designer (216)

unstructured decisions (212)

Problems, Activities, and Discussions

1. What are some guidelines for designing an MSS?

2. How do you avoid failures in designing an EIS?

3. What tools are typically incorporated into a groupware package?

4. The MicroStrategy (*www.microstrategy.com*) platform provides a full range of scorecard and dashboard capabilities. Log on to the company's Web site and explore its offerings. In which situations are scorecard and dashboards helpful? Which management level would benefit the most from these capabilities?

5. Culver's, a Wisconsin-based quick-service restaurant chain, uses ESRI's GIS software to help determine franchise locations. Read the information on the link below and write a one-page paper that summarizes how Culver's has benefited from GIS.

   www.esri.com/library/reprints/pdfs/arcwatch_culvers.pdf

6. Police in Memphis are getting big results from using predictive analytics to fight crime.

   Read the information on the link below and write a one-page paper that summarizes this decision-making capability.

   http://www.computerworld.com/s/article/350967/Police_Use_Analytics_to_Deploy_Officers

7. Decision support systems can provide vital information to health care providers and health care management. Visit the following Web page, consult other sources, and write a one-page paper that summarizes the applications of DSS in health care.

   www.kmworld.com/Articles/Editorial/Feature/Decision-support-systems-prove-vital-to-healthcare-19133.aspx

8. Watch the video on the link below, and identify five GIS applications.

   www.youtube.com/watch?v=kEaMzPo1Q7Q&feature=related

9. Which of the following isn't a main type of decision?

   a. Structured

   b. Semistructured

   c. Unique

   d. Unstructured

10. Cost saving is an advantage of a GSS. True or False?

casestudy

COLLABORATION SYSTEMS AT ISUZU AUSTRALIA LIMITED

Isuzu Australia Limited (IAL) is responsible for marketing and distributing Isuzu trucks in Australia. With just 65 employees, IAL depends on its national dealer network to maintain its market position. In the past, all information was distributed to dealers manually, which was expensive, time consuming, and prone to errors. IAL chose IBM WebSphere Portal and Workplace Web Content Management as an online portal and content management system to replace the existing manual system. With the new system, employees can publish information themselves and use the system's instant-messaging features to collaborate on

© Andreas Pollok/Getty Images

decisions. Since the system's implementation in 2006, IAL has seen major cost savings, reduced errors, and faster, more efficient communication. Collaboration and knowledge sharing have also increased with use of the system's wiki feature.[20]

Answer the following questions:

1. What prompted IAL to use collaboration technologies?

2. What are some applications of collaboration technologies at IAL?

3. Which platform was chosen by IAL as an online portal and content management system?

INTELLIGENT INFORMATION SYSTEMS

this chapter covers intelligent information systems, beginning with artificial intelligence (AI) and how its technologies are used in decision making; this section includes an overview of robotics as one of the earliest applications of AI. Next, you learn about expert systems and their components and see how these systems are used. Case-based reasoning and intelligent agents are also discussed as applications of AI. Finally, you review fuzzy logic, artificial neural networks, genetic algorithms, and natural-language processing systems and learn about the advantages of integrating AI technologies into decision support systems.

learning outcomes

After studying this chapter, you should be able to:

LO1 Define artificial intelligence and explain how these technologies support decision making.

LO2 Explain an expert system, its applications, and its components.

LO3 Describe case-based reasoning.

LO4 Summarize types of intelligent agents and how they're used.

LO5 Describe fuzzy logic and its uses.

LO6 Explain artificial neural networks.

LO7 Describe how genetic algorithms are used.

LO8 Explain natural-language processing and its advantages and disadvantages.

LO9 Summarize the advantages of integrating AI technologies into decision support systems.

1 What Is Artificial Intelligence?

artificial intelligence (AI) consists of related technologies that try to simulate and reproduce human thought behavior, including thinking, speaking, feeling, and reasoning. AI technologies apply computers to areas that require knowledge, perception, reasoning, understanding, and cognitive abilities.[1] To achieve these capabilities, computers must be able to do the following:

- Understand common sense.
- Understand facts and manipulate qualitative data.
- Deal with exceptions and discontinuity.
- Understand relationships between facts.
- Interact with humans in their own language.
- Be able to deal with new situations based on previous learning.

Information systems are concerned with storing, retrieving, and working with data, but AI technologies are concerned with generating and displaying knowledge and facts. In the information systems field, as you've learned, programmers and systems analysts

> **Artificial intelligence (AI)** consists of related technologies that try to simulate and reproduce human thought behavior, including thinking, speaking, feeling, and reasoning. AI technologies apply computers to areas that require knowledge, perception, reasoning, understanding, and cognitive abilities.

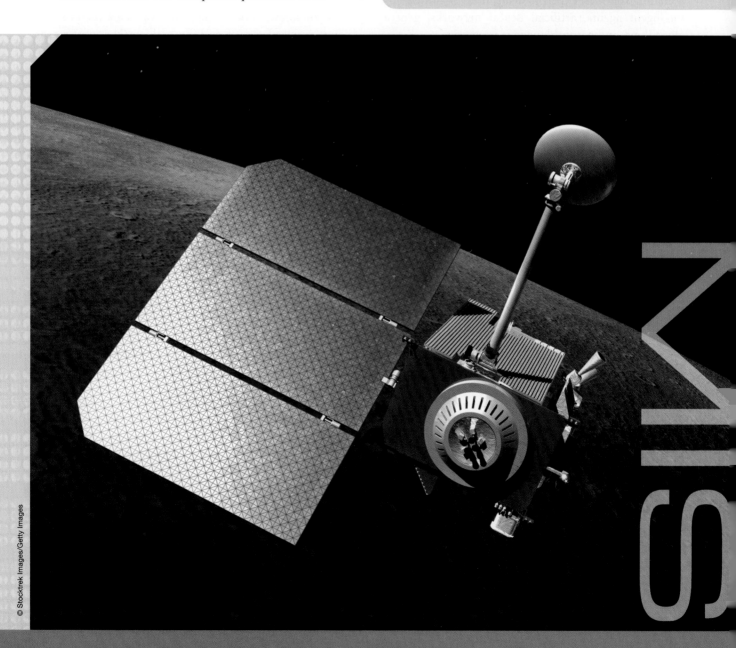

© Stocktrek Images/Getty Images

> Information systems are concerned with storing, retrieving, and working with data, but AI technologies are concerned with generating and displaying knowledge and facts.

design systems that help decision makers by providing timely, relevant, accurate, and integrated information. In the AI field, knowledge engineers try to discover "rules of thumb" that enable computers to perform tasks usually handled by humans. Rules used in the AI field come from a diverse group of experts in areas such as mathematics, psychology, economics, anthropology, medicine, engineering, and physics. AI encompasses several related technologies discussed in this chapter, including robotics, expert systems, fuzzy logic systems, intelligent agents, artificial neural networks, genetic algorithms, and natural-language processing.

Although these applications and technologies may not offer true human intelligence, they're certainly more intelligent than traditional information systems.

Over the years, the capabilities of these systems have improved in an attempt to close the gap between artificial intelligence and human intelligence. The following section discusses possibilities for using AI technologies in decision-making processes.

1.1 AI Technologies Supporting Decision Making

As you know, information technologies are used to support many phases of decision making. The most recent developments in AI technologies promise new areas of decision-making support. Table 13.1 lists some applications of AI-related technologies in several organizations.[2,3,4]

Table 13.1
Applications of AI technologies

Field	Organization	Applications
Energy	Arco and Tenneco Oil Company	Neural networks used to help pinpoint oil and gas deposits
Government	Internal Revenue Service	Software being tested to read tax returns and spot fraud
Human services	Merced County, California	Expert systems used to decide if applicants should receive welfare benefits
Marketing	Spiegel	Neural networks used to determine most likely buyers from a long list
Telecommunications	BT Group	Heuristic search used for a scheduling application that provides work schedules of more than 20,000 engineers
Transportation	American Airlines	Expert systems used to schedule the routine maintenance of airplanes
Inventory/forecasting	Hyundai Motors	Neural networks and expert systems used to reduce delivery time by 20 percent and increase inventory turnover from 3 to 3.4
Inventory/forecasting	SCI Systems	Neural networks and expert systems used to reduce on-hand inventory by 15 percent, resulting in $180 million in annual savings
Inventory/forecasting	Reynolds Aluminum	Neural networks and expert systems used to reduce forecasting errors by 2 percent, resulting in an inventory reduction of 1 million pounds
Inventory/forecasting	Unilever	Neural networks and expert systems used to reduce forecasting errors from 40 percent to 25 percent, resulting in a multimillion-dollar savings

Decision makers use information technologies in the following types of decision-making analyses:[5]

- *What-is*—This analysis is commonly used in transaction processing systems and management information systems. For example, you enter a customer account number, and the system displays the customer's current balance. However, these systems lack the capability to report real-time information or predict what could happen in the future. For example, reports generated by accounting information systems that show past performance over the preceding fiscal quarter consist of past events, so decision makers can't do much with this information.

- *What-if*—This analysis is used in decision support systems. Decision makers use it to monitor the effect of a change in one or more variables. It is available in spreadsheet programs, such as Microsoft Excel.

In addition to these types of analyses, decision makers often need to answer the following questions about information: Why? What does it mean? What should be done? When should it be done? AI technologies have the potential to help decision makers address these questions.

1.2 Robotics

Robots and robotics are some of the most successful applications of AI. You're probably familiar with robots used in factories or ones you've seen on the news. They're far from being intelligent, but progress has been steady. They perform well at simple, repetitive tasks and can be used to free workers from tedious or hazardous jobs. Robots are currently used mainly on assembly lines in Japan and the United States as part

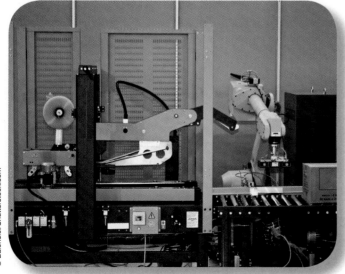

of computer-integrated manufacturing, but they're also used in the military, aerospace, and medical industries as well as for performing services such as delivering mail to employees.

The cost of industrial robots can range from $100,000 to $250,000 or more. Typically, their mobility is limited. For example, they might have only a fixed arm that moves objects from one point to another. Some robots have limited vision that's useful for locating and picking up objects, as long as the objects are isolated from other objects. A robot's operation is controlled by a computer program that includes commands such as when and how far to reach, which direction to go or turn, when to grasp an object, and how much pressure to apply. Some examples of programming languages for controlling robots are Variable Assembly Language (VAL), Functional Robotics (FROB), and A Manufacturing Language (AML). These languages are usually proprietary, meaning they're specific to a robot manufacturer.

One of the most advanced and most popular robots is Honda's Advanced Step in Innovative Mobility (ASIMO) (*http://world.honda.com/ASIMO/*). Honda's intelligence technologies enable ASIMO to work with other robots in coordination. It recognizes moving objects, sound, gesture, multiple environments, different faces, and postures. ASIMO is also able to choose between stepping back and yielding the right-of-way or continuing to walk based on the predicted movement of oncoming people. It is able to automatically charge its battery when its charge falls below a certain level.

Personal robots have attracted a lot of attention recently. These robots have limited mobility, limited vision, and some speech capabilities. Currently, they're used mostly as prototypes designed to perform useful functions like helping the elderly, bringing breakfast to your table, cooking for you, and opening doors and carrying trays and drinks. Examples of these robots include Twendy-One Robot, Motoman Robot, and ApriAttenda Robot.

PR2 is one of the most successful and advanced personal robots on the market today. It performs many ordinary tasks around the home and office. Robots have some unique advantages in the workplace compared with humans:

- They don't fall in love with coworkers, get insulted, or call in sick.

> **Robots** and robotics are some of the most successful applications of AI. They perform well at simple, repetitive tasks and can be used to free workers from tedious or hazardous jobs.

2 Expert Systems

© Angelo Gilardelli/Shutterstock.com

expert systems have been one of the most successful AI-related technologies and have been around since the 1960s. They mimic human expertise in a particular field to solve a problem in a well-defined area. For the purposes of this book, an expert system consists of programs that mimic human thought behavior in a specific area that human experts have solved successfully. The first expert system, called DENDRAL, was developed in the mid-1960s at Stanford University to determine the chemical structure of molecules. For expert systems to be successful, they must be applied to an activity that human experts have already handled, such as tasks in medicine, geology, education, and oil exploration. PortBlue (PortBlue Corporation) is an example of an expert system that can be applied to various financial applications, including examination of complex financial structures, foreign exchange risk management, and more. COGITO (by Italian-based Expert System) is used for monitoring consumer sentiments in blogs, comment sections, message boards, and Web-based articles. It is also used in search engines to better understand users' queries.[6]

Decision support systems generate information by using data, models, and well-defined algorithms, but expert systems work with heuristic data. Heuristics consist of common sense, rules of thumb, educated guesses, and instinctive judgments, and using heuristic data encourages applying knowledge based on experience to solve or describe a problem. In other words, heuristic data isn't formal knowledge, but it helps in finding a solution to a problem without following a rigorous algorithm.

- They're consistent.
- They can be used in environments that are hazardous to humans, such as working with radioactive materials.
- They don't spy for competitors, ask for a raise, or lobby for longer breaks.

Developments in AI-related fields, such as expert systems and natural-language processing, will affect the future development of the robotics industry. For example, natural-language processing will make it easier to communicate with robots in human languages.

Expert systems mimic human expertise in a particular field to solve a problem in a well-defined area.

A **knowledge acquisition facility** is a software package with manual or automated methods for acquiring and incorporating new rules and facts so that the expert system is capable of growth.

A **knowledge base** is similar to a database, but in addition to storing facts and figures it keeps track of rules and explanations associated with facts.

2.1 Components of an Expert System

A typical expert system includes the components described in the following list, which are shown in Exhibit 13.1:

- *Knowledge acquisition facility*—A **knowledge acquisition facility** is a software package with manual or automated methods for acquiring and incorporating new rules and facts so that the expert system is capable of growth. This component works with the KBMS (described later in this list) to ensure that the knowledge base is as up to date as possible.
- *Knowledge base*—A **knowledge base** is similar to a database, but in addition to storing facts and figures

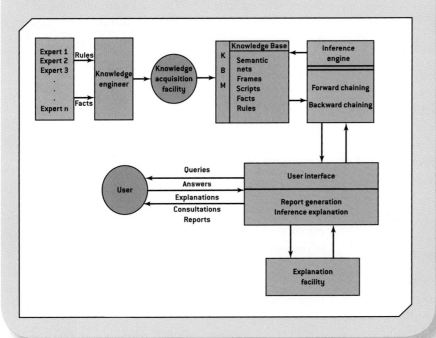

Exhibit 13.1 *An expert system configuration*

it keeps track of rules and explanations associated with facts. For example, a financial expert system's knowledge base might keep track of all figures constituting current assets, including cash, deposits, and accounts receivable. It might also keep track of the fact that current assets can be converted to cash within one year. An expert system in an academic environment might include facts about all graduate students, such as GMAT scores and GPAs, as well as a rule specifying that classified graduate students must have a GMAT of 650 or better and a GPA of 3.4 or better. To be considered part of a true expert system, the knowledge base component must include the following types of knowledge:

- *Factual knowledge*: Facts related to a specific discipline, subject, or problem. For example, facts related to kidney problems might include kidney size, blood levels of certain enzymes, and duration and location of pain.

- *Heuristic knowledge*: Rules related to a problem or discipline. For example, the general rules indicating that a patient has a kidney problem could include severe pain in the lower left or lower right of the back and high levels of creatinine and blood urea nitrogen.

- *Meta-knowledge*: Meta-knowledge is knowledge about knowledge. It enables an expert system to learn from experience and examine and extract relevant facts to determine the path to a solution. For example,

knowing how an expert system makes decisions is considered meta-knowledge. Although this type of knowledge isn't currently available in expert systems, integrating neural networks into expert systems is one possibility for acquiring meta-knowledge.

- *Knowledge base management system*—A **knowledge base management system (KBMS)**, similar to a DBMS, is used to keep the knowledge base updated, with changes to facts, figures, and rules.

- *User interface*—This is the same as the dialog management component of a decision support system. It provides user-friendly access to the expert system. Although GUIs have improved this component, one goal of AI technology is to provide a natural language for the user interface (discussed later in the chapter).

- *Explanation facility*—An **explanation facility** performs tasks similar to what a human expert does by explaining to end users how recommendations are derived. For example, in a loan evaluation expert system, the explanation facility states why an applicant was approved or rejected. In a medical expert system, it explains why the system concluded that a patient has a kidney stone, for instance. This component is important because it helps give users confidence in the system's results.

- *Inference engine*—An **inference engine** is similar to the model base component of a decision support system (discussed in Chapter 12). By using different techniques, such as forward and backward chaining (discussed in the following paragraphs), an inference engine manipulates a series of rules. Some inference engines work from a matrix of facts that includes several rows of conditions and rules, similar to a decision table. In this case, rules are evaluated one at a time, then advice is provided. Some inference engines also learn from doing.

A **knowledge base management system (KBMS)**, similar to a DBMS, is used to keep the knowledge base updated, with changes to facts, figures, and rules.

An **explanation facility** performs tasks similar to what a human expert does by explaining to end users how recommendations are derived.

An **inference engine** is similar to the model base component of a decision support system. By using different techniques, such as forward and backward chaining, it manipulates a series of rules.

In **forward chaining**, a series of "if-then-else" condition pairs is performed. The "if" condition is evaluated first, then the corresponding "then-else" action is carried out. For example, "if" the temperature is less than 80° and the grass is 3 inches long, "then" cut the grass or "else" wait. In a medical diagnostic expert system, the system could evaluate a problem as follows:

- If the patient's temperature is over 101° and
- If the patient has a headache
- Then it's very likely (a 95 percent chance) that the patient has the flu, or else search for other diseases.

In **backward chaining**, the expert system starts with the goal first—the "then" part—and backtracks to find the right solution. In other words, to achieve this goal, what conditions must be met? To understand the differences between these two techniques, consider the following example. In an expert system that provides financial investment advice for investors, the system might use forward chaining and ask 50 questions to determine which of five investment categories—oil-gas, bonds, common stocks, public utilities, and transportation—is more suitable for an investor.[7] In addition, each investor is in a specific tax bracket, and each investment solution provides a different tax shelter. In forward chaining, the system evaluates all the if-then-else conditions before making the final recommendation. In backward chaining, the system might start with the public utilities

category, specified by the investor, and go through all the "if" conditions to see whether this investor qualifies for this investment category. The backward chaining technique can be faster in some situations because it doesn't have to consider irrelevant rules, but the solution the system recommends might not be the best one.

Other techniques are used for representing knowledge in the expert system's knowledge base, such as semantic (associative) networks that represent information as links and nodes, frames that store conditions or objects in hierarchical order, and scripts that describe a sequence of events. For a child's birthday party, for example, events might include buying cake, inviting friends, serving the cake, and lighting the candles. A script for generating a purchase order might include events such as identifying the quantity to order, identifying the supplier and gathering data, generating the purchase order and sending it to the supplier, updating accounts payable, and informing the receiving department that an order has been placed.

2.2 Uses of Expert Systems

Many companies are engaged in research and development of expert systems, and these systems are now used in areas such as the following:

- *Airline industry*—American Airlines developed an expert system to manage frequent flier transactions.
- *Forensics lab work*—Expert systems are used to review DNA samples from crime scenes and generate results quickly and accurately. They help reduce the backlog in labs and get data entered in national crime databases faster.[8]
- *Banking and finance*—JPMorgan Chase developed a foreign currency trade expert system to assess historical trends, new events, and buying and selling factors.
- *Education*—Arizona State University developed an expert system to teach math and evaluate students' math skills.
- *Food industry*—Campbell's Soup Company developed an expert system to capture expertise that a highly specialized, long-time employee had about plant operations and sterilizing techniques.
- *Personnel management*—IBM developed an expert system to assist in training technicians; it has reduced training time.
- *Security*—Canada Trust Bank (now part of TD Canada Trust) developed an expert system to track credit card holders' purchasing trends and report deviations, such as unusual activity on a card.

© Katielittle/Shutterstock.com

- *U.S. government*—Expert systems have been developed to monitor nuclear power plants and assist departments such as the IRS, INS, U.S. Postal Service, Department of Transportation, Department of Energy, and Department of Defense in decision-making processes.
- *Agriculture*—The National Institute of Agricultural Extension Management has designed an expert system to diagnose pests and diseases in rice crops and suggest preventive measures.[10]

The information box above highlights a real-life application of an expert system, this one in burglary and crime detection. The system reduced the time and expenses involved in police operations.

2.3 Criteria for Using Expert Systems

An expert system should be used if one or more of the following conditions exists:

- A lot of human expertise is needed but a single expert can't tackle the problem on his or her own. (An expert system can integrate the experience and knowledge of several experts more easily.)
- The knowledge that's needed can be represented as rules or heuristics; a well-defined algorithm isn't available.
- The decision or task has already been handled successfully by human experts, allowing the expert system to mimic human expertise.
- The decision or task requires consistency and standardization. (Because computers are more accurate at following standard procedures, an expert system can be preferable to humans in this situation.)
- The subject domain is limited. (Expert systems work better if the problem under investigation is narrow.)
- The decision or task involves many rules (typically between 100 and 10,000) and complex logic.

- There's a scarcity of experts in the organization, or key experts are retiring. (An expert system can be used to capture the knowledge and expertise of a long-time employee who's retiring.)

2.4 Criteria for Not Using Expert Systems

The following types of problems are unsuitable to expert systems:

- There are very few rules (less than 10). (Human experts are more effective at solving these problems.)
- There are too many rules (usually more than 10,000). (Processing is slowed down to unacceptable levels.)
- Well-structured numerical problems (such as payroll processing) are involved, which means that standard transaction-processing methods can handle it more quickly and economically.
- The problems are in areas that cover a broad range of topics, and there are not many rules involved. (Expert systems work better in deep and narrow problem areas.)
- There's a lot of disagreement among experts.
- The problems require human experts—for instance, they require some combination of the five senses, such as taste and smell. Selecting a perfume is a problem better solved by human experts.

2.5 Advantages of Expert Systems

Using an expert system instead of humans for some tasks can have the following advantages, among others:

- An expert system never becomes distracted, forgetful, or tired. Therefore, it's particularly suitable for monotonous tasks that human workers might object to.

Case-based reasoning (CBR) is a problem-solving technique that matches a new case (problem) with a previously solved case and its solution, both stored in a database. After searching for a match, the CBR system offers a solution; if no match can be found, even after supplying more information, the human expert must solve the problem.

Intelligent agents are an application of artificial intelligence that are becoming more popular, particularly in e-commerce. They consist of software capable of reasoning and following rule-based processes.

- An expert system duplicates and preserves the expertise of scarce experts and can incorporate the expertise of many experts.

- An expert system can preserve the expertise of employees who are retiring or leaving an organization.

- An expert system creates consistency in decision making and improves the decision-making skills of nonexperts.

3 Case-Based Reasoning

xpert systems solve a problem by going through a series of if-then-else rules, but **case-based reasoning (CBR)** is a problem-solving technique that matches a new case (problem) with a previously solved case and its solution, both stored in a database. Each case in the database is stored with a description and keywords that identify it. If there's no exact match between the new case and cases stored in the database, the system can query the user for clarification or more information. After finding a match, the CBR system offers a solution; if no match can be found, even after supplying more information, the human expert must solve the problem. The new case and its solution are then added to the database.

Hewlett-Packard uses CBR to assist users of its printers; this system performs the role of a help desk operator. Users' complaints and difficulties over the past several years have been stored in a database as cases and solutions. This information is used in dealing with new users, who likely have the same problems as users in the past. In the long term, these systems can improve customer service and save money by reducing the number of help desk employees.

As another example, some banks use a CBR system to qualify customers for loans, using parameters from past customers stored in a database. These parameters include gross income, number of dependents, total assets, net worth, and amount requested for the loan. The database also stores the final decision on each application (acceptance or rejection). When a new customer applies for a loan, the CBR system can compare the application with past applications and provide a response. The new application and its outcome then become part of the database for future use.

 ## 4 Intelligent Agents

ntelligent agents, or bots (short for robots), are applications of artificial intelligence and are becoming more popular, particularly in e-commerce. They consist of software capable of reasoning and following rule-based processes. Some important characteristics of a sophisticated intelligent agent are as follows:[11]

- *Adaptability*—Able to learn from previous knowledge and go beyond information given previously. In other words, the system can make adjustments.

- *Autonomy*—Able to operate with minimum input. The system can respond to environmental stimuli, make a decision without users telling it to do so, and take preemptive action, if needed.

- *Collaborative behavior*—Able to work and cooperate with other agents to achieve a common objective.

- *Humanlike interface*—Able to interact with users in a more natural language.

- *Mobility*—Able to migrate from one platform to another with a minimum of human intervention.

- *Reactivity*—Able to select problems or situations that need attention and act on them. An agent with this capability typically responds to environmental stimuli.

Intelligent agents can collect information about customers, such as items purchased, demographic information, and expressed and implied preferences.

Most intelligent agents today fall short of these capabilities, but improvement is expected in the near future. One important application of intelligent agents that's already available is Web marketing. Intelligent agents can collect information about customers, such as items purchased, demographic information, and expressed and implied preferences. E-commerce sites then use this information to better market their products and services to customers. Other agents, called product-brokering agents, can alert customers to new products and services. Amazon.com has used these agents successfully.

Intelligent agents are also used for smart or interactive catalogs, called "virtual catalogs." A virtual catalog displays product descriptions based on customers' previous experiences and preferences. Currently, Stanford's Center for Information Technology (CIT) is working with IBM, Hewlett-Packard, and National Semiconductor to develop smart catalogs. The CIT project's goals include dynamic creation and updates for catalogs, the ability to search them by content instead of navigating via links, and cross-referencing so that users can find items meeting their needs in multiple catalogs.

Several categories of intelligent agents, or bots, are available, such as the following (discussed in the following sections):

- Shopping and information agents
- Personal agents
- Data-mining agents
- Monitoring and surveillance agents

You can check out additional categories of intelligent agents at BotSpot.com (*www.botspot.com*).

4.1 Shopping and Information Agents

Shopping and information agents help users navigate through the vast resources available on the Web and provide better results in finding information. These agents can navigate the Web much faster than humans and gather more consistent, detailed information. They can serve as search engines, site reminders, or personal surfing assistants. PriceScan (*www.pricescan.com*) is a commercial shopping agent that finds the lowest price for many items and displays all competitive prices. Another example is BestBookBuys.com (*www.bestbookbuys.com*). Using this agent, you identify a book by its title, author, or ISBN, then the agent finds all online booksellers carrying this book and organizes

them in a list from least expensive to most expensive. Another comparison-shopping agent is available at *www.mysimon.com*.

Usenet and newsgroup agents have sorting and filtering features for finding information. For example, DogPile (*www.dogpile.com*) searches the Web by using several search engines, including Google, Yahoo!, Lycos, and Excite, to find information for users. DogPile can remove duplicate results and analyze the results to sort them with the most relevant results at the top.

4.2 Personal Agents

Personal agents perform specific tasks for a user, such as remembering information for filling out Web forms or completing e-mail addresses after the first few characters are typed. An e-mail personal agent can usually perform the following tasks:

- Generate auto-response messages
- Forward incoming messages
- Create e-mail replies based on the content of incoming messages

4.3 Data-Mining Agents

Data-mining agents work with a data warehouse and can detect trend changes and discover new information and relationships among data items that aren't readily apparent. Volkswagen Group, for example, uses a data-mining agent that acts as an early-warning system about market conditions. For example, the data-mining agent might detect a problem that could cause worsening economic conditions and, therefore, delay payments from customers. Having this information early enables decision makers to come up with a solution that minimizes the negative effects of the problem.

Shopping and information agents help users navigate through the vast resources available on the Web and provide better results in finding information. These agents can navigate the Web much faster than humans and gather more consistent, detailed information. They can serve as search engines, site reminders, or personal surfing assistants.

Personal agents perform specific tasks for a user, such as remembering information for filling out Web forms or completing e-mail addresses after the first few characters are typed.

Data-mining agents work with a data warehouse and can detect trend changes and discover new information and relationships among data items that aren't readily apparent.

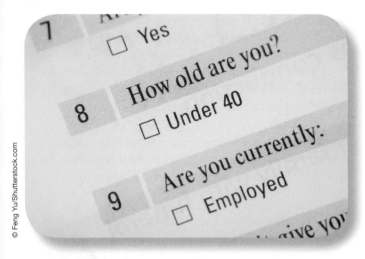

© Feng Yu/Shutterstock.com

4.4 Monitoring and Surveillance Agents

Monitoring and surveillance agents usually track and report on computer equipment and network systems to predict when a system crash or failure might occur. NASA's Jet Propulsion Laboratory, for example, has an agent that monitors inventory, planning, and the scheduling of equipment to keep costs down.[12]

5 Fuzzy Logic

have you ever been given a questionnaire that asks ambiguous questions but expects *yes* or *no* responses? Although you might be tempted to use words such as *usually*, *sometimes*, *only if*, and the like, you know the software used to analyze responses simply can't deal with anything but clear-cut *yes* and *no*

Monitoring and surveillance agents usually track and report on computer equipment and network systems to predict when a system crash or failure might occur.

Fuzzy logic allows a smooth, gradual transition between human and computer vocabularies and deals with variations in linguistic terms by using a degree of membership.

responses. However, with the development of fuzzy logic, a wide variety of responses is possible in questionnaires and other survey tools. **Fuzzy logic** allows a smooth, gradual transition between human and computer vocabularies and deals with variations in linguistic terms by using a degree of membership. A degree of membership shows how relevant an item or object is to a set. A higher number indicates it is more relevant, and a lower number shows it is less. For example, when heating water, as the temperature changes from 50°C to 75°C, you might say the water is warm. What about when the water's temperature reaches 85°C? You can describe it as warmer, but at what point do you describe the water as hot? Describing varying degrees of warmness and assigning them membership in certain categories of warmness involves a lot of vagueness.

Fuzzy logic is designed to help computers simulate vagueness and uncertainty in common situations. Lotfi A. Zadeh developed the fuzzy logic theory in the mid-1960s by using a mathematical method called "fuzzy sets" for handling imprecise or subjective information.[13] Fuzzy logic allows computers to reason in a fashion similar to humans and makes it possible to use approximations and vague data yet produce clear and definable answers.

Fuzzy logic works based on the degree of membership in a set (a collection of objects). For example, heights of 4 feet, 5 feet, and 6 feet could constitute a set of heights for a population. Fuzzy sets have values between 0 and 1, indicating the degree to which an element has membership in the set. At 0, the element has no membership; at 1, it has full membership.

In a conventional set (sometimes called a "crisp" set), membership is defined in a black-or-white fashion; there's no room for gray. For instance, if 90 percent or higher means a "Pass" grade in a course, getting 89.99 percent doesn't give you membership in the "Pass" area of this crisp set. Therefore, despite getting 89.99 percent, you have failed the course. In this example, there's a very small difference between the two grades (0.01), but it means the difference between passing and failing. In other words, a small difference has a huge impact. This doesn't happen in a fuzzy logic environment. To help you understand the membership

Fuzzy logic has been used in search engines, chip design, database management systems, software development, and other areas.

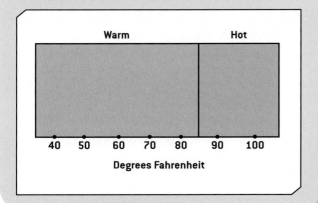

Exhibit 13.2 *An example of a conventional set*

- Dryers that convert information on load size, fabric type, and flow of hot air into drying times and conditions
- Refrigerators that set defrosting and cooling times based on usage patterns
- Shower systems that suppress variations in water temperature
- TVs that adjust screen color and texture for each frame and stabilize the volume based on the viewer's location in the room
- Video camcorders that eliminate shakiness in images (common with handheld video cameras) and adjust focus and lighting automatically

function better, Exhibit 13.2 shows an example of a conventional set. In this example, 84.9° F is considered warm and 85.1° F is considered hot. This small change in temperature can cause a large response in the system.

Exhibit 13.3 shows the same set but with fuzzy logic conventions. For example, 80° F has a membership of 30 percent in the fuzzy set "Warm" and 40 percent in the fuzzy set "Hot." All temperatures from 40° to 100° F make up the membership set.

5.1 Uses of Fuzzy Logic

Fuzzy logic has been used in search engines, chip design, database management systems, software development, and other areas.[14] You might be more familiar with its uses in appliances, as shown in the following examples:

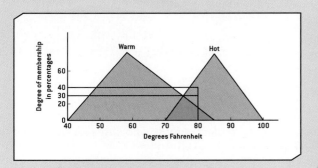

Exhibit 13.3 *Degree of membership in a fuzzy system*

6 Artificial Neural Networks

rtificial neural networks (ANNs) are networks that learn and are capable of performing tasks that are difficult with conventional computers, such as playing chess, recognizing patterns in faces and objects, and filtering spam e-mail. Like expert systems, ANNs are used for poorly structured problems—when data is fuzzy and uncertainty is involved. Unlike an expert system, an ANN can't supply an explanation for the solution it finds because an ANN uses patterns instead of the if-then-else rules that expert systems use.

An ANN creates a model based on input and output. For example, in a loan application problem, input data consists of income, assets, number of dependents, job history, and residential status. The output data is acceptance or rejection of the loan application. After processing many loan applications, an ANN can establish a pattern that determines whether an application is approved or rejected.

As shown in Exhibit 13.4, an ANN has an output layer, an input layer, and a middle (hidden) layer where learning takes place. If you're using an ANN for approving loans in a bank, the middle layer is trained by using past data (from old loan applications, in which the decisions are known) that includes both accepted and rejected applica-

Artificial neural networks (ANNs) are networks that learn and are capable of performing tasks that are difficult with conventional computers, such as playing chess, recognizing patterns in faces and objects, and filtering spam e-mail.

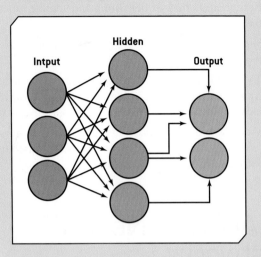

Exhibit 13.4 *An artificial neural network configuration*

Intput

Hidden

Output

and when organizational policies change, the network needs to be retrained so that it can mimic the new policies.

ANNs are used for many tasks, including the following:

- Bankruptcy prediction
- Credit rating
- Investment analysis
- Oil and gas exploration
- Target marketing

The information box below discusses real-life applications of neural networks.

7 Genetic Algorithms

lthough they are not as widely accepted, **genetic algorithms (GAs)** have become more recognized as a form of artificial intelligence. They're used mostly in techniques to find solutions to optimization and search problems. John Holland developed genetic algorithms in the 1940s at MIT, and the term applies to adaptive procedures in a computer system that are based on Darwin's ideas about natural selection and survival of the fittest.[17]

Genetic algorithms are used for optimization problems that deal with many input variables, such as jet engine design, portfolio development, and network

Genetic algorithms (GAs) are used mostly in techniques to find solutions to optimization and search problems.

tions. Based on the pattern of data entered in the input layer—applicant's information, loan amount, credit rating, and so on—and the results in the output layer (the accept or reject decision), nodes in the middle layer are assigned different weights. These weights determine how the nodes react to a new set of input data and mimic decisions based on what they've learned. Every ANN has to be trained,

---

Neural Networks in Microsoft and the Chicago Police Department

Microsoft is using neural network software to maximize the returns on direct mail. Each year, Microsoft sends out approximately 40 million pieces of direct mail to 8.5 million registered customers. The goal of these mailings is to encourage customers to upgrade their software or buy other Microsoft products. The first mailing goes out to all the customers in the company's database. The second mailing goes out only to those customers who are most likely to respond, and neural network software is used to cull the latter from the former. According to Microsoft spokesman Jim Minervino, BrainMaker, the neural network software, has increased the rate of response from 4.9 percent to 8.2 percent. This has resulted in a significant savings for the company—the same revenue at 35 percent less cost.[15]

The Chicago Police Department has used neural network software to predict which police officers are likely to engage in misconduct. Here, BrainMaker has compared the conduct of current officers with the conduct of those who have previously been terminated for disciplinary reasons, and this has produced a list of officers that might be "at risk."[16]

Several other real-life applications of neural networks are posted at: www.calsci.com/BrainIndex.html.

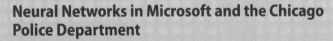

design. They find the combination of inputs that generates the most desirable outputs, such as the stock portfolio with the highest return or the network configuration with the lowest cost. Genetic algorithms can examine complex problems without any assumptions of what the correct solution should be. In a GA system, the following techniques are used:

- *Selection or survival of the fittest*—Gives preference or a higher weight to better outcomes.
- *Crossover*—Combines good portions of different outcomes to achieve a better outcome.
- *Mutation*—Tries combinations of different inputs randomly and evaluates the results.

Genetic algorithms are already used with neural networks and fuzzy logic systems to solve scheduling, engineering design, and marketing problems, among others. For example, a docking algorithm uses a neural network and fuzzy functions with a GA to find the best and shortest route for a robot to take to a docking bay.[18] In addition, researchers at General Electric and Rensselaer Polytechnic Institute have used a GA to design a jet engine turbine in one fourth the time it took to develop a model manually. It improved the design by 50 percent and kept up with the many variables involved better than an expert system did.

Some hybrid products that combine AI technologies use genetic algorithms, such as GeneHunter (*www.wardsystems.com/genehunter.asp*) and NeuroDimension (*www.nd.com/genetic/*).

You can find more information on the uses of GAs in robotics, telecommunications, computer games, and other fields at *http://brainz.org/15-real-world%20-applications-genetic-algorithms*.

8 Natural-Language Processing

espite constant efforts to make information systems user friendly, they still require a certain degree of computer literacy and skills. As mentioned in Chapter 2, **natural-language processing (NLP)** was developed so that users could communicate with computers in their own language. Although GUI elements, such as menus and icons, have helped with communication problems between humans and com-

Table 13.2
NLP systems

NLP system	Use
Dragon Speech Recognition	
Nuance Communications, Inc., (*www.nuance.com/naturallyspeaking*)	Business data retrieval, legal document processing, medical and ER applications, professional dictation systems, respectively
AT&T Natural Voices (*www.wizzardsoftware.com/att_nv_landing.php*)	Creates speech from computer-readable text
e-Speaking software (*www.e-speaking.com*)	Voice and speech recognition for Windows computers

puters, GUIs still involve some training, can be cumbersome to use, and often differ depending on the OS or application. An NLP system provides a question-and-answer setting that's more natural and easier for people to use. It's particularly useful with databases. Table 13.2 lists some currently available NLP systems.

At the time of this writing, these products aren't capable of a dialogue that compares with conversations between humans. The size and complexity of the human language has made developing NLP systems difficult. However, progress has been steady, and NLP systems for tasks such as call routing, stock and bond trading, and banking by phone, among others, are already available.

NLP systems are generally divided into the following categories:[18]

- Interface to databases
- Machine translation, such as translating from French to English
- Text scanning and intelligent indexing programs for summarizing large amounts of text
- Generating text for automated production of standard documents
- Speech systems for voice interaction with computers

Natural-language processing (NLP) was developed so that users could communicate with computers in their own language.

NLP systems usually perform two types of activities. The first is interfacing: accepting human language as input, carrying out the corresponding command, and generating the necessary output. The second is knowledge acquisition: using the computer to read large amounts of text and understand the information well enough to summarize important points and store information so that the system can respond to inquiries about the content.

9 Integrating AI Technologies into Decision Support Systems

AI-related technologies, such as expert systems, natural-language processing, and artificial neural networks, can improve the quality of decision support systems (DSSs). They can add explanation capabilities (by integrating expert systems) and learning capabilities (by integrating ANNs) and create an interface that's easier to use (by integrating an NLP system). These systems are sometimes called integrated DSSs (IDSSs), and the result is a more efficient, powerful DSS.[19] AI technologies, particularly expert systems and natural-language processing, can be integrated into the database, model base, and user interface components of a DSS.

The benefits of integrating an expert system into the database component of a DSS are as follows:[21]

- Adding deductive reasoning to traditional DBMS functions
- Improving access speed
- Improving the creation and maintenance of databases
- Adding the capability to handle uncertainty and fuzzy logic
- Simplifying query operations with heuristic search algorithms

Similarly, you can add AI technologies to a DSS's model base component. For example, expert systems can be added to provide reasons and explanations for output from the model base, to include heuristics in the model base's analysis capabilities, to incorporate fuzzy sets in the model-building process, to reduce the time and cost of calculating data for models, and to select the best model for the problem.[21]

In addition, integrating expert system capabilities into the user interface component can improve the quality and user friendliness of a DSS. This integration can add features such as an explanation capability (explaining responses in more nontechnical terms). Integrating NLP can improve the effectiveness of an interface, too, by making it easier to use, particularly for decision makers who aren't computer savvy.

The Industry Connection highlights Alyuda Research Company, a leading developer of neural network and trading software for business and personal use.

Industry Connection

ALYUDA RESEARCH *

Alyuda Research is a major developer of neural networks and trading software for businesses and individuals. Its products and services include the following:

- *Tradecision*—Provides tools to help investors and brokers make better decisions, such as advanced charting and automated trading. Includes modules for building models, strategies, alerts, and indicators and for managing simulations and data analysis.

- *Scorto Credit Decision*—Offers several methods for developing models for credit scoring, such as decision trees, neural networks, and fuzzy logic, and includes software for loan portfolio analysis.

- *NeuroIntelligence*—Used to analyze and process data sets; find the best neural network architecture; train, test, and optimize a neural network; and apply the network to new data sets.

* This information has been gathered from the company Web site (*www.alyuda.com*) and other promotional materials. For more information and updates, visit the Web site.

10 Chapter Summary

In this chapter, you've learned about intelligent information systems, including how AI technologies are used to support decision-making processes. Next, you explored expert systems, which mimic human expertise to solve problems in well-defined areas. An expert system consists of a knowledge acquisition facility, a knowledge base (with a knowledge base management system), a user interface, an explanation facility, and an inference engine. Another system for problem solving uses case-based reasoning, a technique that matches a new case (problem) with a previously solved case and its solution stored in a database. You also learned about intelligent agents (bots) as another application of AI technologies; they're used for many purposes, including shopping, information gathering, and data mining. Other techniques used in decision making and problem solving include artificial neural networks, fuzzy logic, and genetic algorithms. In addition, natural-language processing was discussed as an intelligent information system that can make interacting with computers easier for users. Finally, you reviewed the advantages of integrating AI technologies into decision support systems.

Key Terms

artificial intelligence (AI) (229)

artificial neural networks (ANNs) (239)

backward chaining (234)

case-based reasoning (CBR) (236)

data-mining agents (237)

expert systems (232)

explanation facility (233)

forward chaining (234)

fuzzy logic (238)

genetic algorithms (GAs) (240)

inference engine (233)

intelligent agents (236)

knowledge acquisition facility (232)

knowledge base (232)

knowledge base management system (KBMS) (233)

monitoring and surveillance agents (238)

natural-language processing (NLP) (241)

personal agents (237)

robots (231)

shopping and information agents (237)

Problems, Activities, and Discussions

1. Give some examples of situations in which using an expert system isn't suitable.

2. Describe the differences and similarities between expert systems and case-based reasoning.

3. What is fuzzy logic?

4. Expert systems could help fight poverty and be used in places where there are shortages of medical personnel. Visit the following Web page, consult other sources, and write a one-page paper that summarizes the applications of expert systems in the medical field.

   www.scientificamerican.com/article. cfm?id=expert-systems-fight-poverty

5. Fuzzy logic applications and other artificial intelligent technologies could be used for banking and loan applications. Visit the following Web page, consult other sources, and write a one-page paper that summarizes the applications of these systems in banking and loan application processing.

   www.artificialintelligence.suite101.com/article.cfm/ fuzzy_logic_applications_for_banking_and_loans

6. Twine (*www.evri.com/*) is a search engine and social networking site that uses semantic Web and artificial intelligent technologies. Based on the information provided on its Web site and information provided on the following link, summarize the applications of this platform. Also highlight the differences between this search technology and a traditional one.

   www.fastcompany.com/magazine/136/digg-this.html

7. Watch the video on the link below and summarize the applications of artificial neural networks.

   www.authorstream.com/Presentation/Group5ISM-277294-11-20-artificial-neural-networks-1-2-entertainment-ppt-powerpoint/

8. The link below provides comprehensive information on intelligent agents. Based on this information, identify five applications of intelligent agents.

   http://teachnet.edb.utexas.edu/~lynda_abbott/ webpage.html

9. AI encompasses which of the following related technologies? (Choose all that apply.)

   a. Electronic data processing

   b. Expert systems

   c. Genetic algorithms

   d. Intelligent agents

10. An expert system consists of programs that attempt to mimic human thought behavior in a specific area that human experts have solved successfully. True or False?

casestudy

GENETIC ALGORITHM AT STAPLES, INC.

Staples, Inc., is the world's largest office products company. In 2007, it decided to overhaul its paper offerings, which accounted for 41.8 percent of company sales. The goal was to find out how customers decided what to buy. In the past, Staples had conducted focus groups, which are time consuming and costly, and the results had not been ideal. So, it sought help from Affinnova, a provider of genetic algorithm (GA) software. Affinnova had assisted many corporations (including Procter & Gamble, Walmart, and Capital One) in reevaluating their product offerings. The main principle behind its software is that, over time, consumer markets evolve, with strong products surviving while weak ones die out. A panel of 750 customers from across the country participated in a 20-minute study of Staples' paper line. Each customer was shown three possible packaging designs on a screen and asked to select his or her favorite. The GA software then analyzed the

© Blaj Gabriel/Shutterstock.com

customers' choices in real time and presented three new designs. Altogether, 22,000 choices were put before Staples' customers. And by looking at those choices over multiple generations, the software identified preference patterns—a certain color or font, for example—as well as the important principles that Staples should use for its paper offerings. As a result of this study, Staples made several changes that the management believes will be good for business, both increasing sales and improving customer satisfaction.[22]

Answer the following questions:

1. What was the main reason for using the genetic algorithm software?

2. How did Staples collect the customer opinions regarding its paper offerings?

3. What was the finding of the study, and how did Staples respond?

MORE BANG
FOR YOUR BUCK!

MIS2 has it all, and so can you! Between the text and online resources, **MIS2** offers everything you need to study:
- In-Text Review Cards
- Online Tutorial Quizzes
- Printable Flash Cards
- Videos
- Interactive Games
- In-Text Cases
- And More!

Visit CourseMate to see all that MIS2 has to offer! Access at login.cengagebrain.com.

EMERGING TRENDS, TECHNOLOGIES, AND APPLICATIONS

t his chapter discusses new trends in software and service distribution, including pull technology, push technology, and application service providers that provide software as a service (SaaS). You also learn about virtual reality components and applications, including CAVE, and see how virtual worlds are becoming a new platform for communication and collaboration. Next, you learn about radio frequency identification (RFID) and its uses before moving on to new uses of biometrics and new trends in networking, including grid, utility, and cloud computing. Finally, you get an overview of how nanotechnology is being used and its future applications.

learning outcomes

After studying this chapter, you should be able to:

LO1 Summarize new trends in software and service distribution.

LO2 Describe virtual reality components and applications.

LO3 Discuss uses of radio frequency identification.

LO4 Summarize new uses of biometrics.

LO5 Explain new trends in networking, including wireless technologies and grid and cloud computing.

LO6 Discuss uses of nanotechnology.

1 Trends in Software and Service Distribution

recent trends in software and service distribution include pull technology, push technology, and application service providers, although pull technology has been around since the Web began. All of these trends are discussed in the following sections.

1.1 Pull and Push Technologies

With **pull technology**, a user states a need before getting information, as when a URL is entered in a Web browser so that the user can go to a certain Web site.

However, for marketing certain products and services and for providing customized information, this technology isn't adequate. People rarely request marketing information, for example. With **push technology**, or Webcasting, a Web server delivers information to users who have signed up for this service instead of waiting for users to request the information be sent to them. Webcasting is supported by many Web browsers and is also available from vendors (described later in this sec-

> With **pull technology**, a user states a need before getting information, as when a URL is entered in a Web browser so that the user can go to a certain Web site.
>
> With **push technology**, or Webcasting, a Web server delivers information to users who have signed up for this service instead of waiting for users to request the information be sent to them.

© Image Source/Getty Images

tion). With push technology, your favorite Web content can be updated in real time and sent to your desktop. Push technology can be effective for B2C and B2B marketing, too. For example, a car manufacturer can send the latest information on new models, prices, features, and related information to all its dealers in real time. Network administrators also use push technology to have antivirus updates downloaded on employees' workstations automatically.

Push technology delivers content to users automatically at set intervals or when a new event occurs. For example, you often see notices such as "A newer version of Adobe Flash is available. Would you like to install it?" In this case, the vendor (Adobe) is pushing the updated product to you as soon as it's available, which is the event triggering the push. Of course, this example assumes you have already downloaded a previous version of Adobe Flash; by doing so, you have signed up for pushed updates. The same process applies to content updates, such as news and movie releases. When users sign up, they specify what content they want (sports, stock prices, political news, etc.) and consent to the "push." They can also specify how often the content should be pushed. For example, if you have subscribed to an online news service and have indicated that you are interested in the latest economic news on China, this online service will send you such news as soon as it becomes available and will do so in the future as well. You do not need to make any further request.

Push technology streamlines the process of users getting software updates and updated content. It benefits vendors, too, because by keeping in constant touch with users, they build customer loyalty. This benefit often outweighs the costs of adding servers and other technology resources needed to use push technology. Push technology also improves business relations and customer service because users get the information they need in a more timely fashion.

Application service providers (ASPs) provide access to software or services for a fee.

Software as a service (SaaS), or on-demand software, is a model for ASPs to deliver software to users for a fee; the software might be for temporary or for long-term use.

Below are examples of push technology in action:

- Research In Motion (RIM) offers a new BlackBerry push API (application programming interface) that allows software developers to push real-time content, as well as alerts, to BlackBerry smartphone users. The content includes news, weather, banking, medical, and games.

- Microsoft Direct Push (AT&T) enables mobile professionals to stay connected to their Microsoft Outlook information while on the go. It helps users work more efficiently with full wireless synchronization of e-mail, calendar, contacts, and tasks on a Windows Mobile-enabled smartphone. Users can receive and respond to e-mail quickly through a Microsoft Office Outlook Mobile interface.

1.2 Application Service Providers

In Chapter 7, you learned about ISPs, which provide access to the Internet for a fee. A more recent business model, called **application service providers (ASPs)**, provides access to software or services for a fee. **Software as a service (SaaS)**, or on-demand software, is a model for ASPs to deliver software to users for a fee; the software might be for temporary or for long-term use. With this delivery model, users don't need to be concerned with new software versions and compatibility problems because the ASP offers the most recent version of the software. Users can also save all application data on the ASP's server so that the software and data are portable. This flexibility is convenient for those who travel or work in different locations, but it can also create privacy and security issues. Saving data on the ASP's servers instead of on users' own workstations might leave this data more exposed to theft or corruption by attackers.

Here's a simple example of how SaaS might work: Say you want to edit a document, Chapter14.doc, and you need word-processing software for this task. With SaaS, you don't need the software installed on your computer. You simply access it from the SaaS provider site. You can then run the software from the provider's server (and not take up your computing resources) or on your computer. The location of the Chapter14.doc file doesn't matter. You make use of the provider's SaaS service to edit the document, which stays on your hard drive (or wherever you had it stored—a flash drive, for example). The word-processing application isn't stored on your computer, so the next time you access the word-processing software from the provider's SaaS site you might get a newer version of the word-processing software. SaaS deals only with software, not with data and document storage or with hardware resources, such as processing power and memory.

The SaaS model can take several forms, such as the following:

- Software services for general use, such as office suite packages
- Offering a specific service, such as credit card processing
- Offering a service in a vertical market, such as software solutions for doctors, accountants, and attorneys

Generally, the advantages of outsourcing, such as being less expensive and delivering information more quickly, apply to the ASP model, too. However, ASPs have some specific advantages, including the following:

- The customer doesn't need to be concerned about whether software is current.
- IS personnel time is freed up to focus on applications (such as customer relationship management and financial information systems) that are more strategically important to the organization.
- Software development costs are spread over several customers, so vendors can absorb some expenses of software development and develop more improved software.
- Software is kept up to date, based on users' requests.
- The ASP contract guarantees a certain level of technical support.
- An organization's software costs can be reduced to a predictable monthly fee.
  Some disadvantages of ASPs are as follows:
- Generally, users must accept applications as provided by ASPs; software customized to users' needs is not offered.

- Because the organization has less control over how applications are developed, there's the risk that applications might not fully meet the organization's needs.
- Integration with the customer's other applications and systems might be challenging.

Google, NetSuite, and Salesforce CRM are three companies that offer software as a service. Google Apps (*www.google.com/apps*) is a service from Google with several Google products. It features several Web applications with similar functionality to traditional office suites, including Gmail, Google Calendar, Talk, Docs, and Sites. The Standard Edition is free. In addition, Basecamp (*http://basecamphq.com*) and Mint.com (*www.mint.com*) offer SaaS. Basecamp is a Web-based project collaboration tool that allows users to share files, set deadlines, assign tasks, and receive feedback. Mint.com is a Web-based personal financial management service. SaaS is also common for human resources applications and has been used in ERP systems with vendors such as Workday (*www.workday.com*).

2 Virtual Reality

he goal of **virtual reality (VR)** is to create an environment in which users can interact and participate as they do in the real world. VR technology uses computer-generated, three-dimensional images to create the illusion of interaction in a real-world environment. It can be integrated with stereo sound and tactile sensations to give users the "feel" of being immersed in a three-dimensional real world. In VR terminology, the everyday physical world is referred to as an "information environment."

> **Virtual reality (VR)** uses computer-generated, three-dimensional images to create the illusion of interaction in a real-world environment.

The distinction between immersion in a VR world and analyzing the same information using blueprints, numbers, or text is the difference between looking at fish in an aquarium and putting on your scuba gear and diving in with them.[1]

Before VR technology, three-dimensional objects were viewed in a two-dimensional computer environment. Even the best graphics programs used a two-dimensional environment to illustrate a three-dimensional object. VR technology has added the third dimension so that users can interact with objects in a way that hasn't been possible before. Thomas Furness, a notable VR pioneer, states, "The distinction between immersion in a VR world and analyzing the same information using blueprints, numbers, or text is the difference between looking at fish in an aquarium and putting on your scuba gear and diving in with them."[1]

Virtual reality began with military flight simulations in the 1960s, but these VR systems were rudimentary compared with today's systems. In the 1990s, Japan's Matsushita built a virtual kitchen that enabled its

customers to change fixtures and appliances and alter the design on a computer and then virtually walk around the kitchen space. A customer's preferences could then become the blueprint for the kitchen's final design. This was the first VR system designed not for games but for general public use.

As you read through the following sections, you'll want to be familiar with these terms:

- *Simulation*—Giving objects in a VR environment texture and shading for a 3-D appearance.
- *Interaction*—Enabling users to act on objects in a VR environment, as by using a data glove to pick up and move objects.
- *Immersion*—Giving users the feeling of being part of an environment by using special hardware and software (such as a CAVE, discussed later in this section). The real world surrounding the VR environment is blocked out so that users can focus their attention on the virtual environment.
- *Telepresence*—Giving users the sense that they're in another location (even one geographically far away) and can manipulate objects as though they're actually in that location. Telepresence systems use a variety of sophisticated hardware, discussed in "Components of a Virtual Reality System."
- *Full-body immersion*—Allowing users to move around freely by combining interactive environments with cameras, monitors, and other devices.
- *Networked communication*—Allowing users in different locations to interact and manipulate the same world at the same time by connecting two or more virtual worlds.

2.1 Types of Virtual Environments

There are two major types of user environments in VR: egocentric and exocentric. In an **egocentric environment**, the user is totally immersed in the VR world. The most common technology used with this environment is a head-mounted display (HMD). Another technology, which uses lasers, is a virtual retinal display (VRD). Exhibit 14.1 shows an example of each of these devices, discussed more in the next section.

In an **exocentric environment**, the user is given a "window view." Data is still rendered in 3-D, but users can only view it on screen. They can't interact with objects, as in an egocentric environment. The main technology used in this environment is 3-D graphics.

In an **egocentric environment**, the user is totally immersed in the VR world.

In an **exocentric environment**, the user is given a "window view." Data is still rendered in 3-D, but users can only view it on screen. They can't interact with objects, as in an egocentric environment.

2.2 Components of a Virtual Reality System

The following are the major components of a VR system:

- *Visual and aural systems*—These components allow users to see and hear the virtual world. HMDs, mentioned earlier, contain two small TV screens, one in front of each eye, along with a magnifying lens to generate the view. Sensing devices on top of the helmet determine the orientation and position of the user's head. The information is then transmitted to the computer, which generates two pictures so that each eye has a slightly different view, just as humans eyes do. HMDs can also incorporate stereo sound into a VR environment to make the environment more convincingly real. With VRDs, a very low-power laser beam carrying an image is projected onto the back of the user's eyes. As with an HMD, users can move their heads in any direction without losing sight of the image.

- *Manual control for navigation*— This component allows the user to navigate in the VR environment and control various objects. The most commonly used device is the data glove (see Exhibit 14.2). With it, users can point to, "grab," and manipulate objects and experience limited tactile sensations, such as

Exhibit 14.1 *Egocentric VR technologies*

© DAJ/Getty Images

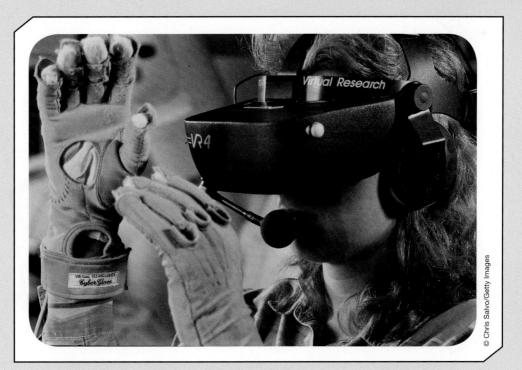

Exhibit 14.2 *VR components*

© Chris Salvo/Getty Images

With data gloves, they can even shake hands, although they might be thousands of miles apart physically.

determining an object's shape, size, and hardness or softness. A data glove can also be used as an input device, much like a mouse. Users can use a data glove with software to open a dialog box or pull down a menu, for example. A data glove is covered with optical sensors that send information to a computer that reconstructs the user's movements graphically. The agent representing the user's hand in the virtual world duplicates the user's hand movements.

- *Central coordinating processor and software system*—This component generates and manipulates high-quality graphics in real time, so it needs a very fast processor. To display images in real time, 3-D image graphics must be rendered rapidly, and the screen's refresh rate has to be extremely fast.

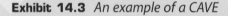

A **cave automatic virtual environment (CAVE)** is a virtual environment consisting of a cube-shaped room in which the walls are rear-projection screens. CAVEs are holographic devices that create, capture, and display images in true 3-D form.

- *Walker*—This input device captures and records movements of the user's feet as the user walks or turns in different directions.

2.3 CAVE

A **cave automatic virtual environment (CAVE)** is a virtual environment consisting of a cube-shaped room in which the walls are rear-projection screens. CAVEs are holographic devices that create, capture, and display images in true 3-D form (see Exhibit 14.3). People can enter CAVEs in other locations, no matter how far away they are geographically. Multiple people in different CAVEs can interact, too. High-speed digital cameras capture one user's presence and movements and then re-create and send these images to users in other CAVEs. In this way, people can carry on a conversation as though they're all in the same room.

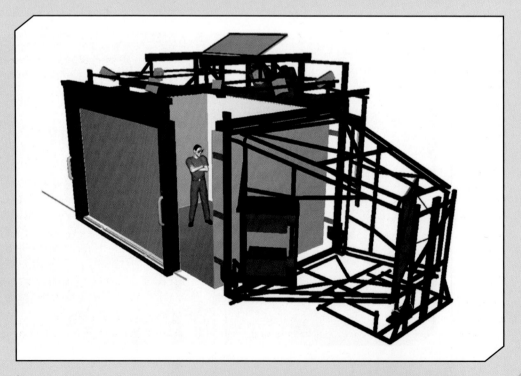

Exhibit 14.3 *An example of a CAVE*

CAVEs are used for research in many fields, including archaeology, architecture, engineering, geology, and physics. Some engineering companies use CAVEs to improve product design and development. With a CAVE, they can create and test prototypes, develop interfaces, and simulate factory layouts and assembly lines, for example, without investing in physical equipment and layouts. Many universities, including Brown University, the University of Illinois at Urbana-Champaign, and Duke University, use CAVEs for geological research, architectural studies, and anatomy studies.

2.4 Virtual Reality Applications

VR systems have been used in military flight simulations, in medicine for "bloodless" surgery, and in the entertainment industry. With further developments, they will one day be used for user interfaces in information systems. You might have seen an example of this in the movie *Minority Report*, in which Tom Cruise uses a 3-D user interface to examine documents, graphics, and video files in crime reports. This technology is here and has been used in many real-life applications.

VR systems can be used for many other business applications, too. For example, a VR system could be used for site selection when a company wants to open a new plant. A simulation model combined with VR capabilities would allow a virtual walk-through of the potential site, a more realistic view than is possible with maps and blueprints. The following are some current applications of VR systems:

- *Assistance for the disabled*—Virtual reality helps extend the capabilities of the disabled. For example, quadriplegics can use an HMD to get a 360-view of their surroundings, or people with cerebral palsy can learn how to operate a motorized wheelchair in a VR environment.[2]
- *Architectural design*—Architects and engineers use VR systems to create blueprints and build prototypes to show to clients. With a VR system, several versions of a design can be created to demonstrate to clients the outcome of modifying different factors. Architects and engineers can also use VR systems to test different conditions, such as wind shear, safely and without the expense of using physical materials.
- *Education*—VR systems are used in educational games and simulations, such as VR "flash cards" for teaching math skills. Incorporating visuals, sound, and touch into a game can help improve the learning process. For example, in a world geography class, a VR globe could be used with touch technology that

displays different facts about a country—language, population, political system, and so forth—when a student touches it.
- *Flight simulation*—Commercial airlines and the military have been using flight simulators for many years. Flight simulators are used for training pilots to handle new equipment or unusual operating conditions. Training in a VR environment is safer and less expensive than training on actual equipment.

VR systems can also be used in videoconferencing and group support systems. Current technologies using TV screens can't fully capture the sense of other people being physically present, and people can't shake hands or engage in direct eye contact effectively. VR systems could help overcome these obstacles by giving participants the impression of being in the same room, which makes achieving true interaction more possible. With data gloves, they can even shake hands, although they might be thousands of miles apart physically. This scenario might sound like science fiction, but the technology already exists. It gives a new meaning to the old AT&T slogan "Reach out and touch someone."

2.5 Obstacles in Using VR Systems

One major obstacle to using VR technology is that not enough fiber-optic cables are currently available to carry the data transmissions needed for a VR environment capable of re-creating a conference. With people in different geographical locations, high-speed transmission capabilities are necessary for participants to interact in real time. Without them, having to wait several seconds every time you want to act in a VR environment would be frustrating.

VR systems have generated a lot of excitement in recent years, but the following problems must be solved before this technology's potential can be realized:

- *Confusion between the VR environment and the real environment*—The possibility of users becoming unable to distinguish reality from virtual reality is a potential danger, particularly if people come to believe that anything they do in the virtual environment is acceptable in the real world. This risk is especially a concern in computer games that allow users to torture or kill others.
- *Mobility and other problems with HMDs*—With current technology, users are "tethered" to a limited area while wearing an HMD and can't switch to performing tasks outside the virtual world without

removing the gear. In addition, refresh rates in HMDs still aren't quite fast enough, so a degree of visual distortion can happen while wearing an HMD.

- *Difficulty representing sound*—Representing sound in a 3-D environment is difficult if the sound needs to move, such as a plane passing overhead. Creating stationary sound is easy, but re-creating the effect of sound fading or getting louder as an object moves away or closer is much harder with current technology.

- *Need for additional computing power*—VR systems require a lot of memory and speed to manipulate large graphics files and provide the instantaneous response needed to give the impression of a real world. Drawing and refreshing frames continuously and rapidly requires extremely fast computers with a lot of memory.

With the rapid pace of technology, however, these problems should be solved in the near future so that VR systems can be used more widely.

A **virtual world** is a simulated environment designed for users to interact via avatars.

An **avatar** is a 2-D or 3-D graphical representation of a person in the virtual world, used in chat rooms and online games.

2.6 Virtual Worlds

A **virtual world** is a simulated environment designed for users to interact via avatars. An **avatar** is a 2-D or 3-D graphical representation of a person in the virtual world, used in chat rooms and online games (see Exhibit 14.4). Strategy Analytics predicts that 640 million people worldwide will inhabit virtual worlds by 2015, an increase of 244 percent over the 186 million who did so in 2009.[3]

Users can manipulate objects in the simulated world and experience a limited telepresence that gives them the feeling of being in another location. Communication between users can take the form of text, graphical icons, and/or sound. For example while you are at Second Life (a virtual world platform developed by Linden Lab) you can shop, and there are fansites, blogs, forums, news sites, and classified ads. Currently, virtual worlds are used most often for gaming, social networking, and entertainment. However, they're beginning to be used in business and education. For example, IBM is using virtual worlds for client training. Other organizations use virtual worlds to conduct a variety of business activities, such as marketing and sales, product development, recruiting, and team meetings. For example, Northrop Grumman Corporation uses Second Life for displaying prototypes, performing simulations,

Exhibit 14.4 *An avatar in Second Life*

© Moneca/Shutterstock.com, Hywit Dimyadi/Shutterstock.com, Moneca/Shutterstock.com

Exhibit 14.5 *RFID tags*

same task as bar codes, universal product codes (UPCs), and magnetic strips on credit and debit cards: It provides a unique identification for the card or the object carrying the tag.

Unlike bar codes and other systems, RFID devices don't have to be in contact with the scanner to be read. Exhibit 14.6 shows an RFID reader. Because of its embedded antenna, it can be read from a distance of about 20 feet. The RFID tag's advantages, along with its decreasing price (less than 10 cents per tag), have made this device more popular with retail and other industries.

and training employees in situations that would be dangerous, expensive, or unfeasible in the physical world. The National Oceanic and Atmospheric Administration (NOAA) also uses Second Life as a marketing channel to reach new customers.[4] Widely used virtual worlds include the following:

- ActiveWorlds (3-D virtual reality platform, *www.activeworlds.com/overview.asp*)
- Club Penguin (online game, *www.clubpenguin.com*)
- EGO (social networking game, *www.ego-city.com*)
- Entropia Universe (multiplayer virtual universe, *www.mindark.com/entropia-universe*)
- Habbo (social networking site, *www.habbo.com*)
- RuneScape (multiplayer role-playing game, *www.runescape.com*)
- Second Life (virtual world platform, *http://secondlife.com*)

A **radio frequency identification (RFID)** tag is a small electronic device consisting of a small chip and an antenna. This device provides a unique identification for the card or the object carrying the tag.

Exhibit 14.6 *An RFID reader*

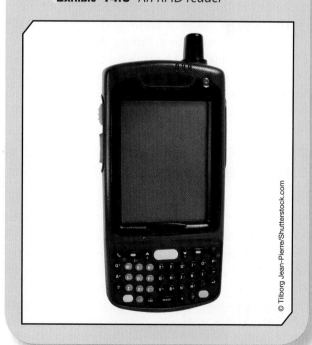

3 Radio Frequency Identification: An Overview

 **radio frequency identification (RFID)** tag is a small electronic device consisting of a small chip and an antenna (see Exhibit 14.5). This device performs the

Table 14.1

RFID applications

Category	Examples
Tracking and identification	Railway cars and shipping containers, livestock and pets, supply-chain management (tracking merchandise from manufacturers to retailers to customers), inventory control, retail checkout and POS systems, recycling and waste disposal
Payment and stored-value systems	Electronic toll systems, contactless credit cards (require no swiping), subway and bus passes, casino tokens, concert tickets
Access control	Building access cards, ski-lift passes, car ignition systems
Anticounterfeiting	Casino tokens, high-denomination currency notes, luxury goods, prescription drugs
Health care	Tracking medical tools and patients (particularly newborns and patients with Alzheimer's), process control, monitoring patient data

There are two types of RFID tags: passive and active. Passive RFID tags have no internal power supply, so they can be very small. Typically, they absorb the signal from the receiver device, convert this signal into energy, and use this energy to respond to the receiver. Passive tags usually last longer than active tags; the best ones have about 10 years of life. Active RFID tags have an internal power source and are usually more reliable than passive tags and can broadcast signals over a much wider range. These tags can also be embedded in a sticker or under the skin (human or animal).

Despite RFID's advantages, there are some technical problems and issues of privacy and security. On the technical level, signals from multiple readers can overlap, signals can be jammed or disrupted, and the tags are difficult to remove. Privacy and security issues include being able to read a tag's contents after an item has left the store, tags being read without the customer's knowledge, and tags with unique serial numbers being linked to credit card numbers.

3.1 RFID Applications

RFID devices have been used by many organizations in both the public and private sectors, including Walmart, the U.S. Department of Defense, Toyota, and The Gap. Table 14.1 lists common applications of RFID, divided into five major categories.[5]

The information box below describes how the Coca-Cola Company uses RFID.

4 Biometrics: A Second Look

Chapter 5 introduced biometrics as a networking security measure. This chapter delves into the topic more, given that these measures have become more

---

Coca-Cola Company Uses RFID-Based Dispensers for Generating BI

Coca-Cola is implementing a system that will offer more than 100 varieties of soda, juice, tea, and flavored water. The system is being tested in California, Georgia, and Utah, and the company has plans to implement it nationwide. The "Freestyle" dispensers not only give customers many choices of soft drinks by allowing them to mix their own flavor combinations, they collect valuable business intelligence that Coca-Cola can use to improve the efficiency and effectiveness of its soft drink production and distribution. The dispensers contain cartridges that are tagged with RFID chips, and each dispenser contains an RFID reader. The system collects data on which drinks customers buy and how much they purchase. This information is then transmitted through a wireless network to a data warehouse system in Atlanta, Georgia. Coca-Cola analyzes the data and generates reports on how new drinks are performing in the marketplace.[6]

widely used since September 11, 2001. Biometrics have become more widespread in forensics and in related law enforcement fields, such as criminal identification, prison security, and airport security. Because biometrics offer a high degree of accuracy that isn't possible with other security measures, they have the potential to be used in many civilian fields, too. They're already used in e-commerce and banking by phone, for example, using voice synthesizers and customers' voices as the biometric element that identifies them remotely. The following are some current and future applications of biometrics:

- *ATM, credit, and debit cards*—Even if users forget their PINs, they can still use their cards if their biometrics attributes are stored. Biometrics make ATM and credit/debit cards more secure, too, because they can't be used by others if they're lost or stolen.

- *Network and computer login security*—Other security measures, such as ID cards and passwords, can be copied or stolen, which isn't likely with biometric security measures. Fingerprint readers, for example, are already available at moderate prices.

- *Web page security*—Biometrics measures could add another layer of security to Web pages to guard against attacks and eliminate or reduce defacing (a kind of electronic graffiti). For example, a stock-trading Web site might require customers to log on with a fingerprint reader.

- *Voting*—Biometrics could be used to make sure people don't vote more than once, for example, and could be useful for authentication purposes in voting via the Internet.

- *Employee time clocks*—Biometrics could uniquely identify each employee and verify clock-in and clock-out times. This technology would also prevent one coworker from checking in for another.

- *Member identification in sport clubs*—Fingerprint scanners are used in some sport clubs to allow members admittance. This technology improves convenience and security.

- *Airport security and fast check-in*—Israeli airports have been using biometrics for this purpose for years. Ben Gurion International Airport in Tel Aviv has a frequent flyer's fast check-in system based on smart cards that store information on users' hand geometry. With this system, travelers can pass through check-in points in less than 20 seconds.[7]

- *Passports and highly secured government ID cards*—A biometrically authenticated passport or ID card can never be copied and used by an unauthorized person. German citizens over the age of 24 can apply for an ePass, a passport containing a chip that stores a digital

photograph and fingerprints. Other biometric identifiers, such as iris scans, could be added.

- *Sporting events*—Germany used biometric technology at the 2004 Summer Olympic Games in Athens, Greece, to protect its athletes. NEC, Inc., created an ID card containing an athlete's fingerprints, which was used for secure access.

- *Cell phones and smart cards*—Biometrics could be used to prevent unauthorized access to cell phones and smart cards and prevent others from using them in case they're lost or stolen.

5 Trends in Networking

t he following sections discuss recent trends in networking technologies. Many are already used in several organizations, such as wireless technologies and grid computing. Others, such as WiMAX and cloud computing, are newer but are attracting a lot of attention.

5.1 Wi-Fi

Wireless Fidelity (Wi-Fi) is a broadband wireless technology. Information can be transmitted over short distances—typically 120 feet indoors and 300 feet outdoors—in the form of radio waves. You can connect computers, mobile phones and smartphones, MP3 players, PDAs, and game consoles to the Internet with Wi-Fi. Some restaurants, coffee shops, and university campuses provide Wi-Fi access; they're called "hotspots." Wi-Fi connections are easy to set up, have fast data transfer rates, and

© Oleksiy Maksymenko/Alamy

offer mobility and flexibility. However, they're susceptible to interference from other devices and to being intercepted, which raises security concerns. In addition, there's a lack of support for high-quality media streaming.

5.2 WiMAX

Worldwide Interoperability for Microwave Access (WiMAX) is a broadband wireless technology based on the IEEE 802.16 standards. It's designed for wireless metropolitan area networks (MANs, discussed in Chapter 6) and usually has a range of about 30 miles for fixed stations and 3 to 10 miles for mobile stations. Compared with Wi-Fi, WiMAX theoretically has faster data transfer rates and a longer range. It can theoretically go up to speeds of 40 Mbps. In addition, it's fast and easy to install and enables devices using the same frequency to communicate. A single station can serve hundreds of users.

Disadvantages of WiMAX include interference from other wireless devices, high costs, and interruptions from weather conditions, such as rain. This technology also requires a lot of power, and when bandwidth is shared among users, transmission speed decreases.

Worldwide Interoperability for Microwave Access (WiMAX) is a broadband wireless technology based on the IEEE 802.16 standards. It's designed for wireless metropolitan area networks and usually has a range of about 30 miles for fixed stations and 3 to 10 miles for mobile stations.

Bluetooth, which can be used to create a personal area network (PAN), is a wireless technology for transferring data over short distances (usually within 30 feet) for fixed and mobile devices.

Grid computing involves combining the processing powers of various computers. With this configuration, users can make use of other computers' resources to solve problems involving large-scale, complex calculations, such as circuit analysis or mechanical design—problems that a single computer isn't capable of solving in a timely manner.

5.3 Bluetooth

Bluetooth, which can be used to create a personal area network (PAN), is a wireless technology for transferring data over short distances (usually within 30 feet) for fixed and mobile devices. The Bluetooth specifications are developed and licensed by the Bluetooth Special Interest Group (SIG, *www.bluetooth.com/bluetooth*). Used with mobile headsets, it's become popular as a safer method of talking on cell phones while driving. Bluetooth uses a radio technology called Frequency Hopping Spread Spectrum (FHSS). It separates data into chunks and can transmit each chunk on a different frequency, if needed. Bluetooth is also used to connect devices such as computers, global positioning systems (GPSs), mobile phones, laptops, printers, and digital cameras. Unlike infrared devices, Bluetooth has no line-of-sight limitations. However, it does have a limited data transfer rate (only 1 MBps), and like other wireless devices, its susceptibility to interception is a security concern. However, Bluetooth 2.1 supports a theoretical speed of 3 Mbps, and Bluetooth 3.0 supports a theoretical speed of 24 Mbps.

Bluetooth is also used in the following ways:

- Wireless controllers are available with video game consoles, such as Nintendo and Sony PlayStation.
- Companies can send short advertisements to Bluetooth-enabled devices. For example, a restaurant can send announcements of dinner specials to nearby devices.
- A wireless device, such as a mouse, keyboard, printer, or scanner, can be connected via Bluetooth.
- Computers that are close together can network via Bluetooth.
- Contact information, to-do lists, appointments, and reminders can be transferred wirelessly between devices with Bluetooth and OBject EXchange (OBEX), a communication protocol for transmitting binary data.

5.4 Grid Computing

Generally, **grid computing** involves combining the processing powers of various computers (see Exhibit 14.7). With this configuration, users can make use of other computers' resources to solve problems involving large-scale, complex calculations, such as circuit analysis or mechanical design—problems that a single computer isn't capable of solving in a timely manner. Each participant in a grid is referred to as a "node." Cost saving

> With utility computing, you can request computing power and memory from the provider. It's like leasing a more powerful computer just for the period of time you need it.

is a major advantage of grid computing because companies don't have to purchase additional equipment. In addition, processing on overused nodes can be switched to idle servers and even desktop systems. Grid computing has already been used in bioinformatics, oil and gas drilling, and financial applications.

Other advantages of grid computing include the following:

- *Improved reliability*—If one node in the grid fails, another node can take over.
- *Parallel processing*—Complex tasks can be performed in parallel, which improves performance. In other words, a large complex task can be split into smaller tasks that run simultaneously on several nodes.
- *Scalability*—If needed, more nodes can be added for additional computing power without affecting the network's operation. Upgrades can also be managed by segmenting the grid and performing the upgrade in stages without any major effect on the grid's performance.

Grid computing does have some drawbacks, however. Some applications can't be spread among nodes, so they aren't suitable for grid computing, and applications requiring extensive memory that a single node can't provide can't be used on a grid. In addition, licensing agreements can be challenging, and synchronizing operations in several different network domains can be difficult and require sophisticated network management tools. Finally, some organizations are resistant to sharing resources, even if doing so benefits them.

5.5 Utility (On-Demand) Computing

Utility (on-demand) computing is similar to the SaaS model and provides IT services on demand. Users pay for computing or storage resources on an as-needed basis, similar to paying for utilities. Convenience and cost savings are two main advantages of utility computing, but this service does have drawbacks in the areas of privacy and security. Because the service is outside the company's location, theft or corruption of data is a concern.

Utility computing can work with the SaaS model you learned about earlier. Returning to the example of editing a Word document, suppose the Chapter 14.doc file is very large because it contains a lot of images. You notice that your computer is running slowly because it has an older CPU and doesn't have enough RAM to handle the file size adequately. With utility computing, you can request computing power and memory from the provider. It's like leasing a more powerful computer just for the period of time you need it. As opposed to SaaS, utility computing handles hardware resources, such as CPU processing and memory, not software.

Utility computing has been made available at universities and research centers that need to run complex programs and don't have the necessary resources. For example, NASA leases its supercomputer for a fee, thereby ensuring that the supercomputer gets used and bringing additional income to NASA. Other organizations, such as Sun Microsystems and IBM, offer this service in the form of storage and virtual servers. Some companies offer virtual data centers with services that enable users to combine memory, storage, and computing capabilities. Liquid Computing's LiquidIQ is one example

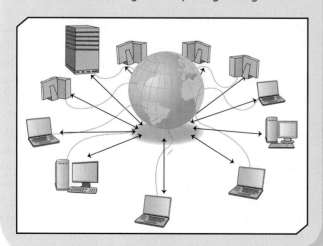

Exhibit 14.7 *A grid computing configuration*

Utility (on-demand) computing is similar to the SaaS model and provides IT services on demand. Users pay for computing or storage resources on an as-needed basis, similar to paying for utilities.

Table 14.2

Cloud computing categories and the top players

Categories	Top Players
Foundations (tools and software that make it possible to build cloud infrastructure)	Vmware, Microsoft, Red Hat
Infrastructure	Amazon, IBM
Network services (the communication components that combine with cloud foundation and infrastructure to form cloud architecture)	Level 3 Computing Services , Amazon, Cisco, Citrix
Platforms	Amazon, IBM
Applications	Google, Salesforce.com, Oracle, DROPBOX
Security	EMC/RSA, Symantec, IBM
Management	IBM, Amazon

(*https://www.liquidweb.com*). ENKI (*www.enkiconsulting.net*), Joyent (*www.joyent.com*), and Layered Tech (*www.layeredtech.com*) are other vendors.

5.6 Cloud Computing

Cloud computing incorporates, under one platform, many recent technologies, including the SaaS model, Web 2.0, grid computing, and utility computing; hence, a variety of resources can be provided to users over the Internet. Business applications are accessed via a Web browser, and data is stored on the providers' servers.[8] In addition, cloud providers, such as Amazon, set up an environment that enables you to subscribe to SaaS, utility, grid, and other services you need and coordinates all these services for you.

Nearly all tech vendors are involved in cloud computing. BTC Logic, an IT consulting firm, has classified cloud computing into seven areas and has identified some of the top players in each category (see Table 14.2).[9]

Going back to the example of editing the Chapter14.doc file, say you're using your iPhone instead of your computer. Clearly, your iPhone doesn't have the storage space to save such a large file, and it doesn't have the necessary computing power or Word

software installed. With cloud computing, you can subscribe to Word at the provider's SaaS site, store the document on an external storage unit provided by the vendor, and run Word on a multiprocessor system the vendor provides. You might even get extra RAM from another computer available in the cloud, and the cloud provider coordinates all these tasks for you. Your iPhone is simply the device for viewing the document while you're editing it, and because it's a mobile device, you can do your work anywhere. In other words, the document, the software, and the computing resources are like a cloud that surrounds you wherever you go and are available whenever you need them.

Generally, cloud computing includes components in the form of infrastructure as a service (IaaS), platform as a service (PaaS), and software as a service (SaaS).

Cloud computing offers many of the advantages and disadvantages of distributed computing. With this platform, users can request services, applications, and storage. For small and medium businesses, it means they don't have to invest in expensive equipment to compete effectively with large companies and can concentrate on the services and products they provide. Cloud computing services typically require a fee, although some are free. Google Apps, which includes Gmail, Google Talk, and Google Docs, provides commonly used applications accessed via a Web browser; software and data are stored on Google's servers, not on the user's computer. The standard edition for personal use is free.[10] The Cloud Computing in Action information box explains the cloud computing service at Amazon.com.

Cloud computing incorporates, under one platform, many recent technologies, including the SaaS model, Web 2.0, grid computing, and utility computing; hence, a variety of resources can be provided to users over the Internet. Business applications are accessed via a Web browser, and data is stored on the providers' servers.

The document, the software, and the computing resources are like a cloud that surrounds you wherever you go and are available whenever you need them.

6 Nanotechnology

nanotechnology incorporates techniques that involve the structure and composition of materials on a nanoscale. A nanometer is one billionth of a meter (10^{-9}). To better understand this scale of measurement, consider that a nanometer is:[14]

- About 1/100,000 the thickness of a human hair
- Almost as wide as a DNA molecule
- Ten times the diameter of a hydrogen atom
- How much your fingernails grow each second
- How far the San Andreas Fault slips in half a second
- The thickness of a drop of water spread over a square meter
- One-tenth the thickness of the metal film on a potato chip bag

Nanotechnology has become an exciting development in many fields. For example, scientists are working on miniature devices that can unclog arteries, detect and eradicate cancer cells, filter water pollution, and more. Nanotechnology is also being developed to make computers much faster and smaller, with more memory. However, the cost of nanotechnology is currently too expensive to justify its use in many applications. Further research and development should reduce its cost in the future.

In the field of information systems, the current technology for making transistors and other components might reach their miniaturization limits in the next decade, so new technologies, including nanotechnology, will be necessary.[15] Nanotechnology might also play a role in the following areas:

- Energy (reduction of energy consumption, increase in the efficiency of energy production, more environmentally friendly energy systems)
- Information and communication (larger and faster storage devices, faster and cheaper computers, display monitors with low energy consumption)
- Heavy industry (aerospace, construction, refineries, vehicle manufacturing)

Some consumer goods incorporating nanotechnology are already on the market. They use what's called "nanomaterials." Nanomaterials have been added to sports gear, such as tennis and golf balls and tennis rackets, to make them more durable and improve their responsiveness; tennis balls incorporating nanomaterials bounce better. Nanomaterials have also been applied as coatings on eyewear for increased comfort and durability, and they are used in clothing and footwear to cut down bacteria growth and improve stain resistance.[16] In addition, IBM has developed the scanning tunneling microscope (STM), which is capable of imaging atoms and incorporating nanomaterial layers into hard disk recording heads and magnetic disk coatings. This technology might also improve electronic circuits and data storage devices.[17]

The Industry Connection highlights Mechdyne Corporation and its virtual reality products.

> **Nanotechnology** incorporates techniques that involve the structure and composition of materials on a nanoscale.

Cloud Computing in Action

Jeff Bezos, the founder of Amazon.com, says, "You don't generate your own electricity. Why generate your own computing?"[11] Amazon.com has established a computing platform that companies can use, regardless of their locations. This platform provides storage and processing power on demand, and companies pay only for the resources they use. By using this service, companies don't have to invest in technology that might become obsolete quickly.[12]

Google Apps, introduced in February 2007, competes with Microsoft's Office Suite, and many companies use it now, including universities such as Arizona State and Northwestern.[13] Other vendors offering cloud computing services include IBM and Salesforce.com.

7 Chapter Summary

I n this chapter, you've learned about new trends in disseminating software and technology services, including software as a service (SaaS). You've also learned about exciting developments in virtual reality, such as using a CAVE or virtual worlds for communication and collaboration. You reviewed how RFID technologies are used as well as new trends in networking, including Wi-Fi, WiMAX, and Bluetooth. Finally, you learned what's available via grid, utility, and cloud computing and some of the future trends in nanotechnology.

Industry Connection

MECHDYNE CORPORATION *

Mechdyne developed the first commercial CAVE system with rear-projection screens and offers a wide variety of VR hardware, software, and services, including the following:

- *CAVELib*—An application programming interface (API) with tools for creating interactive three-dimensional applications.
- *Trackd*—Enhanced VR software for the immersive displays industry; capable of incorporating input from a variety of devices.

- *Conduit for Google Earth Pro*—Includes features such as stereoscopic viewing for true depth perception to emulate real-life settings and real-time changes in viewing perspectives as users navigate through the environment to simulate a real-world experience.

* This information has been gathered from the company Web site (*www.mechdyne.com*) and other promotional materials. For more information and updates, visit the Web site.

Key Terms

application service providers (ASPs) (248)

avatar (254)

Bluetooth (258)

cave automatic virtual environment (CAVE) (252)

cloud computing (260)

egocentric environment (250)

exocentric environment (250)

grid computing (258)

nanotechnology (261)

pull technology (247)

push technology (247)

radio frequency identification (RFID) (255)

software as a service (SaaS) (248)

utility (on-demand) computing (259)

virtual reality (VR) (249)

virtual world (254)

Wireless Fidelity (Wi-Fi) (257)

Worldwide Interoperability for Microwave Access (WiMAX) (258)

Problems, Activities, and Discussions

1. What are application service providers?

2. Why have biometrics increased in use recently?

3. Explain the advantages of utility computing.

4. What are the seven cloud categories outlined in the chapter? Who are some of the major players in these categories?

5. Amazon is a major player in cloud and high-end computing. After reading the information on the following links, write a one-page paper that summarizes Amazon's offerings in these areas.

   *www.infoworld.com/d/cloud-computing/
   amazon-and-ibm-are-the-clouds-biggest-players-
   484?source=IFWNLE_nlt_wrapup_2010-07-16*

   *www.infoworld.com/d/cloud-computing/
   amazoncoms-big-move-high-end-computing-
   438?source=IFWNLE_nlt_daily_2010-07-16*

6. Some experts believe that cloud computing could save the economy. Visit the following Web page, consult other sources, and write a one-page paper that summarizes why this claim might be true.

   *www.infoworld.com/d/cloud-computing/
   will-cloud-computing-save-the-economy-
   794?source=IFWNLE_nlt_blogs_2010-07-27*

7. Nanotechnology will make computers faster. Visit the following Web page, consult other sources, and write a one-page paper that summarizes why this claim might be true.

*www.sciencedaily.com/releases/2010/05/
100531082857.htm*

8. IBM has designed 3D World for business meetings. Visit the following Web page, consult other sources, and write a one-page paper that summarizes the advantages and disadvantages of this platform.

   *www.informationweek.com/news/internet/web2.0/
   showArticle.jhtml?articleID=218101365*

9. Which of the following is not a problem with RFID technology?

   a. Tags can be linked to customers' credit card numbers.

   b. Signals from readers can overlap.

   c. Signals can collide if too many tags are in a small area.

   d. Tags are too easy to remove.

10. Nanotechnology has no prospective applications for storage devices. True or False?

casestudy

CLOUD COMPUTING HELPS UNIVERSITIES CUT COSTS

Australia's Queensland University of Technology implemented cloud computing in order to provide enterprise software to more than 140 universities in the Asia Pacific. According to Glenn Stewart, Professor of Information Systems at the university, a cloud computing platform has not only reduced costs, it has provided greater reliability and scalability. Professor Stewart is in charge of the SAP University Competence Centre (UCC), which provides the SAP suite of business software to over 800 academics and 42,000 students from 140 universities in the Asia Pacific and Japan. If a university chooses to run this suite of software without the help of UCC, it has to invest in hardware, software, backup facilities, and so forth. The upfront investment would be over $300,000, which is a major undertaking for any university. By migrating the services into a private cloud, each university pays $6,760 for that same package, which

© Alphaspirit/Shutterstock.com

is more than 433 percent reduction in cost.[17] However, security could be a concern when using a cloud computing platform, and users play an important role in its success. An organization needs to provide end-user education, force software updates, and work with the cloud computing provider to spot unusual activities.[18]

Answer the following questions:

1. How does Queensland University of Technology save money by using cloud computing?

2. Which computing costs are reduced when cloud computing is used?

3. How could end-users improve the security of cloud computing?

© Andersen Ross/Getty Images

LEARN YOUR WAY!

We know that no two students are alike. You come from different walks of life and with many different preferences. You need to study just about anytime and anywhere. **MIS2** was developed to help you learn Management Information Systems in a way that works for you.

Not only is the format fresh and contemporary, it's also concise and focused. And, **MIS2** is loaded with a variety of study tools, like in-text review cards, printable flash cards, and more.

Go to CourseMate for MIS2 to find plenty of resources to help you study—no matter what learning style you like best! Access at login.cengagebrain.com.

Endnotes

Chapter 1

1 Reinhard CG. YouTube brands: 5 outstanding leaders in YouTube marketing. http://mashable.com/2009/06/01/youtube-brands. Accessed July 5, 2010.

2 Top 10 YouTube Applications. http://hyveup.tv/2009/07/top-10-youtube-applications.html. Accessed July 5, 2010.

3 Third Most Populated Country in the World – The United States of FaceBook. http://techchunks.com/social-media/third-most-populated-country-in-the-world-united-states-of-facebook-infographic. Accessed July 5, 2010.

4 Gaudin S. Half of social networkers post risky information, study finds. *InfoWorld*. 2010; May 4. www.infoworld.com/d/security-central/half-social-networkers-post-risky-information-study-finds-623?source=IFWNLE_nlt_wrapup_2010-05-04. Accessed July 5, 2010.

5 O'Leary M. Putting Hertz executives in the driver's seat. CIO. February 1990;62-69.

6 Bidgoli H. *Handbook of Management Information Systems: A Managerial Perspective*. San Diego, CA: Academic Press; 1999.

7 Bednarz A. The Home Depot's latest project: XML, Web services. *NetworkWorld*. 2002; January 21. www.networkworld.com/news/2002/129295_01-21-2002.html. Accessed October 20, 2010.

8 Songini ML. Home Depot kicks off big data warehouse effort. *Computerworld*. September 30, 2002. http://www.computerworld.com/action/article.do?command=viewArticleBasic&articleId=74751. Accessed October 20, 2010.

9 Operational technology. UPS pressroom. www.pressroom.ups.com/mediakits/etech/technology/0,1370,,00.html. Accessed July 3, 2010.

10 UPS changes the delivery game with new intercept service. UPS pressroom. 2007; March 26. www.pressroom.ups.com/mediakits/pressrelease/0,2300,4877,00.html. Accessed July 3, 2010.

11 Kotler P, Armstrong G. *Principles of Marketing*, 12th ed. Upper Saddle River, NJ: Prentice Hall; 2007.

12 Porter M. How competitive forces shape strategy. *Harvard Business Review*. 1979; March-April, 57.

13 Cole W. *I Robot*. *Boeing Frontiers*. 2005; December. www.boeing.com/news/frontiers/archive/2005/december/ts_sf09.html. Accessed July 3, 2010.

14 Linthicum D. The cloud job market: A golden opportunity for IT pros. 2010; July 1. www.infoworld.com/d/cloud-computing/the-cloud-job-market-golden-opportunity-it-pros-817?source=IFWNLE_nlt_blogs_2010-07-01. Accessed July 3, 2010.

15 Kennerson G. IT job titles. http://education.yahoo.net/articles/it_job_titles.htm. Accessed July 3, 2010.

16 Babcock C. FedEx locks in customers by tying shipping data to back-office apps. *InformationWeek*. 2006; September 12. www.informationweek.com/news/management/showArticle.jhtml?articleID=192700339. Accessed July 3, 2010.

17 Gage D. FedEx: Personal touch. *Baseline*. 2005; January 13. www.baselinemag.com/c/a/Projects-Supply-Chain/FedEx-Personal-Touch. Accessed July 3, 2010.

Chapter 2

1 IBM's 'Millipede' project demonstrates trillion-bit data storage density. *IBM*. 2002; June 11. http://domino.research.ibm.com/comm/pr.nsf/pages/news.20020611_millipede.html. Accessed July 31, 2010.

2 Shah A. SanDisk SD card can store data for 100 years. *InfoWorld*. 2010; June 23. www.infoworld.com/d/storage/sandisk-sd-card-can-store-data-100-years-161?source=IFWNLE_nlt_storage_2010-06-30. Accessed July 31, 2010.

3 Apple sells three million iPads in 80 days. www.apple.com/pr/library/2010/06/22ipad.html. Accessed July 5, 2010.

4 Fleck C. Top 10 business uses for the iPad. 2010; April 1. http://community.citrix.com/display/ocb/2010/04/01/Top+10+Business+Uses+For+the+iPad. Accessed July 5, 2010.

5 Google Docs Tour. www.google.com/google-d-s/tour1.html.

6 Kirk J. Security analyst spots three flaws in Google Docs. *IT World*. 2009; March 27. www.itworld.com/saas/65211/security-analyst-spots-three-flaws-google-docs. Accessed July 31, 2010.

Chapter 3

1 Smalltree H. Business intelligence case study: Gartner lauds police for crime-fighting BI. *Data Management*. 2007; April 5. http://search-datamanagement.techtarget.com/news/article/0,289142,sid91_gci1250435,00.htm. Accessed July 5, 2010.

2 Kim SH, Yu B, Chang J. Zoned-partitioning of tree-like access methods. *J Inform Systems*. 2008;33(3):315-331.

3 Storey VC, Goldstein RC. Knowledge-based approaches to database design. *MIS Quarterly*. 1993; March:25-32.

4 Rauch-Hindin W. True distributed DBMSs presage big dividends. *Mini-Micro Systems*. 1987; May:65-73.

5 Brueggen D, Lee S. Distributed database systems: Accessing data more efficiently. *Inform Systems Manage*. 1995; Spring:15-20.

6 Moad J. What IBM says about client/server. *Datamation*. 1991; February:53-58.

7 Inmon WH. *Building the Data Warehouse*, 4th ed. New York, NY: Wiley; 2005.

8 Thomas TD. An IBM solution for: KeyCorp. *Inform Manage Mag*. 2000; January 1. www.information-management.com/issues/20000101/1775-1.html. Accessed July 5, 2010.

9 Janies L. Room to grow, IHG boosts information delivery performance with a data warehouse appliance. *Teradata Magazine Online*. 2010; Q2. www.teradatamagazine.com/Article.aspx?id=14186. Accessed July 5, 2010.

10 Tremblay MC, Fuller R, Berndt D, et al. Doing more with more information: Changing healthcare planning with OLAP tools. *Decision Support Systems*. 2007;43(4):1305-1320.

11 Harrah's Entertainment. 2009, February 27. www-01.ibm.com/software/success/cssdb.nsf/CS/LWIS-7PNLEC?OpenDocument&Site=dmmain&cty=en_us. Accessed July 23, 2010.

Chapter 4

1 Prince B. FBI: Internet fraud cost $559 million in 2009. 2010; March 15. www.eweek.com/c/a/Security/FBI-Internet-Fraud-Cost-559-Million-in-2009-538939. Accessed August 6, 2010.

2 A MySpace Photo Costs a Student a Teaching Certificate. *Chronicle of Higher Education*. 2007; April 27. http://chronicle.com/wiredcampus/index.php?id=2029. Accessed August 6, 2010.

3 Office of the Privacy Commissioner. Online Privacy Tools. www.privacy.gov.au/internet/tools/#disc. Accessed August 6, 2010.

4 Internet 2009 in numbers. Pingdom blog. 2010; January 22. http://royal.pingdom.com/2010/01/22/internet-2009-in-numbers. Accessed August 6, 2010.

5 Chattopadhyay D. Where ethics rule: A global ranking of the world's most ethical companies could be a source of worry for Asian firms. *Business Today*. 2009; March 8.

6 Kallman EA, Grillo JP. *Ethical Decision Making and Information Technology*. San Francisco, CA: McGraw-Hill; 1993.

7   Identity theft statistics 2010. 2010; February 18. www.identitytheftlabs
    .com/identity-theft/identity-theft-statistics-2010. Accessed August
    6, 2010.

8   What is intellectual property? World Intellectual Property Organization.
    www.wipo.int/about-ip/en. Accessed August 6, 2010.

9   What is copyright protection? WhatIsCopyright.org. 2007; March 4.
    www.whatiscopyright.org. Accessed August 6, 2010.

10 Jones & Askew, LLP. Patent protection for e-commerce business
    models. 1999. www.versaggi.net/ecommerce/articles/e-business-
    models/patnets-for-models.pdf. Accessed August 6, 2010.

11 Tetzeli R. Getting your company's Internet strategy right. *Fortune*. 1996;
    March 18:72-78.

12 Verizon wins $33 Million in suit over domain names. *Bloomberg News*.
    2008; December 24. www.nytimes.com/2008/12/25/technology/
    companies/25verizon.html. Accessed August 6, 2010.

13 Milone Jr. MN, Salpeter J. Technology and equity issues. *Technology &
    Learning*. 1996; January:39-47.

14 Harvey DA. Health and safety first. *Byte*. 1991; October:119-128.

15 Klein MM. The virtue of being a virtual corporation. *Best Review*. 1994;
    October:88-94.

16 Becker D. When games stop being fun. *CNET News*. 2002; April 12.
    http://news.cnet.com/2100-1040-881673.html. Accessed August 6,
    2010.

17 Explaining green computing. ExplainingComputers.com. www
    .youtube.com/watch?v=350Rb2sOc3U. Accessed July 4, 2010.

18 Murugesan S. Harnessing green IT: Principles and practices.
    IT Professional. 2008;10(1):24-33.

19 IBM Project Big Green. www-03.ibm.com/press/us/en/photo/
    21514.wss. Accessed July 4, 2010.

20 Behar R. Never Heard of Acxiom…? *Fortune*. 2004; February 23.
    http://money.cnn.com/magazines/fortune/fortune_archive/
    2004/02/23/362182/index.htm. Accessed August 6, 2010.

Chapter 5

1   McMillan R. 1.5 million stolen Facebook IDs up for sale. April 23, 2010.
    http://www.infoworld.com/d/the-industry-standard/15-million-
    stolen-facebook-ids-sale-645. Accessed July 8, 2010.

2   Sanders S. Putting a lock on corporate data. *Data Communications*.
    1996; January:78-80.

3   McCumber J. *Assessing and managing security risk in IT systems*.
    Boca Raton, FL: Auerbach; 2004.

4   Richards J. Number of computer viruses tops one million. *Times Online*.
    2008; April 10. http://technology.timesonline.co.uk/tol/news/tech_
    and_web/article3721556.ece. Accessed August 10, 2010.

5   Grimes RA. Fighting today's malware. *InfoWorld*. 2010; August 5.
    www.infoworld.com/d/security-central/fighting-todays-malware-869.
    Accessed August 10, 2010.

6   Prevent data theft using removable devices. GetSafeOnline.org. 2009.
    http://www.getsafeonline.org/nqcontent.cfm?a_id=1103. Accessed
    August 10, 2010.

7   Anderson H. Case study: The motivation for biometrics. June 24, 2010.
    http://www.healthcareinfosecurity.com/articles.php?art_id=2686.
    Accessed July 13, 2010.

8   Bueb F, Fife P. Line of defense: Simple, complex securities measures
    help prevent lost and stolen laptops. California Society of Certified
    Public Accountant and Gale Group; 2006. http://www.thefreelibrary
    .com/Line+of+defense:+simple,+complex+security+measures+help+
    prevent+lost...-a0155477162. Accessed August 10, 2010.

9   The Sarbanes-Oxley Act of 2002. http://www.soxlaw.com. Accessed
    August 10, 2010.

10 Bidgoli H, ed. *Global perspectives in information security: Legal, social and
    international issues*. Hoboken, NJ: John Wiley & Sons; 2008.

11 Goodman MD, Brenner SW. The emerging consensus on criminal
    conduct in cyberspace. *Inter J Law Inform Technol*. 2002;10(2):139-223.

Chapter 6

1   Keizer G. Forrester: iPad and tablet rivals will kill netbooks. 2010; June
    17. www.infoworld.com/d/mobilize/forrester-ipad-and-tablet-rivals-
    will-kill-netbooks-633. Accessed July 31, 2010.

2   Apple - iPhone 4 - This changes everything. Again. www.youtube.com/
    watch?v=FHngLJ0RINg&feature=youtube. Accessed July 17, 2010.

3   iPhone in business. www.apple.com/iphone/business/features/.
    Accessed July 17, 2010.

4   Cramer S. Telepresence: The business case. *MIS Asia*. 2008,
    October 20. www.mis-asia.com/magazines/cio_asia/october-2008/
    telepresence-the-business-case. Accessed August 10, 2010.

5   Janah M. Wal-Mart links to suppliers. *InformationWeek*. 1998; October 5.
    www.informationweek.com/703/03iuwal.htm;jsessionid=O3CRMFURN
    JGDMQSNDLPCKH0CJUNN2JVN. Accessed August 10, 2010.

6   Wal-Mart takes on industrial vehicle management. *Food Manufactur-
    ing*. 2009; September 29. www.foodmanufacturing.com/scripts/
    ShowPR~RID~12487.asp. Accessed August 10, 2010.

Chapter 7

1   Gaudin S. Internet hits major milestone, surpassing 1 billion monthly
    users. *Computerworld*. 2009; January 26. www.computerworld.com/
    action/article.do?command=viewArticleBasic&articleId=9126796.
    Accessed August 10, 2010.

2   Cross R. Internet: The missing marketing medium found. *Direct
    Marketing*. 1994;57(6):20-23.

3   Hayes M. Online shopping for software. *Information Week*. 1995;
    January 2:23-24.

4   Anonymous. Health care on the information superhighway poses ad-
    vantages and challenges. *Employee Benefit Review*. 1994; August:24-29.

5   Bazzolo F. Putting patients at the center. *Internet Health Care Magazine*.
    2000; May:42-51.

6   Marsan CD. 10 fool-proof predictions for the Internet in 2020. 2010;
    January 4. www.networkworld.com/news/2010/010410-outlook-
    vision-predictions.html?page=1. Accessed July 13, 2010.

7   Bidgoli H. *Electronic Commerce: Principles and Practice*. San Diego, CA:
    Academic Press; 2002.

8   Fletcher T. Intranet pays dividends in time and efficiency for
    investment giant. *InfoWorld*. 1997; September 29:84.

9   Jones K. Copier strategy as yet unduplicated. *Interactive Week*.
    1998;5(5):41.

10 Maloff J. Extranets: Stretching the Net to boost efficiency. *NetGuide*.
    1997; August:62-68.

11 O'Reilly T. What is Web 2.0? Design patterns and business models for
    the next generation of software? 2005; September 30. www.oreillynet
    .com/pub/a/oreilly/tim/news/2005/09/30/what-is-web-20.html.
    Accessed August 10, 2010.

12 Cho A. What is Web 3.0? The next generation Web: Search context for
    online information. 2008, July 22. http://internet.suite101.com/article
    .cfm/what_is_web_30. Accessed August 10, 2010.

13 King R. No rest for the Wiki. *BusinessWeek*. 2007; March 12. www
    .businessweek.com/technology/content/mar2007/tc20070312_
    740461.htm. Accessed August 10, 2010.

14 King R. How companies use Twitter to bolster their brands. *Business-Week*. 2008; September 6. www.businessweek.com/technology/content/sep2008/tc2008095_320491.htm. Accessed August 10, 2010.

15 IBM's intranet one of the world's top ten. *IBM*. 2006; January 26. www.03.ibm.com/press/us/en/pressrelease/19156.wss. Accessed August 10, 2010.

Chapter 8

1 Porter ME. *Competitive advantage: Creating and sustaining superior performance*. New York, NY: Free Press; 1985.

2 Zetlin M. Use Twitter to find customers. 2009; July 24. www.inc.com/news/articles/2009/07/twitter.html. Accessed July 14, 2010.

3 Afuah A, Tucci CL. *Internet business models and strategies: Text and cases*, 2nd ed. Boston, MA: McGraw-Hill/Irwin; 2002.

4 Ovans A. E-procurement at Schlumberger. *Harvard Business Review*. 2000; May-June:21-22.

5 Blankenhorn D. GE's e-commerce network opens up to other marketers. *NetMarketing*. 1987; May 1.

6 Caswell S. Voice-based e-commerce looms large. *E-Commerce Times*. 2000; March 28. www.ecommercetimes.com/story/2838.html. Accessed July 14, 2010.

7 Greenberg PA. Get ready for wireless e-commerce. *E-Commerce Times*. 1988. www.ecommercetimes.com/story/2059.html. Accessed July 14, 2010.

8 Blodget H. Google developing micropayments and subscription system to save newspapers. *Business Insider*. 2009; September 9. www.businessinsider.com/henry-blodget-google-launching-micropayments-and-subscription-system-to-save-newspapers-2009-9. Accessed July 14, 2010.

9 Smith B. Yahoo's FareChase: The stealth disruptor? SearchEngineWatch.com. 2006; April 27. http://searchenginewatch.com/3601971. Accessed July 14, 2010.

Chapter 9

1 Ives B, Jarvenpaa S. Application of global information technology: Key issues for management. *Management Information Systems Quarterly*. 1991;15:33-49.

2 Eom SB. Transnational management systems: An emerging tool for global strategic management. *Advanced Management Journal*. 1994;59(2):22-27.

3 Lucas H. *Information Systems Concepts for Managers*. San Francisco, CA: McGraw-Hill; 1994:137-152, 289-315.

4 Harrington LH. The information challenge: The dream of global information visibility becomes a reality. *IndustryWeek*. 1997; April 7:97-100.

5 Internet World Stats. Miniwatts Marketing Group. 2009. www.internetworldstats.com/stats.htm. Accessed August 17, 2010.

6 Edwards R, Ahmad A, Moss S. Subsidiary autonomy: The case of multinational subsidiaries in Malaysia. *J Inter Business Studies*. 2002;33.

7 Karimi J, Konsynski BR. Globalization and information management strategies. *J Manage Inform Systems*. 1991;7(4):7-26.

8 Bar F, Borrus M. Information networks and competitive advantage: Issues for government policy and corporate strategy. *Intern J Technol Manage*. 1992;7(6-8):398-408.

9 Yadav V, Adya M, Sridhar V, et al. Flexible global software development (GSD): Antecedents of success in requirements analysis. *J Global Inform Manag*. 2009;17(1)1-30.

10 Haeckel S, Nolan R. Managing by wire (enterprise models driving strategic information technology). *Harvard Business Review*. 1993;71:122-132.

11 Passino J, Severance D. Harnessing the potential of information technology for support of the new global organization. *Human Resource Management*. 1990;57:69-76.

12 Schatz W. Scatter-shot systems: There are lots of ways to buy and manage technology in a decentralized environment. *Computerworld*. 1993; August 30:75-79.

13 Kettinger W, Varun G, Subashish G. Strategic information systems revisited: A study in sustainability and performance. *Manag Inform Systems Q*. 1994;18:31-58.

14 Fleenor D. The coming and going of the global corporation. *Columbia Journal of World Business*. 1993;28:6-10.

15 Laudon K, Laudon J. *Management Information Systems: Organization and Technology*. Upper Saddle River, New Jersey: Prentice Hall; 1996:668-692.

16 Dubie D. Gartner: Top offshore locations for 2008. 2008; September 3. www.itbusiness.ca/it/client/en/home/News.asp?id=49768. Accessed August 17, 2010.

17 Klein KE. Globalization, small biz-style. 2008; January 23. www.businessweek.com/smallbiz/content/jan2008/sb20080123_487864.htm. Accessed July 14, 2010.

18 Biehl M. Success factors for implementing global information systems. *Communications of the ACM*. 2007;50(1).

19 Gelfand MJ, Erez M, Aycan Z. Cross-cultural organizational behavior. *Annual Rev Psychol*. 2007;58:479-514.

20 Huff S. Managing global information technology. *Business Q*. 1991;56:71-75.

21 Ambrosio J. Global software: When does it make sense to share software with offshore units? *Computerworld*. 1993; August 2:74-77.

22 Spiegler RM. Globalization: Easier said than done. *Industry Standard*. 2000; October:136-155.

23 Aune SP. Two Persian Gulf nations planning to block features of BlackBerry phones. 2010; August 2. http://tech.blorge.com/Structure:%20/2010/08/02/two-persian-gulf-nations-planning-to-block-features-of-blackberry-phones. Accessed August 17, 2010.

24 *Yahoo v. LICRA*, U.S. Court of Appeals (9th Cir., 2004, D.C. No. CV-00-21275-JF).

25 Business Software Alliance. Fourth annual BSA and IDC global software piracy study. 2007. http://global.bsa.org/idcglobalstudy2007/studies/2007_global_piracy_study.pdf. Accessed August 17, 2010.

26 Beck E. Translating the Web. 2007; October 12. www.businessweek.com/innovate/content/oct2007/id20071012_139633.htm. Accessed July 24, 2010.

Chapter 10

1 Wood J, Silver D. *Joint Application Design*. New York, NY: John Wiley and Sons; 1989.

2 Bensaou M, Earl M. The right mind-set for managing information technology. *Harvard Business Review*. 1998;76(5):109.

3 Duvall M. Airline reservation system hits turbulence. *Baseline*. 2007; July 25. www.baselinemag.com/c/a/Intelligence/Airline-=Reservation-System-Hits-Turbulence. Accessed July 16, 2010.

4 Doke RE. An industry survey of emerging prototyping methodologies. *Information & Management*. 1990;18:169-176.

5 Benson DH. A field study of end user computing: Findings and issues. *MIS Quarterly*. 1983; December:35–40.

6 Tayntor CB. New challenges or the end of EUC. *Information Systems Management*. 1994; Summer:86-88.

7 McKendrick J. Ten examples of SOA at work, right now. 2006; January 3. www.zdnet.com/blog/service-oriented/ten-examples-of-soa-at-work-right-now/508. Accessed July 15, 2010.

8 Goldstein A. Hand-to-hand teamwork. Sabre Airline Solutions. 2003; February 23. www.sabreairlinesolutions.com/news/030223_Hand-to-Hand_Teamwork.htm; reprinted with permission from the *Dallas Morning News*).

9 Principles behind the Agile Manifesto.www.agilemanifesto.org/principles.html. Accessed July 16, 2010.

10 Who we are. www.overstock.com/about. Accessed July 16, 2010.

11 Weier MH. Overstock.com divulges secret to its cyber Monday success. 2007; November 28. www.informationweek.com/news/software/soa/showArticle.jhtml?articleID=204300504. Accessed July 16, 2010.

12 SEB Latvia builds a comprehensive business platform on IBM Lotus Notes and Domino. IBM. 2008; December 16. www-01.ibm.com/software/success/cssdb.nsf/CS/STRD-7MDN5V?OpenDocument&Site=software&cty=en_us. Accessed July 16, 2010.

Chapter 11

1 Payne A. *Handbook of CRM: Achieving excellence in customer management*. Oxford, MA: Butterworth-Heinemann; 2005.

2 Wailgum T. CRM definition and solutions. CIO.com. www.cio.com/article/40295/Customer_Relationship_Management. Accessed July 16, 2010.

3 Team productivity increases up to 10% after Time Warner cable business class implements salesforce. Salesforce.com. www.salesforce.com/au/customers/communications-media/timewarnercable.jsp. Accessed July 16, 2010.

4 Allen C. Personalization vs. customization. ClickZ.1999; July 13. www.clickz.com/814811.

5 Selling with personalization: Why Amazon.com succeeds and you can too. 2008; August 2. Smallbiztechnology.com. smallbiztechnology.com/archive/2008/08/selling-with-personalization-w.html. Accessed July 16, 2010.

6 Sherman C. Google personalized search leaves Google Labs. Search EngineWatch.com. 2005; November 10. http://searchenginewatch.com/3563036.

7 Aaronson J. Personalization technologies: A primer. August 24, 2007. ClickZ.com. http://www.clickz.com/3626837.

8 Levinson M. Knowledge management definition and solutions. CIO.com. http://www.cio.com/article/40343/Knowledge_Management_Definition_and_Solutions. Accessed July 16, 2010.

9 Saudi-based auto and electronics distributor streamlines operations. http://www.vai.net/company/success-stories/naghi-group.html. Accessed July 17, 2010.

10 Jabil Circuit optimizes plant processes on three continents with a global ERP solution. IBM. November 30, 2004. http://www-01.ibm.com/software/success/cssdb.nsf/CS/JSTS-66YM6A?OpenDocument&Site=gicss67sap&cty=en_us. Accessed July 16, 2010.

11 Johns Hopkins unites its businesses with SAP and IBM. IBM. 2008; June 26. www-01.ibm.com/software/success/cssdb.nsf/CS/STRD-7FYDM6?OpenDocument&Site=gicss67sap&cty=en_us. Accessed July 16, 2010.

Chapter 12

1 Simon H. *The New Science of Management Decision*. Englewood Cliffs, NJ: Prentice-Hall; 1977.

2 Sprague R Jr, Carlson ED. *Building Effective Decision Support Systems*. Englewood Cliffs, NJ: Prentice-Hall; 1982.

3 Keen PG. Value analysis: Justifying decision support systems. *MIS Quarterly*. 1981;5(1):1-15.

4 Alter SL. *Decision Support Systems Current Practice and Continuing Challenges*. Reading, MA: Addison-Wesley; 1980.

5 Watson HJ, Rainer RK Jr. A manager's guide to executive support systems. *Business Horizons*. 1991; March–April:44-50.

6 Glover H, Watson H, Rainer R. 20 ways to waste an EIS investment. *Information Strategy: The Executive's Journal*. 1992; Winter:11-17.

7 Watson HJ, Satzinger J. Guidelines for designing EIS interfaces. *Information Systems Management*. 1994; Fall:46-52.

8 Cronk R. EISs mind your data. *Byte*. 1993; June:121-128.

9 Overton K, Frolick MN, Wilkes RB. Politics of implementing EISs. *Information Systems Management*. 1996; Summer: 50-57.

10 Kibbe DC. Why clinical groupware may be the next big thing in health IT. 2009; February 8. http://e-caremanagement.com/why-clinical-groupware-may-be-the-next-big-thing-in-health-it. Accessed July 18, 2010.

11 Covaleski JM. Software help carriers avoid perilous areas. *Best's Review*. 1994; February:82-83.

12 Gilbert E. GUS guards against costly classification errors. *National Underwriter*. 1993; October 4:5.

13 Lewis R. Putting sales on the map. *Sales & Marketing*. 1992; August:76-80.

14 Thom J, Walters L. A map for marketing. *Sales & Marketing*. 1992; July:102-104.

15 Aalberts RJ, Bible DS. Geographic information system: Application for the study of real estate. *Appraisal Journal*. 1992; October:483-492.

16 Weber BR. Application of geographic information systems to real estate market analysis and appraisal. *Appraisal Journal*. 1990; January:127-131.

17 Smith B, Eglowstein H. Putting your data on the map. *Byte*. 1993; January:188-200.

18 O'Cionnaith F. GIS on the front lines in fighting disease. *GPS World*. 2007; December 19. www.gpsworld.com/gis/health-and-education/gis-front-lines-fighting-disease-5442. Accessed July 16, 2010.

19 Monitor the swine flu real-time with Google maps. GIS Lounge. 2009; April 27. http://gislounge.com/monitor-the-swine-flu-real-time-with-google-maps. Accessed July 16, 2010.

20 Isuzu Australia takes the road to collaboration success with IBM WebSphere and Lotus technologies. IBM. 2008; December 9. www-01.ibm.com/software/success/cssdb.nsf/CS/EMAU-7JC232?OpenDocument&Site=gicss67retl&cty=en_us. Accessed July 16, 2010.

Chapter 13

1 McCarthy J. What is artificial intelligence? Stanford University. 2007; November 12. www.formal.stanford.edu/jmc/whatisai/whatisai.html. Accessed July 16, 2010.

2 Smart programs go to work. *BusinessWeek*. 1992; March 2:96-101.

3 Lesaint D, Voudouris C, Azarmi N. Dynamic workforce scheduling for British telecommunications. *Interfaces*. 2000;30(1):45-56.

4 Hall O. Artificial intelligence techniques enhance business forecasts. *Graziadio Business Report*. http://gbr.pepperdine.edu/022/intelligence.html. Accessed November 15, 2009.

5 Bidgoli H. Integration of technologies: An ultimate decision-making aid. *Industrial Management & Data Systems*. 1993;1:10-17.

6 Ives B. Expert system puts more meaning into semantic search. 2008; June 2. www.theappgap.com/expert-system-puts-more-meaning-into-semantic-search.html. Accessed July 16, 2010.

7 Kneale D. How Coopers & Lybrand put expertise into its computers. *Wall Street Journal*. 1986; November 14: 33.

8 Roby RK. Expert systems help labs process DNA samples. *NIJ Journal*. 2008; July. www.ojp.usdoj.gov/nij/journals/260/expert-systems .htm. Accessed July 16, 2010.

9 Wai KS, Latif AB, Rahman A, Zaiyadi MF, et al. Expert system in real world applications. www.generation5.org/content/2005/Expert_ System.asp. Accessed July 16, 2010.

10 Adderley RW, Musgrove P. Police crime recording and investigation systems—A user's view. *Policing: Inter J Police Strat Manag*. 2001;24(1):100-114.

11 Lewis DJ. Intelligent agents and the semantic Web. 2008, October 28. www.ibm.com/developerworks/web/library/wa-intelligentage. Accessed July 16, 2010.

12 Boehle S, Goldwasser DG, Stamps J. The return of artificial intelligence. *Training*. 2000; November:26.

13 Zadeh LA. Yes, no and relatively part 2. *Chemtech*. 1987; July:406-410.

14 Feng G. A survey on analysis and design of model-based fuzzy control systems. *IEEE Transactions on Fuzzy Systems*. 2006;14(5):676-697.

15 Maximize returns on direct mail with BrainMaker neural network software. www.calsci.com/DirectMail.html. Accessed July 20, 2010.

16 Neural network red-flags police officers with potential for misconduct. www.calsci.com/Police.html. Accessed July 20, 2010.

17 Holland JH. Genetic algorithms. *Scientific American*. 1992; July:66-72.

18 Wiggins R. Docking a truck: A genetic fuzzy approach. *AI Expert*. 1992; May:28-35.

19 Obermeir KK. Natural language processing. *Byte*. 1987; December: 225-231.

20 Liu SF, Duffy AHB, Whitfield RI, Boyle IM. Integration of decision support systems to improve decision support performance. In *Knowledge and Information Systems*. London, UK: Springer; 2009.

21 Turban E, Watkins PR. Integrating expert systems and decision support systems. *MIS Quarterly*. 1986; June:121-136.

22 Scanlon J. Staples' evolution. 2008, December 29. www.businessweek .com/innovate/content/dec2008/id20081229_162381.htm. Accessed July 25, 2010.

Chapter 14

1 Virtual reality: How a computer-generated world could change the real world. *BusinessWeek*. www.businessweek.com/1989-94/pre88/ b328653.htm. Accessed July 16, 2010.

2 Virtual reality, tools for the disabled. 1994; April 13. www.nytimes .com/1994/04/13/garden/in-virtual-reality-tools-for-the-disabled.html. Accessed July 16, 2010.

3 Hayes HB. The next frontier—The vGov Project will provide agencies the security they need to explore and inhabit virtual worlds. 2010; July 29. http://fedtechmagazine.com/article.asp?item_id=819&sv=related. Accessed August 21, 2010.

4 Second Life Case Studies. www.secondlifegrid.net/casestudies. Accessed August 21, 2010.

5 Weis SA. RFID: Technical considerations, in H. Bidgoli (ed.), *The Handbook of Technology Management*. Hoboken, NJ: John Wiley & Sons: 2010.

6 Weier MH. Coke's RFID-based dispensers redefine business intelligence. 2009; June 6. www.informationweek.com/news/mobility/RFID/ showArticle.jhtml?articleID=217701971. Accessed October 7, 2010.

7 Mesenbrink J. Biometrics plays big role with airport security. *Security*. 2002; February 4.

8 Knorr E, Gruman G. What cloud computing really means. InfoWorld. 2008; April 7. www.infoworld.com/d/cloud-computing/what-cloud-computing-really-means-031.

9 Brodkin J. Amazon and IBM are the cloud's biggest players. 2010; July 15. http://www.infoworld.com/d/cloud-computing/amazon-and-ibm-are-the-clouds-biggest-players-484?page=0,0&source=IFWNLE_nlt_ wrapup_2010-07-16. Accessed July 31, 2010.

10 Knorr E, Gruman G. What cloud computing really means. InfoWorld. 2008; April 7. www.infoworld.com/article/08/04/07/15FE-cloud-computing-reality_1.html. Accessed August 21, 2010.

11 McFedries P. The cloud is the computer. *IEEE Spectrum*. 2008; August. www.spectrum.ieee.org/computing/hardware/the-cloud-is-the-computer. Accessed August 21, 2010.

12 Dignan L. Amazon's cloud computing will surpass its retailing business. 2008; April 14. http://blogs.zdnet.com/BTL/?p=8471. Accessed August 21, 2010.

13 Brodkin J. 10 cloud computing companies to watch. *Network World*. 2009; May 18. www.networkworld.com/supp/2009/ndc3/051809-cloud-companies-to-watch.html?page=1. Accessed August 21, 2010.

14 Nanotech defined. http://domino.watson.ibm.com/comm/research .nsf/pages/r.nanotech.defined.html. Accessed August 21, 2010.

16 Wood S, Jones R, Geldart A. Commercial applications of nanotechnology in computing and information technology. Report given at ESRC: The Social and Economic Challenges of Nanotechnology. 2003; July. www.azonano.com/details.asp?ArticleID=1057. Accessed August 21, 2010.

16 New report on nanotechnology in consumer goods. AZoM.com. 2009; May 26. www.azom.com/news.asp?NewsID=17263. Accessed August 21, 2010.

Index

Page numbers in *italic* refer to figures or information boxes.

review

Learning Outcomes

Here you'll find a summary of the key points for each Learning Outcome.

Learning Outcomes

LO¹ Discuss common applications of computers and information systems.

Organizations use computers and information systems to reduce costs and gain a competitive advantage in the marketplace.

As a knowledge worker of the future, computers and information technology will help you be more effective and productive, no matter what profession you choose.

LO² Explain the differences between computer literacy and information literacy.

In the 21st century, knowledge workers need two types of knowledge to be competitive in the workplace: com... Computer literacy is having skills in using productivi... ...preadsheet, database management systems, and prese... ...owledge of hardware and software, the Internet, and c... ...ormation literacy, on the other hand, is understanding th... ...using business intelligence.

How to use this Card:

1. Look over the card to preview the new concepts you'll be introduced to in the chapter.

2. Read the chapter to fully understand the material.

3. Go to class (and pay attention).

4. Review the card one more time to make sure you've registered the key concepts.

5. Don't forget—this card is only one of many MIS learning tools available to help you succeed in your management information systems course.

LO³ Define tr...

For the past 60 yea... ...Ss) have been applied to structured tasks, such as reco... ...and inventory control. TPSs focus on data collection... ...for using them is cost reduction.

LO⁴ Define m...

A management inf... ...integration of hardware and software technolo... ...ents designed to produce timely, integrated, relevan... ...decision-making purposes.

In designing an MI... ...s objectives clearly. Second, data must be collected... ...t be provided in a useful format for decision-making p...

LO⁵ Describe... ...information system.

In addition to hardw... ..., an information system includes four major components:

- Data
- Database
- Process
- Information

LO⁶ Discuss the differences between data and information.

The data component of an information system is considered the input to the system. An information system should collect data from external and internal sources.

Data consists of raw facts and by itself is difficult to use for decision making. Information—the output of an information system—consists of facts that have been analyzed by the process component and, therefore, are more useful for decision making.

LO⁷ Explain the importance and applications of information systems in functional areas of a business.

Information is the second most important resource (after the human element) in any organization. Timely, relevant, and accurate information is a critical tool for enhancing a company's competitive position in the marketplace and managing the four Ms of resources: manpower, machinery, materials, and money.

Key Terms

The key terms and definitions are organized by Learning Outcomes.

LO²

...e skills ...ch as ...atabase ...ntation software, and having a basic knowledge of hardware and software, the Internet, and collaboration tools and technologies.

Information literacy is understanding the role of information in generating and using business intelligence.

Business intelligence (BI) provides historical, current, and predictive views of business operations and environments and gives organizations a competitive advantage in the marketplace.

LO³

Transaction processing systems (TPSs) focus on data collection and processing, and the major reason for using them is cost reduction.

LO⁴

A **management information system (MIS)** is an organized integration of hardware and software technologies, data, processes, and human elements designed to produce timely, integrated, relevant, accurate, and useful information for decision-making purposes.

LO⁵

Data consists of raw facts and is a component of an information system.

A **database** is a collection of all relevant data organized in a series of integrated files.

The **process** component of an information system generates the most useful type of information for decision making, including transaction-processing reports and models for decision analysis.

Information consists of facts that have been analyzed by the process component and is an output of an information system.

Additional content is available on the CourseMate for MIS Web site. Login at **www.cengagebrain.com**

Chapter 1: Information Systems: An Overview

Key Terms (con't)

LO⁸

Information technologies support information systems and come in the form of the Internet, computer networks, database systems, POS systems, and radio frequency identification (RFID) tags.

LO⁹

Michael Porter's **Five Forces Model** analyzes an organization, its position in the marketplace, and how information systems could be used to make it more competitive. The five forces include buyer power, supplier power, threat of substitute products or services, threat of new entrants, and rivalry among existing competitors.

Exhibits

Key exhibits from the chapter are included on the Review Card to make it easier for you to study.

To manage these resources, different types of information systems have been developed. Although all have the major components (data, database, process, and information), they vary in the kind of data they collect and the analyses they perform. Major functional information systems include HRIS, LIS, MFIS, FIS, and MKIS.

LO⁸ Discuss how information technologies are used to gain a competitive advantage

Timely, relevant, and accurate information is a critical tool for enhancing a company's competitive position in the marketplace and managing the four Ms of resources:

manpower, machinery, materials, and money.

Three strategies for competing in the marketplace successfully include:

- Overall cost leadership
- Differentiation
- Focus

Information systems can help organizations reduce the cost of products and services and assist with differentiation and focus strategies.

Information systems can help bottom-line and top-line strategies.

LO⁹ Explain the Five Forces Model and strategies for gaining a competitive advantage.

Developed by Michael Porter, the Five Forces model is a comprehensive framework for analyzing an organization, distinguishing its position in the marketplace, and identifying information systems that could be used to make the organization more competitive.

The five forces are:

- Buyer power
- Supplier power
- Threat of substitute products or services
- Threat of new entrants
- Rivalry among existing competitors

LO¹⁰ Review the IT job market.

In the last decade the IT job market has been one of the fastest growing. Broadly speaking, IT jobs fall into the following categories: operations and help desk, programming, systems design, Web design and Web hosting, network design and maintenance, database design and maintenance, and robotics and artificial intelligence.

LO¹¹ Summarize the future outlook of information systems.

- Hardware and software costs will continue to decline making it affordable for all organizations, regardless of size, to utilize information systems.
- Artificial intelligence will continue to improve and expand.
- Computer literacy will improve as computer basics are taught in schools.
- Compatibility issues between networks will become more manageable .
- Personal computers will continue to improve in power and quality. They should become more affordable.
- Computer criminals will become more sophisticated, and protecting personal identity information will become more difficult.

Chapter 1: Information Systems: An Overview

Additional content is available on the CourseMate for MIS Web site. Login at **www.cengagebrain.com**

Learning Outcomes

LO¹ Discuss common applications of computers and information systems.

Organizations use computers and information systems to reduce costs and gain a competitive advantage in the marketplace.

As a knowledge worker of the future, computers and information technology will help you be more effective and productive, no matter what profession you choose.

LO² Explain the differences between computer literacy and information literacy.

In the 21st century, knowledge workers need two types of knowledge to be competitive in the workplace: computer literacy and information literacy. Computer literacy is having skills in using productivity software, such as word processing, spreadsheet, database management systems, and presentation software, and having a basic knowledge of hardware and software, the Internet, and collaboration tools and technologies. Information literacy, on the other hand, is understanding the role of information in generating and using business intelligence.

LO³ Define transaction processing systems.

For the past 60 years, transaction processing systems (TPSs) have been applied to structured tasks, such as record keeping, simple clerical operations, and inventory control. TPSs focus on data collection and processing, and the major reason for using them is cost reduction.

LO⁴ Define management information systems.

A management information system (MIS) is an organized integration of hardware and software technologies, data, processes, and human elements designed to produce timely, integrated, relevant, accurate, and useful information for decision-making purposes.

In designing an MIS, the first task is to define the system's objectives clearly. Second, data must be collected and analyzed. Finally, information must be provided in a useful format for decision-making purposes.

LO⁵ Describe the four major components of an information system.

In addition to hardware, software, and human elements, an information system includes four major components:

- Data
- Database
- Process
- Information

LO⁶ Discuss the differences between data and information.

The data component of an information system is considered the input to the system. An information system should collect data from external and internal sources.

Data consists of raw facts and by itself is difficult to use for decision making. Information—the output of an information system—consists of facts that have been analyzed by the process component and, therefore, are more useful for decision making.

LO⁷ Explain the importance and applications of information systems in functional areas of a business.

Information is the second most important resource (after the human element) in any organization. Timely, relevant, and accurate information is a critical tool for enhancing a company's competitive position in the marketplace and managing the four Ms of resources: manpower, machinery, materials, and money.

Key Terms

LO²

Computer literacy is having the skills in using productivity software, such as word processing, spreadsheet, database management systems, and presentation software, and having a basic knowledge of hardware and software, the Internet, and collaboration tools and technologies.

Information literacy is understanding the role of information in generating and using business intelligence.

Business intelligence (BI) provides historical, current, and predictive views of business operations and environments and gives organizations a competitive advantage in the marketplace.

LO³

Transaction processing systems (TPSs) focus on data collection and processing, and the major reason for using them is cost reduction.

LO⁴

A **management information system (MIS)** is an organized integration of hardware and software technologies, data, processes, and human elements designed to produce timely, integrated, relevant, accurate, and useful information for decision-making purposes.

LO⁵

Data consists of raw facts and is a component of an information system.

A **database** is a collection of all relevant data organized in a series of integrated files.

The **process** component of an information system generates the most useful type of information for decision making, including transaction-processing reports and models for decision analysis.

Information consists of facts that have been analyzed by the process component and is an output of an information system.

Additional content is available on the CourseMate for MIS Web site. Login at **www.cengagebrain.com**

Chapter 1: Information Systems: An Overview

LO⁸

Information technologies support information systems and come in the form of the Internet, computer networks, database systems, POS systems, and radio frequency identification (RFID) tags.

LO⁹

Michael Porter's **Five Forces Model** analyzes an organization, its position in the marketplace, and how information systems could be used to make it more competitive. The five forces include buyer power, supplier power, threat of substitute products or services, threat of new entrants, and rivalry among existing competitors.

To manage these resources, different types of information systems have been developed. Although all have the major components (data, database, process, and information), they vary in the kind of data they collect and the analyses they perform. Major functional information systems include HRIS, LIS, MFIS, FIS, and MKIS.

LO⁸ Discuss how information technologies are used to gain a competitive advantage

Timely, relevant, and accurate information is a critical tool for enhancing a company's competitive position in the marketplace and managing the four Ms of resources:

manpower, machinery, materials, and money.

Three strategies for competing in the marketplace successfully include:

- Overall cost leadership
- Differentiation
- Focus

Information systems can help organizations reduce the cost of products and services and assist with differentiation and focus strategies.

Information systems can help bottom-line and top-line strategies.

LO⁹ Explain the Five Forces Model and strategies for gaining a competitive advantage.

Developed by Michael Porter, the Five Forces model is a comprehensive framework for analyzing an organization, distinguishing its position in the marketplace, and identifying information systems that could be used to make the organization more competitive.

The five forces are:

- Buyer power
- Supplier power
- Threat of substitute products or services
- Threat of new entrants
- Rivalry among existing competitors

LO¹⁰ Review the IT job market.

In the last decade the IT job market has been one of the fastest growing. Broadly speaking, IT jobs fall into the following categories: operations and help desk, programming, systems design, Web design and Web hosting, network design and maintenance, database design and maintenance, and robotics and artificial intelligence.

LO¹¹ Summarize the future outlook of information systems.

- Hardware and software costs will continue to decline making it affordable for all organizations, regardless of size, to utilize information systems.
- Artificial intelligence will continue to improve and expand.
- Computer literacy will improve as computer basics are taught in schools.
- Compatibility issues between networks will become more manageable .
- Personal computers will continue to improve in power and quality. They should become more affordable.
- Computer criminals will become more sophisticated, and protecting personal identity information will become more difficult.

Additional content is available on the CourseMate for MIS Web site. Login at **www.cengagebrain.com**

review card / CHAPTER 2
COMPUTERS: THE MACHINES BEHIND COMPUTING

Learning Outcomes

LO¹ Define a computer system and describe its components.

A computer is a machine that accepts data as input, processes data without human intervention by using stored instructions, and outputs information. The instructions, also called a "program," are step-by-step directions for performing a specific task, written in a language the computer can understand.

A computer system consists of hardware and software. Hardware components are physical devices, such as keyboards, monitors, and processing units. The software component consists of programs written in computer languages.

LO² Discuss the history of computer hardware and software.

Major developments in hardware have taken place over the past 60 years. To understand these developments better, computers are often categorized into "generations" to mark technological breakthroughs.

LO³ Explain the factors distinguishing computing power of computers.

Computers draw their power from three factors that far exceed human capacities: speed, accuracy, and storage and retrieval capabilities.

Every character, number, or symbol on the keyboard is represented as a binary number in computer memory. A binary system consists of 0s and 1s, with a 1 representing "on" and a 0 representing "off," similar to a light switch.

Computers and communication systems use data codes to represent and transfer data between computers and network systems.

Three data codes are: ASCII, Extended ASCII, and Unicode

LO⁴ Describe the major operations of computers.

Computers can perform three basic tasks: arithmetic operations, logical operations, and storage and retrieval operations. All other tasks are carried out by one or a combination of these operations.

Computers can add, subtract, multiply, divide, and raise numbers to a power. They also perform comparison operations.

Computers can store massive amounts of data in very small spaces and locate a particular item quickly.

LO⁵ Discuss the types of input, output, and memory devices.

Input devices send data and information to the computer. Examples include keyboards, a mouse, touch screens, light pens, a trackball, data tablet, bar code reader, optical character readers, magnetic ink character recognition, and optical mark recognition.

Output devices are available for both mainframe and personal computers. Output devices are either soft copy (plasma display, LCD) or hard copy (printers).

Memory is considered either main or secondary.

LO⁶ Explain how computers are classified.

Computers are classified based on cost, amount of memory, speed, and sophistication.

Computers can be classified as: subnotebook; notebook; personal computer; minicomputer; mainframes; or supercomputer.

Key Terms

LO¹

A **computer** is a machine that accepts data as input, processes data without human intervention by using stored instructions, and outputs information.

The **central processing unit (CPU)** is the heart of a computer. It's divided into two components: the arithmetic logic unit and the control unit.

The **arithmetic logic unit (ALU)** performs arithmetic operations (addition, subtraction, multiplication, and division) and logical operations, such as comparing numbers.

The **control unit** tells the computer what to do, such as instructing the computer which device to read or send output to.

A **bus** is a link between devices connected to the computer. It can be parallel or serial or internal (local) or external.

A **disk drive** is a peripheral devise for recording, storing, and retrieving information.

A **CPU case** is also known as a computer chassis or tower. It is the enclosure containing the computer's main components.

A **motherboard** is the main circuit board containing connectors for attaching additional boards. It usually contains the CPU, Basic Input/Output System (BIOS), memory, storage interfaces, serial and parallel ports, expansion slots, and all the controllers for standard peripheral devices, such as the display monitor, disk drive, and keyboard.

LO⁵

Input devices send data and information to the computer. Examples include keyboards and mouses.

An **output device** is capable of representing information from a computer. The form of this output might be visual, audio, or digital and examples include printers, display monitors, and plotters.

Main memory stores data and information and is usually volatile.

Additional content is available on the CourseMate for MIS Web site. Login at **www.cengagebrain.com**

Chapter 2: Computers: The Machines Behind Computing

Secondary memory, which is nonvolatile, serves mostly as archival storage, as for backups.

Random access memory (RAM) is volatile memory, in which data can read from and be written to "read-write memory."

A special type of RAM, called **cache RAM**, resides on the processor. Because memory access from main RAM storage generally takes several clock cycles (a few nanoseconds), cache RAM stores recently accessed memory so that the processor isn't waiting for the memory transfer.

Nonvolatile memory is called **read-only memory (ROM)**, so data can't be written to ROM.

Magnetic tape, made of a plastic material, resembles a cassette tape and stores data sequentially.

A **magnetic disk** made of mylar or metal is used for random-access processing. In other words, data can be accessed in any order, regardless of its order on the surface.

Optical discs use laser beams to access and store data. Examples include CD-ROMs, WORM discs, and erasable optical discs.

A **redundant array of independent disks (RAID) system** is a collection of disk drives used for fault tolerance and improved performance, typically in large network systems.

A **storage area network (SAN)** is a dedicated high-speed network consisting of both hardware and software, used to connect and manage shared storage devices, such as disk arrays, tape libraries, and optical storage devices.

Network-attached storage (NAS) is essentially a network-connected computer dedicated to providing file-based data storage services to other network devices.

LO⁶

A **server** is a computer and all the software for managing network resources and offering services to a network.

LO⁷   Describe the two major types of software.

Software is all of the programs that run a computer system.

The two types of software include: system software and application software.

System software is run by operating systems (OS). Examples of OS include Microsoft Windows, Mac OS, and Linux.

Application software is used to perform a variety of tasks on a personal computer, including word processing, spreadsheet, database, presentation, graphics, etc. Examples of application software suites include Microsoft Office, OpenOffice, and Corel.

LO⁸   List the generations of computer languages.

There are four generations of computer language with a fifth generation in development.

- First: Machine Language
- Second: Assembly Language
- Third: High-level Language
- Fourth: Fourth-Generation Language
- Fifth: Natural Language Processing

LO⁷

An **operating system (OS)** is a set of programs for controlling and managing computer hardware and software. An OS provides an interface between a computer and the user and increases computer efficiency by helping users share computer resources and performing repetitive tasks for users.

Application software can be commercial software or software developed in-house and is used to perform a variety of tasks on a personal computer.

LO⁸

Machine language consists of a series of 0s and 1s representing data or instructions and was part of the first generation of computer languages. It depends on the machine, so a code written for one type of computer does not work on another type of computer.

Assembly language, the second generation of computer languages, is a higher-level language than machine language but is also machine dependent. It uses a series of short codes, or mnemonics, to represent data or instructions.

High-level languages are machine independent and part of the third-generation computer language. Many options are available and each is designed for a specific purpose.

Fourth-generation languages (4GLs) use macro codes that can take the place of several lines of programming. The commands are powerful and easy to learn, particularly for people with little computer training.

Fifth-generation languages, also called natural language processing (NLP), are the ideal computer languages for people with minimal computer training. These languages are designed to facilitate natural conversations between you and the computer.

Additional content is available on the CourseMate for MIS Web site. Login at **www.cengagebrain.com**

Learning Outcomes

LO¹ Define a database and a database management system.

A database is a critical component of information systems because any type of analysis that's done is based on data available in the database. To make using databases more efficient, a DBMS is used.

A user issues a request, and the DBMS searches the database and returns the information to the user. See Exhibit 3.2.

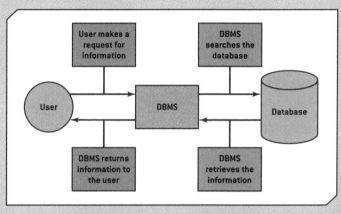

Exhibit 3.2 *Interaction between the user, DBMS, and database*

- User makes a request for information
- DBMS searches the database
- User
- DBMS
- Database
- DBMS returns information to the user
- DBMS retrieves the information

Files are accessed by using a sequential, random, or indexed sequential method.

LO² Explain logical database design and the relational database model.

The first step in database design is defining a data model (which includes considering data structure, operations, and integrity rules).

Consider which view you'll use: physical or logical.

Many data models (including, object-oriented, hierarchical, and network) are used but the most common is the relational model.

LO³ Define the components of a database management system.

DBMS software includes these components:

- Database engine: heart of DBMS software. Responsible for data storage, manipulation, and retrieval.
- Data definition: creates and maintains the data dictionary and defines the structure of files in a database.
- Data manipulation: adds, deletes, modifies, and retrieves records from a database.
- Application generation: designs elements of an application.
- Data administration: used for backup and recovery, security, and change management. Also determines CRUD.

LO⁴ Summarize recent trends in database design and use.

Trends in database design and use include data-driven Web sites, natural language processing, distributed databases, client/server databases, and object-oriented databases. In addition to

Key Terms

LO¹

A **database** is a collection of related data that can be stored in a central location or in multiple locations.

Data hierarchy is the structure and organization of data, which involves fields, records, and files.

A **database management system (DBMS)** is software for creating, storing, maintaining, and accessing database files. A DBMS makes using databases more efficient.

In a **sequential file structure**, records in files are organized and processed in numerical or sequential order, typically the order in which they were entered.

In a **random access file structure**, records can be accessed in any order, regardless of their physical location in storage media. This method of access is fast and very effective when a small number of records need to be processed daily or weekly.

With the **indexed sequential access method (ISAM)**, records can be accessed sequentially or randomly, depending on the number being accessed. For a small number, random access is used, and for a large number, sequential access is used.

LO²

The **physical view** involves how data is stored on and retrieved from storage media, such as hard disks, magnetic tapes, or CDs.

The **logical view** involves how information appears to users and how it can be organized and retrieved.

A **data model** determines how data is created, represented, organized and maintained. It usually contains data structure, operations, and integrity rules.

In a **hierarchical model**, the relationships between records form a treelike structure (hierarchy). Records are called nodes, and relationships between records are called branches. The node at the top is called the root, and every other node (called a child) has a parent. Nodes with the same parents are called twins or siblings.

Additional content is available on the CourseMate for MIS Web site. Login at **www.cengagebrain.com**

Chapter 3: Database Systems, Data Warehouses, and Data Marts

The **network model** is similar to the hierarchical model, but records are organized differently. Unlike the hierarchical model, each record in the network model can have multiple parent and child records.

A **relational model** uses a two-dimensional table of rows and columns of data. Rows are records (also called "tuples"), and columns are fields (also referred to as "attributes").

The **data dictionary** stores definitions, such as data types for fields, default values, and validation rules for data in each field.

Primary keys uniquely identify every record in a relational database. Examples include Student ID numbers, account numbers, Social Security numbers, and invoice numbers.

A **foreign key** is a field in a relational table that matches the primary key column of another table. It can be used to cross-reference tables.

Normalization improves database efficiency by eliminating redundant data and ensuring that only related data is stored in a table.

LO³

Structured Query Language (SQL) is a standard fourth-generation query language used by many DBMS packages, such as Oracle 11g and Microsoft SQL Server. SQL consists of several keywords specifying actions to take.

With **query by example (QBE)**, you request data from a database by constructing a statement made up of query forms. With current graphical databases, you simply click to select query forms instead of having to remember keywords, as you do with SQL. You can add AND, OR, and NOT operators to the QBE form to fine-tune the query.

Data administration is used to determine who has permission to perform certain functions, often summarized as **create, read, update, and delete (CRUD)**

Database administrators (DBA), found in large organizations, design and set up databases, establish security measures, develop recovery procedures, evaluate database performance, and add and fine-tune database functions.

LO⁴

A **data-driven Web site** acts as an interface to a database, retrieving data for users and allowing users to enter data in the database.

A **distributed database** stores data on multiple servers throughout an organization.

The **fragmentation** approach to a distributed DBMS addresses how tables are divided among multiple locations. There are three variations: horizontal, vertical, and mixed.

The **replication** approach to distributed DBMS has each site store a copy of data of the organization's database.

The **allocation** approach to a distributed DBMS combines fragmentation and replication, with each site storing the data it uses most often.

In a **client/server database**, users' workstations (client) are linked in a local area network (LAN) to share the services of a single server.

In **object-oriented databases** both data and their relationships are contained in a single object. An object consists of attributes and methods that can be performed on the object's data.

Grouping objects along with their attributes and methods into a class is called **encapsulation**, which essentially means grouping similar items into a single unit. It helps handle more complex types of data, such as images and graphs.

Inheritance means new objects can be created faster and more easily by entering new data in attributes.

LO⁵

A **data warehouse** is a collection of data from a variety of sources used to support decision-making applications and generate business intelligence.

Extraction, transformation, and loading (ETL) describe the processes used in a data warehouse. It includes extracting data from outside sources, transforming it to fit operational needs, and loading it into the end target (database or data warehouse).

Online analytical processing (OLAP), generates business intelligence. It uses multiple sources of information and provides multidimensional analysis, such as viewing data based on time, product, and location.

Data-mining analysis is used to discover patterns and relationships.

LO⁶

A **data mart** is usually a smaller version of a data warehouse, used by a single department or function.

---

these trends, advances in artificial intelligence and natural language processing will have an impact on database design and use, such as improving user interfaces.

LO⁵   Explain the components and functions of a data warehouse.

Data warehouses support decision-making applications and generate business intelligence. Data in a data warehouse has the following characteristics: subject oriented, integrated, time variant, aggregate data, and purpose.

LO⁶   Describe the functions of a data mart.

A data mart is usually a smaller version of a data warehouse, used by a single department or function. Despite being smaller, data marts can usually perform the same types of analysis as a data warehouse. Advantages include: faster access, improved response time, less expensive, easier to create, and targeted to users' needs.

Learning Outcomes

LO¹ Describe information technologies that could be used in computer crimes.

Information technologies can be misused to invade users' privacy and commit computer crimes.

- Cookies
- Spyware and adware
- Phishing
- Keyloggers
- Sniffing and Spoofing
- Computer crime and fraud

You can minimize or prevent many of these risks by installing operating system updates regularly, using antivirus software, and using e-mail security features.

Sniffing is capturing and recording network traffic to intercept information.

Spoofing is an attempt to gain access to a network by posing as an authorized user to find sensitive information, such as passwords.

Computer fraud is the unauthorized use of computer data for personal gain, such as charging purchases to someone else's account.

LO² Review privacy issues and methods for improving privacy of information.

Despite the benefits of information technologies, they've created some concerns about privacy in the workplace. Employers can check social networking sites during the hiring process. They can also monitor employees' performance, accuracy, and time spent away from their computers.

Personal information is stored on databases, and misuse of extremely sensitive information, such as medical records, could prevent someone from getting employment, health insurance, or housing.

Information in databases can be cross-matched to create profiles of people and even predict their behavior, based on transactions with educational, financial, government, and other institutions. This information is often used for direct marketing and credit checks on potential borrowers or renters.

Several federal laws now regulate collecting and using information on people and corporations, but they're narrow in scope and contain some loopholes.

To better understand the legal and privacy issues of the Internet and networks, three important concepts should be understood: acceptable use policies, accountability, and nonrepudiation.

Although e-mail is widely used, it presents some serious privacy issues including exposure to spam and ease of access.

Two commonly used technologies for data collection are log files and cookies. Sometimes users give incorrect information on purpose. If the information collected isn't accurate, the result could be identity misrepresentation. Therefore, data collected on the Internet must be used and interpreted with caution.

LO³ Discuss ethical issues of information technology.

Ethics and ethical decision making involve the moral guidelines people or organizations follow in dealing with others. In essence, ethics means doing the right thing, and its meaning can vary in different cultures and even from person to person. Determining

Key Terms

LO¹

Cookies are small text files with a unique ID tag which is embedded in a Web browser and are saved on the user's hard drive.

Spyware is software that secretly gathers information about users while they browse the Web.

Adware is a form of spyware that collects information about the user (without the user's consent) to display advertisements in the Web browser, based on information it collects from the user's browsing patterns.

Phishing is sending fraudulent e-mails that seem to come from legitimate sources, such as a bank or university, for the purpose of capturing private information, such as bank account numbers or Social Security numbers.

Keyloggers monitor and record keystrokes and can be software or hardware devices.

LO²

An **acceptable use policy** is a set of rules specifying legal and ethical system use and activities and the consequences of noncompliance.

Spam is unsolicited e-mail sent for advertising purposes.

Log files, which are generated by Web server software, record a user's actions on a Web site.

LO³

Intellectual property is a legal umbrella covering protections that involve copyrights, trademarks, trade secrets, and patents for "creations of the mind" developed by people or businesses.

Cybersquatting is registering, selling, or using a domain name to profit from someone else's trademark.

Additional content is available on the CourseMate for MIS Web site. Login at **www.cengagebrain.com**

Chapter 4: Personal, Legal, Ethical, and Organizational Issues of Information Systems

Information technology and the Internet have created a **digital divide**. Computers still aren't affordable for many people. The digital divide has implications for education.

LO⁴

Virtual organizations are networks of independent companies, suppliers, customers, and manufacturers connected via information technologies so that they can share skills and costs and have access to each other's markets.

what's legal or illegal is usually clear, but drawing a line between what's ethical and unethical is more difficult. See Exhibit 4.1.

Exhibit 4.1 *Ethical versus legal grid*

	Legal	Illegal
Ethical	I	II
Unethical	III	IV

Cybercrime, cyberfraud, identity theft, and intellectual property theft are on the rise. Many experts believe management can reduce employees' unethical behavior by developing and enforcing codes of ethics. Many associations promote ethically responsible use of information systems and technologies and have developed codes of ethics for their members.

Two types of information are available on the Web: public and private. Public information, posted by an organization or public agency, could be censored for public policy reasons. Private information—what's posted by a person—can't be subject to censorship because of freedom of expression.

Parents are concerned about what their children are exposed to on the Web. Programs such as CyberPatrol prevent children's access to certain sites.

Intellectual property can be divided into two categories: industrial properties (inventions, trademarks, logos, etc.) and copyrighted material (literary and artistic works).

Copyright laws cover online materials, including Web pages, HTML code, and computer graphics, as long as the content can be printed or saved on a hard drive. They give only the creator exclusive rights, meaning no one else can reproduce, distribute, or perform the work without permission.

Other intellectual property protections include trademarks, patents, software piracy, and cybersquatting.

LO⁴ **Describe the impact of information technology in the workplace.**

Information technologies have a direct effect on the nature of jobs. Telecommuting or virtual work, for example, has enabled some people to perform their jobs from home. With telecommunications technology, a worker can send and receive data to and from the main office, and organizations can use the best and most cost-effective human resources in a large geographical region.

Information technology is also creating virtual organizations, which are networks of independent companies, suppliers, customers, and manufacturers. These networks are connected via information technology and can share skills and costs.

Ergonomics experts believe using better-designed furniture as well as flexible keyboards, correct lighting, special monitors for workers with vision problems, and so forth can resolve many health-related problems.

LO⁵ **Discuss green computing and ways it can help improve the quality of the environment.**

Green computing promotes a sustainable environment and consumes the least amount of energy. Information and communications technology generates approximately 2% of the world's carbon dioxide emissions, roughly the same as the aviation industry.

Green computing involves the design, manufacture, use, and disposal of computers, servers, and computing devices in such a way that there is minimal impact on the environment.

Learning Outcomes

LO¹   Describe basic safeguards in computer and network security.

A comprehensive security system includes hardware, software, procedures, and personnel that collectively protect information resources and keep intruders and hackers at bay. There are three important aspects of computer and network security: confidentiality, integrity, and availability.

When planning a comprehensive security system, the first step is designing fault-tolerant systems, which have a combination of hardware and software for improving reliability. See Exhibit 5.1.

Exhibit 5.1   *The McCumber cube*

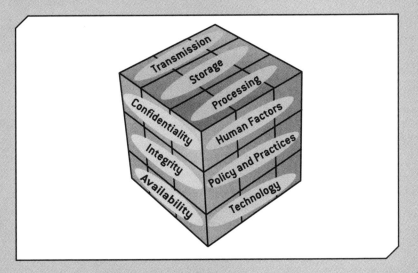

LO²   Explain the major security threats.

Threats can be categorized by whether they're unintentional or intentional. Intentional threats include:

- Viruses
- Worms
- Trojan programs
- Logic bombs
- Backdoors
- Blended threats (e.g., worm launched by Trojan)
- Rootkits
- Denial-of-service attacks
- Social engineering

Unintentional threats include:

- Natural disasters
- User's accidental deletion of files
- Structural failure

Key Terms

LO¹

Confidentiality means a system must not allow the disclosure of information to anyone who isn't authorized to access it.

Integrity ensures the accuracy of information resources in an organization.

Availability ensures that computers and networks are operating, and authorized users can access the information they need.

Fault-tolerant systems ensure availability in case of a system failure by using a combination of hardware and software.

LO²

A **virus** consists of self-propagating program code that's triggered by a specified time or event.

A **worm** travels from computer to computer in a network. Worms are independent programs that can spread themselves without having to be attached to a host program.

A **Trojan program** contains code intended to disrupt a computer, network, or Web site and is usually hidden inside a popular program.

A **logic bomb** is a type of Trojan program used to release a virus, worm, or other destructive code. Logic bombs are triggered at a certain time or by an event, such as a user pressing Enter or running a specific program.

A **backdoor** (also called a "trapdoor") is a programming routine built into a system by its designer or programmer. This routine enables the designer or programmer to bypass system security and sneak back into the system later to access programs or files.

A **blended threat** is a security threat that combines the characteristics of computer viruses, worms, and other malicious codes with vulnerabilities found on public and private networks.

A **denial-of-service (DoS) attack** floods a network or server with service requests to prevent legitimate users' access to the system.

Additional content is available on the CourseMate for MIS Web site. Login at **www.cengagebrain.com**

Chapter 5: Protecting Information Resources

In the context of security, **social engineering** means using "people skills"—such as being a good listener and assuming a friendly, unthreatening air—to trick others into revealing private information.

LO³

Biometric security measures use a physiological element to enhance security measures. These elements are unique to a person and can't be stolen, lost, copied, or passed on to others.

A **callback** modem is used to verify whether a user's access is valid by logging the user off and then calling the user back at a predetermined number.

A **firewall** is a combination of hardware and software that acts as a filter or barrier between a private network and external computers or networks, including the Internet.

An **intrusion detection system (IDS)** can protect against both external and internal access. They're usually placed in front of a firewall and can identify attack signatures, trace patterns, generate alarms for the network administrator, and cause routers to terminate connections with suspicious sources.

Physical security measures primarily control access to computers and networks and include devices for securing computers and peripherals from theft.

Access controls are designed to protect systems from unauthorized access to preserve data integrity.

A **virtual private network (VPN)** provides a secure "tunnel" through the Internet for transmitting messages and data via a private network.

Data encryption transforms data, called "plaintext" or "cleartext," into a scrambled form called "ciphertext" that can't be read by others.

Secure Sockets Layers (SSL) is a commonly used encryption protocol that manages transmission security on the Internet.

LO³ Describe security and enforcement measures.

In addition to backing up data and storing it securely, organizations can take many other steps to guard against threats. A comprehensive security system should include the following:

- Biometric security measures
- Nonbiometric security measures (See Exhibit 5.3)

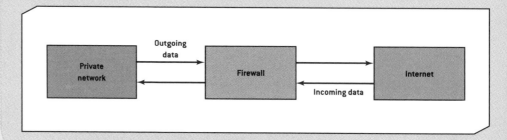

Exhibit 5.3 *A basic firewall configuration*

- Physical security measures
- Access controls
- Virtual private networks
- Data encryption
- E-commerce transaction security measures
- Computer Emergency Response Team

LO⁴ Summarize the guidelines for a comprehensive security system, including business continuity planning.

An organization's employees are an essential part of the success of any security system. When developing a comprehensive security plan focus on both user-related security issues such as posting the security policy, and revoking terminated employees' authorization immediately, and procedural and technical security such as reviewing logs, making sure fire protection systems are up to date, and installing firewalls, and IDS.

To lessen the effects of an attack or intrusion, planning for disaster recovery is important. This planning should include business continuity planning, which outlines procedures for keeping an organization operational.

Transport Layer Security (TLS) is a cryptographic protocol that ensures data security and integrity over public networks, such as the Internet.

Asymmetric encryption uses two keys: a public key known to everyone and a private or secret key known only to the recipient.

In **symmetric encryption** (also called "secret key encryption"), the same key is used to encrypt and decrypt the message.

LO⁴

Business continuity planning outlines procedures for keeping an organization operational in the event of a natural disaster or network attack.

reviewcard/

CHAPTER 6
DATA COMMUNICATION: DELIVERING INFORMATION ANYWHERE AND ANYTIME

Learning Outcomes

LO¹ Describe major applications of a data communication system.

Data communication is the electronic transfer of data from one location to another. Data communication applications enhance decision makers' efficiency and effectiveness.

LO² Explain the major components of a data communication system.

Typical data communication systems include the following components:

- Sender and receiver devices
- Modems or routers
- Communication medium (channel) (see Exhibit 6.1)

Exhibit 6.1 *Types of communication media*

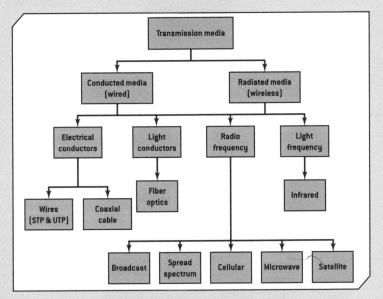

LO³ Describe the major types of processing configurations.

Data communication systems can be used in several different configurations, including centralized, decentralized, and distributed processing systems.

LO⁴ Explain the three types of networks.

The three major types of networks are local area networks, wide area networks, and metropolitan area networks.

LO⁵ Describe the main network topologies.

The five common topologies are star, ring, bus, hierarchical, and mesh.

LO⁶ Explain important networking concepts, such as bandwidth, routing, routers, and the client/server model.

Protocols are agreed-on methods and rules that electronic devices use to exchange information. Transmission Control Protocol/Internet Protocol (TCP/IP) is an industry-standard suite of communication protocols.

Key Terms

LO¹

Data communication is the electronic transfer of data from one location to another.

LO²

Bandwidth is the amount of data that can be transferred in a certain time period.

Attenuation is the loss of power in a signal.

In **broadband** data transmission, multiple pieces of data are sent simultaneously to increase the transmission rate.

Narrowband is a voice-grade transmission channel.

Protocols are rules that govern data communication.

A **modem** is a device that connects a user to the Internet.

Digital subscriber line (DSL) is a high-speed service that uses ordinary phone lines.

Communication media connect sender and receiver devices.

Conducted media provide a physical path along which signals are transmitted.

Radiated media use an antenna for transmitting data.

LO³

In a **centralized** processing system, all processing is done at one central computer.

In **decentralized processing**, each user, department, or division has its own computer for performing processing tasks.

Distributed processing maintains centralized control and decentralizes operations.

The **Open Systems Interconnection (OSI) model** is a seven-layer architecture for defining how data is transmitted in a network.

LO⁴

A **network interface card (NIC)** enables computers to communicate over a network.

A **local area network (LAN)** connects workstations and peripheral devices that are in close proximity.

Additional content is available on the CourseMate for MIS Web site. Login at **www.cengagebrain.com**

A **wide area network (WAN)** can span several cities, states, or even countries.

A **metropolitan area network (MAN)**, is designed to handle data communication for multiple organizations in a city.

LO⁵

A **network topology** represents a network's physical layout.

The **star topology** usually consists of a central computer and a series of nodes.

In a **ring topology** each computer manages its own connectivity.

The **bus topology** connects nodes along a network segment.

A **hierarchical topology** combines computers with different processing strengths in different organizational levels.

LO⁶

A **controller** controls data transfer from a computer to a peripheral device and vice versa.

A **multiplexer** allows several nodes to share one communication channel.

In a **mesh topology,** every node is connected to every other node.

Transmission Control Protocol/ Internet Protocol (TCP/IP) is an industry-standard suite of communication protocols that enables interoperability.

A **packet** is a collection of binary digits, sent from computer to computer over a network.

Routing is the process of deciding which path to take on a network.

A **routing table** is used to determine the best possible route for the packet.

In **centralized routing,** one node is in charge of selecting the path for all packets.

Distributed routing relies on each node to calculate its own best possible route.

A **router** is a network connection device that connects network systems and controls traffic flow between them.

A **static router** requires the network routing manager to give it information about which addresses are on which network.

A **dynamic** router can build tables that identify addresses on each network.

The process of deciding which path data takes is called routing.

In the client/server model, software runs on the local computer (the client) and communicates with the remote server to request information or services.

LO⁷ Describe wireless and mobile technologies and networks.

A wireless network is a network that uses wireless instead of wired technology. A mobile network is a network operating on a radio frequency (RF). See Exhibit 6.10.

LO⁸ Discuss the importance of wireless security and the techniques used.

Wireless security is especially important since anyone within walking or driving distance of an access point can gain use of a wireless network.

There are several techniques for improving the security of wireless networks:
- Service set identifier (SSI)
- Wired equivalent privacy (WEP)
- Extensible authentication protocol (EAP)
- Wi-Fi protected access (WPA)
- WPA2 or 802.11i

LO⁹ Summarize the convergence phenomenon and its applications for business and personal use.

In data communication, convergence refers to integrating voice, video, and data so that multimedia information can be used for decision making.

In the **client/server model,** software runs on the local computer and communicates with the remote server to request information or services.

In the **two-tier architecture**, a client communicates directly with the server.

An *n*-tier architecture places application processing on a middle-tier server.

LO⁷

A **wireless network** is a network that uses wireless instead of wired technology.

A **mobile network** is a network operating on a radio frequency (RF).

Throughput is the amount of data transferred or processed in a specified time.

Time Division Multiple Access (TDMA) divides each channel into six time slots. Each user is allocated two slots: one for transmission and one for reception.

Code Division Multiple Access (CDMA) transmits multiple encoded messages over a wide frequency and then decodes them at the receiving end.

LO⁸

Convergence refers to integrating voice, video, and data so that multimedia information can be used for decision making.

Learning Outcomes

LO¹ Describe the makeup of the Internet and the World Wide Web.

The Internet backbone is a foundation network linked with fiber-optic cables that can support very high bandwidth.

The World Wide Web changed the Internet by introducing a graphical interface to the largely text-based Internet in 1989.

When information is transferred from one network to another, domain names are converted to IP addresses by the protocol Domain Name System (DNS). You see domain names used in uniform resource locators (URLs), also called "universal resource locators," to identify a Web page.

There are several methods for connecting to a network, including the Internet. These methods include dial-up and cable modems as well as Digital Subscriber Line (DSL).

LO² Discuss navigational tools, search engines, and directories.

Navigational tools are used to travel from site-to-site or to "surf" the Internet.

Search engines give you an easy way to look up information and resources on the Internet by entering key words related to your topic of interest.

Directories are indexes of information based on keywords in documents and make it possible for search engines to find what you're looking for. Some Web sites, such as Yahoo!, also use directories to organize content into categories.

LO³ Describe common Internet services.

Many services are available via the Internet, and most are made possible by the TCP suite of protocols in the Application layer. These include:

- E-mail
- Newsgroups and discussion groups
- Internet Relay Chat (IRC) and instant messaging
- Internet telephony

LO⁴ Summarize widely used Web applications.

Several service industries use the Internet and its supporting technologies to offer services and products to a wide range of customers at more competitive prices and with increased convenience.

LO⁵ Explain the purpose of intranets.

An intranet is a network within an organization that uses Internet protocols and technologies for collecting, storing, and disseminating useful information that supports business activities. See Table 7.2 and Table 7.3.

LO⁶ Explain the purpose of extranets.

An extranet is a secure network that uses the Internet and Web technologies to connect intranets of business partners, so communication between organizations or between consumers is possible.

Key Terms

LO¹

The **Internet** is a worldwide collection of millions of computers and networks of all sizes.

The **Advanced Research Projects Agency Network (ARPANET)** was the project started in 1969 by the U.S. Department of Defense that was the beginning of the Internet.

The **Internet backbone** is a foundation network linked with fiber-optic cables that can support very high bandwidth.

With **hypermedia**, documents can include embedded references to audio, text, images, video, and other documents.

Hypertext consists of links users can click to follow a particular thread (topic).

Domain Name System (DNS) protocol is used to convert domain names to IP addresses.

Uniform resource locators (URLs), also called "universal resource locators," identify a Web page.

LO²

Hypertext Markup Language (HTML) is the language used to create Web pages.

Navigational tools are used to travel from site-to-site or to "surf" the Internet.

Directories are indexes of information based on keywords in documents.

A **search engine** is an information system that enables users to retrieve data from the Web by searching for information using search terms.

Table 7.2 *The Internet versus intranets*

Key feature	Internet	Intranet
User	Anybody	Approved users only
Geographical scope	Unlimited	Limited to unlimited
Speed	Slower than an intranet	Faster than the Internet
Security	Less than an intranet's	More than the Internet's; user access is restricted more

Additional content is available on the CourseMate for MIS Web site. Login at **www.cengagebrain.com**

Chapter 7: The Internet, Intranets, and Extranets

Table 7.3 *Comparison of the Internet, intranets, and extranets*

	Internet	Intranet	Extranet
Access	Public	Private	Private
Information	General	Typically confidential	Typically confidential
Users	Everybody	Members of an organization	Groups of closely related companies, users, or organizations

Key Terms (con't)

LO³

Discussion groups are usually formed for people to exchange opinions and ideas on a specific topic, usually of a technical or scholarly nature.

Newsgroups can be established for any topic; they allow people to get together for fun or for business purposes.

Internet Relay Chat (IRC) enables users in chat rooms to exchange text messages in real time.

Instant messaging (IM) is a service for communicating with others via the Internet in a private "chat room."

Internet telephony is using the Internet to exchange spoken conversations.

Voice over Internet Protocol (VoIP) is the protocol used for Internet telephony.

LO⁵

An **intranet** is a network within an organization that uses Internet protocols and technologies for collecting, storing, and disseminating useful information that supports business activities.

LO⁶

An **extranet** is a secure network that uses the Internet and Web technologies to connect intranets of business partners.

LO⁷

Web 2.0 describes the trend of Web applications that are more interactive than traditional Web applications.

A **blog** is a journal or newsletter that's updated frequently and intended for the general public.

LO⁷ Summarize new trends in the Web 2.0 and 3.0 eras.

Web 2.0 describes the trend of Web applications that are more interactive than traditional Web applications. Some of these applications include blogs, wikis, social networking sites, RSS, and podcasts. See Table 7.4.

The Internet2 (I2) is a collaborative effort involving more than 200 U.S. universities and corporations to develop advanced Internet technologies and applications for higher education and academic research.

Most experts agree that Web 3.0, or the Semantic Web, provides personalization that allows users to access the Web more intelligently.

Table 7.4 *Web 1.0 versus Web 2.0*

Web 1.0	Web 2.0
DoubleClick (used for online marketing)	Google AdSense
Ofoto (sharing digital photos)	Flickr
Akamai (streaming media services)	BitTorrent
mp3.com	iTunes
Britannica Online	Wikipedia
Personal Web sites	Blogging
eVite	Upcoming.org and Events and Venues Database (EVBD), a type of wiki for event planning
Domain name speculation	Search engine optimization
Page views	Cost per click
Content management systems	Wikis
ERoom and Groove (collaboration software)	Collaboration portals, such as IBM Quickr and Microsoft Sharepoint
Posting a movie file on a personal Web page	YouTube

A **wiki** is a type of Web site that allows users to add, delete, and sometimes modify content.

Social networking refers to Web sites and services that allow users to connect with friends, family, and colleagues.

Really Simple Syndication (RSS) feeds are a fast, easy way to distribute Web content in Extensible Markup Language (XML) format.

A **podcast** is an electronic audio file, such as an MP3 file, that's posted on the Web for users to download to their mobile devices.

Internet2 (I2) is a collaborative effort involving U.S. universities and corporations to develop advanced Internet technologies and applications for higher education and academic research.

A **gigapop** is a local connection point-of-presence.

Additional content is available on the CourseMate for MIS Web site. Login at **www.cengagebrain.com**

Learning Outcomes

LO¹ **Define e-commerce and describe its advantages, disadvantages, and business models.**

E-commerce is buying and selling goods and services over the Internet. E-commerce builds on traditional commerce by adding the flexibility that networks offer and the availability of the Internet. One way to examine e-commerce and its role in the business world is through value chain analysis. See Table 8.1.

Table 8.1 *E-commerce versus traditional commerce*

Activity	Traditional commerce	E-commerce
Product information	Magazines, flyers	Web sites, online catalogs
Business communication	Regular mail, phone calls	E-mail
Check product availability	Phone calls, faxes, letters	E-mail, Web sites, and extranets
Order generation	Printed forms	E-mail, Web sites
Product acknowledgements	Phone calls, faxes	E-mail, Web sites, and EDI
Invoice generation	Printed forms	Web sites

If e-commerce is based on a sound business model (discussed in the next section), its advantages outweigh its disadvantages.

The most widely used business models in e-commerce are:

- Merchant
- Brokerage
- Advertising
- Mixed
- Informediary
- Subscription

LO² **Explain the major categories of e-commerce.**

The major categories of e-commerce are:

- Business-to-consumer (B2C): companies selling directly to consumers
- Business-to-business (B2B): electronic transactions between businesses
- Consumer-to-consumer (C2C): business transactions between users
- Consumer-to-business (C2B): people selling products or services to businesses
- E-government: government or other nonbusiness organizations using e-commerce applications
- Organizational or intrabusiness e-commerce: e-commerce activities within an organization

LO³ **Describe the business-to-consumer e-commerce cycle.**

There are five major activities involved in conducting B2C e-commerce:

Key Terms

LO¹

E-business encompasses all activities a company performs for selling and buying products and services, using computers and communication technologies.

E-commerce is buying and selling goods and services over the Internet.

A **value chain** is a series of activities designed to meet business needs by adding value (or cost) in each phase of the e-commerce process.

Click-and-brick e-commerce mixes traditional commerce and e-commerce.

The **merchant model** transfers the old retail model to the e-commerce world by using the medium of the Internet.

Using the **brokerage model** brings sellers and buyers together on the Web and collects commissions on transactions between these parties.

The **advertising model** is an extension of traditional advertising media, such as radio and television.

The **mixed model** generates revenue from more than one source.

E-commerce sites that use the **informediary model** collect information on consumers and businesses and then sell this information to other companies.

Using the **subscription model**, an e-commerce site sells digital products or services to customers.

Additional content is available on the CourseMate for MIS Web site. Login at **www.cengagebrain.com**

Chapter 8: E-Commerce

LO²

Business-to-consumer (B2C) companies sell directly to consumers.

Business-to-business (B2B) e-commerce involves electronic transactions between businesses.

Consumer-to-consumer (C2C) e-commerce involves business transactions between users.

Consumer-to-business (C2B) e-commerce involves people selling products or services to businesses.

E-government applications can include government-to-citizen, government-to-business, government-to-government, and government-to-employee.

Organizational or intrabusiness e-commerce involves e-commerce activities that take place inside an organization, typically via the organization's intranet.

LO⁴

In the **seller-side marketplace** model, sellers who cater to specialized markets come together to create a common marketplace for buyers.

E-procurement enables employees in an organization to order and receive supplies and services directly from suppliers.

In a **buyer-side marketplace** model, a buyer, or a group of buyers, opens an electronic marketplace and invites sellers to bid on announced products or requests for quotation (RFQs).

The **third-party exchange marketplace** is controlled by a third party, and the marketplace generates revenue from the fees charged for matching buyers and sellers.

A **vertical market** concentrates on a specific industry or market.

A **horizontal market** concentrates on a specific function or business process.

Trading partner agreements automate negotiating processes and enforce contracts between participating businesses.

- Information sharing
- Ordering
- Payment
- Fulfillment
- Service and support

LO⁴ Summarize the major models of business-to-business e-commerce.

There are three major types of B2B e-commerce models, based on who controls the marketplace: seller, buyer, or intermediary (third-party). As a result, the following marketplace models have been created: seller-side marketplace, buyer-side marketplace, and third-party exchange marketplace. A fourth model, called trading partner agreements, facilitates contracts and negotiations among business partners and is gaining popularity.

LO⁵ Describe mobile- and voice-based e-commerce.

M-commerce is using handheld devices, such as smart phones or PDAs, to conduct business transactions, such as making stock trades on an online brokerage firm. Many telecommunication companies offer Web-ready cell phones. A wide variety of m-commerce applications are available.

You can already use a mobile phone to access a Web site and order a product. The next step is voice-based e-commerce, which will rely on voice recognition and text-to-speech technologies that have improved dramatically in the past decade.

LO⁶ Explain two supporting technologies for e-commerce.

A number of technologies and applications support e-commerce activities. The following sections explain these widely used supporting technologies: electronic payment systems, Web marketing, and search engine optimization.

Electronic payment systems include smart cards, e-cash, e-checks, e-wallets, PayPal, and micropayment processors.

Web marketing uses the Web and its supporting technologies to promote goods and services.

Search engine optimization (SEO) is a method for improving the volume or quality of traffic to a Web site.

LO⁵

Mobile commerce (m-commerce) is using handheld devices to conduct business transactions.

Voice-based e-commerce relies on voice recognition and text-to-speech technologies.

LO⁶

Electronic payment refers to money or scrip which is exchanged only electronically.

A **smart card** is about the size of a credit card and contains an embedded microprocessor chip storing important financial and personal information.

E-cash is a secure and convenient alternative to bills and coins.

An **e-check** is the electronic version of a paper check.

E-wallets offer a secure, convenient, and portable tool for online shopping.

PayPal is a popular online payment system used in many online auction sites.

Micropayments are used for very small payments on the Web.

Web marketing uses the Web and its supporting technologies to promote goods and services.

Search engine optimization (SEO) is a method for improving the volume or quality of traffic to a Web site.

Learning Outcomes

LO¹ Discuss reasons for globalization and using global information systems, including e-business and Internet growth.

The global economy is creating customers who demand integrated worldwide services, and the expansion of global markets is a major factor in developing global information systems to handle these integrated services. Many companies have become international. In 2008, for example, the Coca-Cola company generated more than 80% of its revenue from outside the United States. Airline reservation systems are considered the first large-scale interactive global system, and hotels, rental car companies, and credit card services also require worldwide databases now to serve their customers more efficiently and effectively.

E-business is a major factor in the widespread use of global information systems. The Internet can simplify communication, change business relationships, and offer new opportunities to both consumers and businesses.

Today, the Internet is a part of daily life in most parts of the world. Growth has been highest in the Middle East and lowest in North America. The number of Internet users worldwide tops 1.5 billion, with Asia having the most.

LO² Describe global information systems and their requirements and components.

A global information system (GIS) is an information system that works across national borders, facilitates communication between headquarters and subsidiaries in other countries, and incorporates all the technologies and applications found in a typical information system to store, manipulate, and transmit data across cultural and geographic boundaries.

Although a GIS can vary quite a bit depending on a company's size and business needs, most GISs have these basic components:

- A network capable of global communication, including transmission equipment and communication media
- A global database
- Information-sharing technologies

A GIS must be capable of supporting complex global decisions. This complexity stems from the global environment in which multinational corporations (MNCs) operate. A GIS, like any information system, can be classified by levels of management affecting different functions: operational, tactical, and strategic. The complexities of global decision making mean that a GIS has some functional requirements that differ from a domestic information system's requirements.

GISs can be categorized in different ways, depending on their function or application. Global marketing information systems, strategic intelligent systems, transnational management support systems, and global competitive intelligent systems are some different names for GISs.

LO³ Explain the types of organizational structures used with global information systems.

The most important factor for effective operation of an MNC is coordination, and a global information system can provide essential information for this task. There are four commonly accepted types of global organizations:

- Multinational
- Global
- International
- Transnational

Key Terms

LO²

A **global information system (GIS)** is an information system that works across national borders, facilitates communication between headquarters and subsidiaries in other countries, and incorporates all the technologies and applications found in a typical information system to store, manipulate, and transmit data across cultural and geographic boundaries.

Transborder data flow (TDF) restricts what type of data can be captured and transmitted in foreign countries.

A **multinational corporation** or enterprise refers to a corporation that has asset and operations in at least one country other than its home country. This cooperation delivers products and services across its national borders and is usually managed centrally from its headquarters.

LO³

In a **multinational structure**, production, sales, and marketing are decentralized, and financial management remains the parent's responsibility.

A **global structure** (or headquarters-driven structure) manages highly centralized information systems. Subsidiaries have little autonomy and rely on headquarters for all process and control decisions as well as system design and implementation.

An organization using an **international structure** operates much like a multinational corporation, but subsidiaries depend on headquarters more for process and production decisions.

In an organization following a **transnational structure**, the parent and all subsidiaries work together in designing policies, procedures, and logistics for delivering products to the right market.

With **offshore outsourcing**, an organization chooses an outsourcing firm in another country that can provide needed services and products.

Additional content is available on the CourseMate for MIS Web site. Login at **www.cengagebrain.com**

Chapter 9: Global Information Systems

In a multinational structure production, sales, and marketing are decentralized, and financial management remains the parent's responsibility. An organization following a global structure manages highly centralized information systems. Subsidiaries have little autonomy and rely on headquarters for all process and control decisions as well as system design and implementation. An organization using an international structure operates much like a multinational corporation, but subsidiaries depend on headquarters more for process and production decisions. In an organization following a transnational structure, the parent and all subsidiaries work together in designing policies, procedures, and logistics for delivering products to the right market.

Offshore outsourcing is an alternative for developing information systems. With this approach, an organization chooses an outsourcing firm in another country that can provide needed services and products. See Table 9.1.

Table 9.1

Top offshore locations for outsourcing in 2008

Americas	Asia/Pacific	Europe, Middle East, and Africa
Argentina	Australia	Czech Republic
Brazil	China	Hungary
Canada	India	Ireland
Chile	Malaysia	Israel
Costa Rica	New Zealand	Northern Ireland
Mexico	Pakistan	Poland
Uruguay	Philippines	Romania
	Singapore	Russia
	Sri Lanka	Slovakia
	Vietnam	South Africa
		Spain
		Turkey
		Ukraine

LO⁴ Discuss obstacles to using global information systems.

The following factors can prevent the success of a GIS:

- Lack of standardization (can also include differences in time zones, taxes, language, work habits, and so forth)
- Cultural differences
- Diverse regulatory practices
- Poor telecommunication infrastructures
- Lack of skilled analysts and programmers

review card/ CHAPTER 10
BUILDING SUCCESSFUL INFORMATION SYSTEMS

Learning Outcomes

LO¹ Describe the systems development life cycle (SDLC) as a method for developing information systems.

The systems development life cycle (SDLC) is a series of well-defined phases performed in sequence. Each phase's output (results) becomes the input for the next phase.

The five phases for the SDLC are (see Exhibit 10.1):

- Planning
- Requirements gathering and analysis
- Design
- Implementation
- Maintenance

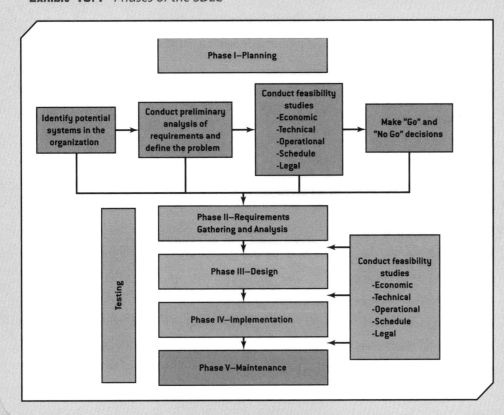

Exhibit 10.1 *Phases of the SDLC*

LO² Explain the tasks involved in the planning phase.

During the planning phase, the systems designer must understand and define the problem the organization faces. After identifying the problem, an analyst or team of analysts assesses the current and future needs.

Feasibility is the measure of how beneficial or practical an information system will be to an organization. The tool used for this purpose is a feasibility study, and it usually has five major dimensions: economic, technical, operational, schedule, and legal.

Key Terms

LO¹

Systems development life cycle (SDLC) is a series of well-defined phases performed in sequence that serve as a framework for developing a system or project.

LO²

During the **planning phase**, the systems designer must understand and define the problem the organization faces.

Internal users are employees who will use the system regularly.

External users include customers, contractors, suppliers, and other business partners.

Joint application design (JAD) centers on a structured workshop where users and system professionals come together to develop an application.

A **feasibility study** analyzes a proposed solution's feasibility and determines how best to present the solution to management.

Economic feasibility assesses a system's costs and benefits.

Technical feasibility is concerned with technology to be used in the system.

Operational feasibility is the measure of how well the proposed solution will work in the organization.

Schedule feasibility is concerned with whether the new system can be completed on time.

Legal feasibility is concerned with legal issues.

Additional content is available on the CourseMate for MIS Web site. Login at **www.cengagebrain.com**

Chapter 10: Building Successful Information Systems

LO³

In the **requirements-gathering and analysis phase**, analysts define the problem and generate alternatives for solving it.

LO⁴

During the **design phase**, analysts choose the solution that's the most realistic and offers the highest payoff for the organization.

Computer-aided systems engineering (CASE) tools automate parts of the application development process.

In **prototyping**, a small-scale version of the system is developed.

A **proof-of-concept prototype** shows users how a particular task that technically wasn't feasible can be done.

A **selling prototype** is used to sell a proposed system to users or management.

LO⁵

During the **implementation phase**, the solution is transferred from paper to action.

In **parallel conversion**, the old and new systems run simultaneously for a short time.

In **phased-in-phased-out conversion**, as each module of the new system is converted, the corresponding part of the old system is retired.

In **plunge (direct cutover) conversion**, the old system is stopped and the new system is implemented.

In **pilot conversion**, the analyst introduces the system in only a limited area of the organization.

A **request for proposal (RFP)** is a written document with detailed specifications, used to request bids from vendors.

A **request for information (RFI)** is a screening document for gathering vendor information and narrowing the list of potential vendors.

Insourcing happens when an organization's team develops the system internally.

Self-sourcing is when end users have been developing their own information systems with little or no formal assistance from the information systems team.

With the **outsourcing** approach, an organization hires an external vendor or consultant who specializes in providing development services.

LO⁶

In the **maintenance phase**, the information system is operating, enhancements and modifications to the system have been developed and tested, and hardware and software components have been added or replaced.

LO⁷

Service-oriented architecture (SOA) is a philosophy and a software and system development methodology that focuses on the development, use, and reuse of small, self-contained blocks of codes to meet the software needs of an organization.

LO³ Explain the tasks involved in the requirements-gathering and analysis phase.

In the requirements-gathering and analysis phase, analysts define the problem and generate alternatives for solving it. The team attempts to understand the requirements for the system, analyzes these requirements, and looks for ways to solve problems.

LO⁴ Explain the tasks involved in the design phase.

During the design phase, analysts choose the solution that's the most realistic and offers the highest payoff for the organization. The output of this phase is a document with exact specifications for implementing the system.

Prototyping has gained popularity in designing information systems because needs can change quickly and lack of specifications for the system can be a problem. Prototyping is usually done in four steps.

LO⁵ Explain the tasks involved in the implementation phase.

During the implementation phase, the solution is transferred from paper to action, and the team configures the system and procures components for it.

LO⁶ Explain the tasks involved in the maintenance phase.

During the maintenance phase, the maintenance team assesses how the system is working and takes steps to keep the system up and running.

LO⁷ Describe new trends in systems analysis and design, including service-oriented architecture, rapid application development, extreme programming, and agile methodology.

The SDLC is not appropriate for all systems development efforts. Alternatives include:

- Service-oriented architecture (SOA)
- Rapid application development (RAD)
- Extreme programming (XP)
- Agile methodology

Rapid application development (RAD) concentrates on user involvement and continuous interaction between users and designers.

Extreme programming (XP) divides a project into smaller functions, and developers can't go on to the next phase until the current phase is finished.

Pair programming is where two programmers participate in one development effort at one workstation.

Agile methodology focuses on an incremental development process and timely delivery of working software.

Additional content is available on the CourseMate for MIS Web site. Login at **www.cengagebrain.com**

Learning Outcomes

LO¹ Explain how supply chain management is used.

A supply chain is an integrated network consisting of an organization, its suppliers, transportation companies, and brokers used to deliver goods and services to customers. Supply chains exist in both service and manufacturing organizations, although the chain's complexity can vary widely in different organizations and industries.

Supply chain management (SCM) is the process of working with suppliers and other partners in the supply chain to improve procedures for delivering products and services. An SCM system coordinates:

- Procuring materials
- Transforming materials into intermediate and finished products or services
- Distributing finished products or services

The following tools help overcome some challenges associated with SCM:

Electronic data interchange (EDI) enables business partners to send and receive information on business transactions.

An e-marketplace is a third-party exchange (a B2B business model) that provides a platform for buyers and sellers to interact with each other and trade more efficiently online.

An online auction brings traditional auctions to customers around the globe and makes it possible to sell far more goods and services than at a traditional auction.

Collaborative planning, forecasting, and replenishment (CPFR) is used to coordinate supply chain members through point-of-sale (POS) data sharing and joint planning. CPFR ensures that inventory and sales data is shared across the supply chain.

LO² Describe customer relationship management systems.

Customer relationship management (CRM) consists of the processes a company uses to track and organize its contacts with customers. The main goal of a CRM system is to improve services offered to customers and use customer contact information for targeted marketing. CRM gives organizations more complete pictures of their customers. Typically, CRM applications are implemented with one of these approaches: on-premise CRM or Web-based CRM.

LO³ Explain knowledge management systems.

Knowledge management (KM) is a technique used to improve CRM systems (and many other systems) by identifying, storing, and disseminating "know-how"—facts about how to perform tasks.

Personalization is the process of satisfying customers' needs, building customer relationships, and increasing profits by designing goods and services that meet customers' preferences better.

Customization, which is somewhat different from personalization, allows customers to modify the standard offering, such as selecting a different home page to be displayed each time you open your Web browser.

Key Terms

LO¹

An **enterprise system** is an application used in all functions of a business that supports decision making throughout the organization.

A **supply chain** is an integrated network consisting of an organization, its suppliers, transportation companies, and brokers used to deliver goods and services to customers.

Supply chain management (SCM) is the process of working with suppliers and other partners in the supply chain to improve procedures for delivering products and services.

Electronic data interchange (EDI) enables business partners to send and receive information on business transactions.

An **e-marketplace** is a third-party exchange (B2B model) that provides a platform for buyers and sellers to interact with each other and trade more efficiently online.

By using the Internet, an **online auction** brings traditional auctions to customers around the globe and makes it possible to sell far more goods and services than at a traditional auction.

A **reverse auction** invites sellers to submit bids for products and services. A one-to-many relationship is represented by one buyer and many sellers. The buyer can choose the seller that offers the service or product at the lowest price.

Collaborative planning, forecasting, and replenishment (CPFR) is used to coordinate supply chain members through point-of-sale (POS) data sharing and joint planning.

LO²

Customer relationship management (CRM) consists of the processes a company uses to track and organize its contacts with customers. It improves services offered to customers and uses customer contact information for targeted marketing.

Additional content is available on the CourseMate for MIS Web site. Login at **www.cengagebrain.com**

Chapter 11: Enterprise Systems

LO³

Knowledge management draws on concepts of organizational learning, organizational culture, and best practices to convert tacit knowledge into explicit knowledge, create a knowledge-sharing culture in an organization, and eliminate obstacles to sharing knowledge.

LO⁴   Describe enterprise resource planning systems.

Enterprise resource planning (ERP) is an integrated system that collects and processes data and manages and coordinates resources, information, and functions throughout an organization. A well-designed ERP system offers many benefits. More than 40 vendors, such as SAP, Oracle, Sage Group, and Microsoft, offer ERP software with varying capabilities. See Table 11.1.

Table 11.1 *ERP components*

Component	Function
Unified database	Collects and analyzes relevant internal and external data and information needed by other functions
Inventory management	Provides inventory status and inventory forecasts
Supply chain	Provides information on supply chain members, including suppliers, manufacturing, distribution, and customers
Manufacturing	Supplies information on production costs and pricing
Human resources	Provides information on assessing job candidates, scheduling and assigning employees, and predicting future personnel needs
CRM	Supplies information on customers and their needs and preferences
Purchasing	Provides information related to the purchasing function, including e-procurement
Accounting	Tracks financial information, such as budget allocations and debits and credits
Vendor integration	Integrates information for vendors, such as offering automated downloads of data on product pricing, specifications, and availability
E-commerce	Provides B2C information related to order status and B2B information related to suppliers and business partners
Sales	Supplies information on sales and marketing

Personalization is the process of satisfying customers' needs, building customer relationships, and increasing profits by designing goods and services that meet customers' preferences better. It involves not only customers' requests, but also the interaction between customers and the company.

Customization allows customers to modify the standard offering, such as selecting a different home page to be displayed each time you open your Web browser.

Collaborative filtering (CF) searches for specific information or patterns, using input from multiple business partners and data sources. It identifies groups of people based on common interests and recommends products or services based on what members of the group purchased or didn't purchase.

LO⁴

Enterprise resource planning (ERP) is an integrated system that collects and processes data and manages and coordinates resources, information, and functions throughout an organization.

Learning Outcomes

LO¹ Define types of decisions and phases of the decision-making process in a typical organization.

In a typical organization, decisions usually fall into one of these categories (see Exhibit 12.1):

- Structured decisions
- Semistructured decisions
- Unstructured decisions

Herbert Simon defines three phases in the decision-making process: intelligence, design, and choice. A fourth phase, implementation, can be added.

Exhibit 12.1 *Organizational levels and types of decisions*

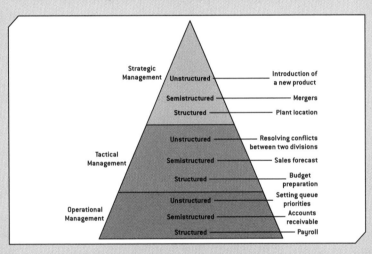

Exhibit 12.2 *Components of a DSS*

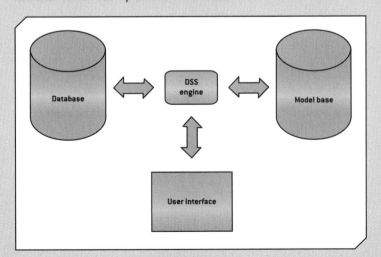

Key Terms

LO¹

Structured decisions can be automated because a well-defined standard operating procedure exists for these types of decisions.

Semistructured decisions include a structured aspect that benefits from information retrieval, analytical models, and information systems technology.

Unstructured decisions are unique, typically one-time decisions, with no standard operating procedure pertaining to them.

Management support systems (MSSs) are the different types of information systems that support certain aspects and types of decisions.

In the **intelligence phase**, a decision maker examines the organization's environment for conditions that need decisions. Data is collected from a variety of sources and processed. From this information, the decision maker can discover ways to approach the problem.

In the **design phase**, the objective is to define criteria for the decision, generate alternatives for meeting the criteria, and define associations between the criteria and the alternatives.

During the **choice phase**, the best and most effective course of action is chosen.

In the **implementation phase**, the organization devises a plan for carrying out the alternative selected in the choice phase and obtains the resources to implement the plan.

LO²

A **decision support system (DSS)** is an interactive information system consisting of hardware, software, data, and models (mathematical and statistical) designed to assist decision makers in an organization.

The **model base** component includes mathematical and statistical models that, along with the database, enable a DSS to analyze information.

Additional content is available on the CourseMate for MIS Web site. Login at **www.cengagebrain.com**

Chapter 12: Managing Support Systems

A **managerial designer** defines the management issues in designing and using a DSS. These issues don't involve the technological aspects of the system; they're related to management's goals and needs.

The **technical designer** focuses on how the DSS is implemented and usually addresses questions about data storage, file structure, user access, response time, and security measures.

A **model builder** is the liaison between users and designers. He or she is responsible for supplying information on what the model does, what data inputs it accepts, how the model's output should be interpreted, and what assumptions go into creating and using the model.

LO³

Executive information systems (EISs), branches of DSSs, are interactive information systems that give executives easy access to internal and external data and typically include "drill-down" features and a digital dashboard for examining and analyzing information.

A **digital dashboard** integrates information from multiple sources and presents it in a unified, understandable format, often charts and graphs. They offer up-to-the-minute snapshots of information and assist decision makers in identifying trends and potential problems.

LO⁴

Group support systems (GSSs) assist decision makers working in groups. These systems use computer and communication technologies to formulate, process, and implement a decision-making task and can be considered a kind of intervention technology that helps overcome the limitations of group interactions.

Groupware assists groups in communicating, collaborating, and coordinating their activities. It is a collection of applications that support decision makers by providing access to a shared environment and information.

Electronic meeting systems enable decision makers in different locations to participate in a group decision-making process.

LO⁵

Geographic information system (GIS) capture, store, process, and display geographic information or information in a geographic context, such as showing the location of all city streetlights on a map.

LO²   Describe a decision support system.

A decision support system (DSS) is an interactive information system consisting of hardware, software, data, and models (mathematical and statistical) designed to assist decision makers in an organization. The emphasis is on semistructured and unstructured tasks.

A DSS includes three major components: a database, a model base, and a user interface. In addition, a fourth component, the DSS engine, manages and coordinates these major components. See Exhibit 12.2.

To design, implement, and use a DSS, several groups, or roles, must be involved. These roles include: user, managerial designer, technical designer, and model builder.

LO³   Explain an executive information system's importance in decision making.

Executive information systems (EISs) are interactive information systems that give executives easy access to internal and external data and typically include "drill-down" features and a digital dashboard for examining and analyzing information. An EIS includes graphical representations of data that help managers make critical decisions.

LO⁴   Describe group support systems, including groupware and electronic meeting systems.

Group support systems (GSSs) are intended to assist decision makers working in groups. Intervention features of a GSS reduce communication barriers and introduce order and efficiency into situations that are inherently unsystematic and inefficient, such as group meetings and brainstorming sessions.

The goal of groupware is to assist groups in communicating, collaborating, and coordinating their activities. It's intended more for teamwork than for decision support.

LO⁵   Summarize uses for a geographic information system.

A geographic information system (GIS) captures, stores, processes, and displays geographic information or information in a geographic context. A GIS uses spatial and non-spatial data and specialized techniques for storing coordinates of complex geographic objects, including networks of lines and reporting zones.

LO⁶   Describe guidelines for designing a management support system.

Before designing any management support system, the system's objectives should be defined clearly, and then standard system development methods can be followed.

reviewcard/ CHAPTER 13
INTELLIGENT INFORMATION SYSTEMS

Learning Outcomes

LO¹ Define artificial intelligence and explain how these technologies support decision making.

Artificial intelligence (AI) consists of related technologies that try to simulate and reproduce human thought behavior, including thinking, speaking, feeling, and reasoning. The most recent developments in AI technologies promise new areas of decision-making support.

Over the years, the capabilities of these systems have improved in an attempt to close the gap between artificial intelligence and human intelligence. See Table 13.1.

Table 13.1
Applications of AI technologies

Field	Example of an organization	Applications
Energy	Arco & Tenneco	Neural networks used to help pinpoint oil and gas deposits
Government	Internal Revenue Service	Testing a software to read tax returns and spot fraud
Human Services	Merced County in California	Expert systems used to decide if applicants should receive welfare benefits
Marketing	Spiegel	Neural networks used to determine most likely buyers from a long list
Telecommunications	BT Group	Heuristic search used for scheduling application that provides the work schedules of more than 20,000 engineers
Transportation	American Airlines	Expert systems used to schedule the routine maintenance of its airplanes
Inventory/forecasting	Hyundai Motors	Used neural nets and expert systems to reduce delivery time by 20% and increased inventory turns from 3 to 3.4
Inventory/forecasting	SCI Systems	Used neural nets and expert systems to reduce on-hand inventory by 15% resulting in $180 million in annual savings
Inventory/forecasting	Reynolds Aluminum	Used neural nets and expert systems to reduce forecasting errors by 2% that resulted in a reduction of one million pounds in inventory
Inventory/forecasting	Unilever	Used neural nets and expert systems to reduce forecasting errors from 40% to 25% yielding resulting in multi-million dollar savings

LO² Explain an expert system, its applications, and its components.

Expert systems mimic human expertise in a field to solve a problem in a well-defined area. A typical expert system includes:

- Knowledge acquisition facility
- Knowledge base
- Factual knowledge
- Heuristic knowledge
- Meta-knowledge
- Knowledge base management system
- User interface
- Explanation facility
- Inference engine

Key Terms

LO¹

Artificial intelligence (AI) consists of related technologies that try to simulate and reproduce human thought behavior, including thinking, speaking, feeling, and reasoning. AI technologies apply computers to areas that require knowledge, perception, reasoning, understanding, and cognitive abilities.

Robots and robotics are some of the most successful applications of AI. They perform well at simple, repetitive tasks and can be used to free workers from tedious or hazardous jobs.

LO²

Expert systems mimic human expertise in a field to solve a problem in a well-defined area.

A **knowledge acquisition facility** is a software package with manual or automated methods for acquiring and incorporating new rules and facts so that the expert system is capable of growth.

Additional content is available on the CourseMate for MIS Web site. Login at **www.cengagebrain.com**

Chapter 13: Intelligent Information Systems

A **knowledge base** is similar to a database, but in addition to storing facts and figures, it keeps track of rules and explanations associated with facts.

A **knowledge base management system (KBMS)**, similar to a DBMS, is used to keep the knowledge base updated with changes to facts, figures, and rules.

An **explanation facility** performs tasks similar to what a human expert does by explaining to end users how recommendations are derived.

An **inference engine** is similar to the model base component of a decision support system. By using different techniques, such as forward and backward chaining, it manipulates a series of rules.

In **forward chaining**, a series of "If-Then Else" condition pairs are performed.

In **backward chaining**, the expert system starts with the goal first—the Then part and backtracks to find the right solution.

LO³

Case-based reasoning (CBR) is a problem-solving technique that matches a new case (problem) with a previously solved case and its solution stored in a database.

LO⁴

Intelligent agents are applications of artificial intelligence and are becoming more popular, particularly in e-commerce. They consist of software capable of reasoning and following rule-based processes.

Shopping and information agents help users navigate through the vast resources available on the Web and provide better results in finding information.

Personal agents perform specific tasks for a user, such as remembering information for filling out Web forms or completing e-mail addresses after the first few characters are typed.

Data-mining agents work with a data warehouse and can detect trend changes and discover new information and relationships among data items that aren't readily apparent.

Monitoring and surveillance agents usually track and report on computer equipment and network systems to predict when a system crash or failure might occur.

LO³ Describe case-based reasoning.

Case-based reasoning (CBR) is a problem-solving technique that matches a new case (problem) with a previously solved case and its solution stored in a database.

LO⁴ Summarize types of intelligent agents and how they're used.

Intelligent agents, or bots, consist of software capable of reasoning and following rule-based processes. These include:

- Shopping and information agents
- Personal agents
- Data-mining agents
- Monitoring and surveillance agents

LO⁵ Describe fuzzy logic and its uses.

Fuzzy logic is designed to help computers simulate vagueness and uncertainty in common situations. Fuzzy logic has been used in search engines, chip design, database management systems, software development, and more.

LO⁶ Explain artificial neural networks.

Artificial neural networks (ANNs) are networks that learn and are capable of performing tasks that are difficult with conventional computers.

LO⁷ Describe how genetic algorithms are used.

Genetic algorithms (GAs) are used mostly in techniques to find solutions to optimization and search problems. Genetic algorithms can examine complex problems without any assumptions of what the correct solution should be.

LO⁸ Explain natural language processing and its advantages and disadvantages.

Natural language processing (NLP) was developed so that users could communicate with computers in their own language. The size and complexity of the human language has made developing NLP systems difficult.

LO⁹ Summarize the advantages of integrating AI technologies into decision support systems.

AI-related technologies add explanation capabilities (by integrating expert systems) and learning capabilities (by integrating ANNs) and create an interface that's easier to use (by integrating an NLP system).

LO⁵

Fuzzy logic allows a smooth, gradual transition between human and computer vocabularies and deals with variations in linguistic terms by using a degree of membership.

LO⁶

Artificial neural networks (ANNs) are networks that learn and are capable of performing tasks that are difficult with conventional computers.

LO⁷

Genetic algorithms (GAs) are used mostly in techniques to find solutions to optimization and search problems.

LO⁸

Natural language processing (NLP) was developed so that users could communicate with computers in their own language.

Learning Outcomes

LO¹ Summarize new trends in software and service distribution.

Recent trends in software and service distribution include pull and push technologies and application service providers. With pull technology, a user states a need before getting information, such as entering a URL in a Web browser to go to a certain Web site. With push technology, or Webcasting, a Web server delivers information to users who have signed up for this service instead of waiting for user requests to send the information.

A recent business model called application service providers (ASPs) provides access to software or services for a fee. Software as a service (SaaS), or on-demand software, is a model for ASPs to deliver software to users for a fee; the software might be for temporary or long-term use. The SaaS model can take several forms.

LO² Describe virtual reality components and applications.

The goal of virtual reality (VR) is to create an environment in which users can interact and participate as they do in the real world. There are two major types of user environments in VR: egocentric and exocentric.

The major components of a VR system are:

- Visual and aural systems
- Manual control for navigation
- Central coordinating processor and software system
- Walker

A cave automatic virtual environment (CAVE) is a virtual environment consisting of a cube-shaped room in which the walls are rear-projection screens. A virtual world is a simulated environment designed for users to interact via avatars. An avatar is a 2-D or 3-D graphical representation of a person in the virtual world, used in chat rooms and online games.

LO³ Discuss uses of radio frequency identification.

A radio frequency identification (RFID) tag is a small electronic device consisting of a small chip and an antenna. The device provides a unique identification for the card or the object carrying the tag. There are two types of RFID tags: passive and active. See Table 14.1.

Key Terms

LO¹

With **pull technology**, a user states a need before getting information.

With **push technology**, or Webcasting, a Web server delivers information to users who have signed up for this service instead of waiting for user requests to send the information.

Application service providers (ASPs) provide access to software or services for a fee.

Software as a service (SaaS), or on demand software, is a model for ASPs to deliver software to users for a fee.

LO²

Virtual reality (VR) uses computer generated, three-dimensional images to create the illusion of interaction in a real-world environment.

In an **egocentric environment** the user is totally immersed in the VR world.

The **exocentric environment** can be called a "window view"; users can only view data onscreen. They can't interact with objects.

A **cave automatic virtual environment (CAVE)** is a virtual environment consisting of a cube-shaped room in which the walls are rear-projection screens. They are holographic devices that create, capture, and display images in true 3-D form.

Table 14.1 *RFID applications*

Category	Examples
Tracking and identification	Railway cars and shipping containers, livestock and pets, supply-chain management (tracking merchandise from manufacturers to retailers to customers), inventory control, retail checkout and POS systems, and recycling and waste disposal
Payment and stored-value systems	Electronic toll systems, contactless credit cards (require no swiping), subway and bus passes, casino tokens, and concert tickets
Access control	Building access cards, ski-lift passes, concert tickets, and car ignition systems
Anticounterfeiting	Casino tokens, high-denomination currency notes, luxury goods, and prescription drugs
Healthcare	Tracking medical tools and patients (particularly newborns and patients with Alzheimer's), process control, and monitoring patient data

Additional content is available on the CourseMate for MIS Web site. Login at **www.cengagebrain.com**

Chapter 14: Emerging Trends, Technologies, and Applications

A **virtual world** is a simulated environment designed for users to interact via avatars.

An **avatar** is a 2-D or 3-D graphical representation of a person in the virtual world.

LO³

A **radio frequency identification (RFID)** tag is a small electronic device consisting of a small chip and an antenna. This device provides a unique identification for the card or the object carrying the tag.

LO⁵

Wireless Fidelity (Wi-Fi) is a broadband wireless technology. Information can be transmitted over short distances, typically 120 feet indoors and 300 feet outdoors, in the form of radio waves.

Worldwide Interoperability for Microwave Access (WiMAX) is a broadband wireless technology based on the IEEE 802.16 standards. It's designed for wireless metropolitan area networks.

Bluetooth, a technology that can be used to create a personal area network (PAN), is a wireless technology for transferring data over short distances.

Grid computing involves connecting all the different computers combining their processing power to solve a particular problem. Users can make use of other computers' resources to solve problems involving large-scale, complex calculations that a single computer isn't capable of solving.

Utility (on-demand) computing is similar to the SaaS model and provides IT services on demand. Users pay for computing or storage resources on an as-needed basis, similar to paying for utilities.

Cloud computing is a platform incorporating many recent technologies under one platform, including the SaaS model, Web 2.0, grid computing, and utility computing, so that a variety of resources can be provided to users over the Internet.

LO⁴ **Summarize new uses of biometrics.**

Biometrics have become more widespread in forensics and in related law enforcement fields, such as criminal identification, prison security, and airport security. Because biometrics offer a high degree of accuracy that isn't possible with other security measures, they have the potential to be used in many civilian fields, too.

LO⁵ **Explain new trends in networking, including wireless technologies and grid and cloud computing.**

Recent trends in networking technologies include:

- Wireless Fidelity (Wi-Fi)
- Worldwide Interoperability for Microwave Access (WiMAX)
- Bluetooth
- Grid computing (see Exhibit 14.7)
- Utility (on-demand) computing
- Cloud computing

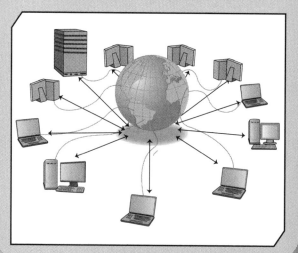

Exhibit 14.7 *A grid computing configuration*

LO⁶ **Discuss uses of nanotechnology.**

Nanotechnology incorporates techniques that involve the structure and composition of materials on a nanoscale. Nanotechnology has become an exciting development in many fields including healthcare, information technology, energy, heavy industry, and consumer goods.

Nanotechnology incorporates techniques that involve the structure and composition of materials on a nanoscale.

Additional content is available on the CourseMate
for MIS Web site. Login at **www.cengagebrain.com**

Notes